Applied Chemistry

O.V. Roussak • H.D. Gesser

Applied Chemistry

A Textbook for Engineers and Technologists

Second Edition

 Springer

O.V. Roussak
Chemistry Department
University of Manitoba
Winnipeg, Manitoba, Canada

H.D. Gesser
Chemistry Department
University of Manitoba
Winnipeg, Manitoba, Canada

ISBN 978-1-4614-4261-5   ISBN 978-1-4614-4262-2 (eBook)
DOI 10.1007/978-1-4614-4262-2
Springer New York Heidelberg Dordrecht London

Library of Congress Control Number: 2012947030

Printed on acid-free paper

Springer is part of Springer Science+Business Media (www.springer.com)

O.V. Roussak: In memory of my father, Roussak Vladimir Alexandrovich, a smart mining engineer, my best friend and teacher.

H.D. Gesser: To Esther, Isaac, Sarah and Avi.

Preface to the Second Edition

The first edition of this book appeared 10 years ago. This book is the result of teaching in the Applied Chemistry (Dr. H.D. Gesser, the Chemistry 2240 course) as well as in the Water Quality Analysis for Civil Engineers (Dr. O.V. Roussak, the CHEM 2560 course) to second year engineering students for many years at the University of Manitoba (Winnipeg, Manitoba, Canada). Much has transpired in science during this period and that includes applied chemistry. The major change in this new edition that becomes obvious is the addition of several (eight) experiments to accompany the book and the course for which it was intended. A new solutions manual is also a valuable asset to the second edition of the book.

Chemistry is primarily an experimental science and the performance of a few experiments to accompany the text was long considered while the course was taught. The choice of experiments we include was determined by the equipment that is usually available (with one or two possible exceptions) and by the expected usefulness of these experiments to the student, who will eventually become a practicing professional, and to the cost that is involved in student time. We welcome any reasonable and inexpensive additional experiments to introduce for the next edition of our book and topics to include in the next edition.

Winnipeg, Manitoba, Canada   O.V. Roussak
September 2012   H.D. Gesser

Preface to the First Edition

This book is the result of teaching a one semester course in applied chemistry (Chemistry 224) to second year engineering students for over 15 years. The contents of the course evolved as the interests and needs of both the students and the engineering faculty changed. All the students had at least one semester of introductory chemistry and it has been assumed in this text that the students have been exposed to thermodynamics, chemical kinetics, solution equilibrium, and organic chemistry. These topics must be discussed either before starting the applied subjects or developed as required if the students are not familiar with these prerequisites.

Engineering students often ask "Why is another chemistry course required for non-chemical engineers?"

There are many answers to this question but foremost is that the professional engineer must know when to consult a chemist and be able to communicate with him. When this is not done, the consequences can be disastrous due to faulty design, poor choice of materials, or inadequate safety factors.

Examples of blunders abound and only a few will be described in an attempt to convince the student to take the subject matter seriously.

The Challenger space shuttle disaster which occurred in January 1986 was attributed to the cold overnight weather which had hardened the O-rings on the booster rockets while the space craft sat on the launchpad. During flight, the O-ring seals failed, causing fuel to leak out and ignite. The use of a material with a lower glass transition temperature (T_g) could have prevented the disaster.

A similar problem may exist in automatic transmissions used in vehicles. The use of silicone rubber O-rings instead of neoprene may add to the cost of the transmission but this would be more than compensated for by an improved and more reliable performance at $-40°C$ where neoprene begins to harden; whereas the silicone rubber is still flexible.

A new asphalt product from Europe incorporates the slow release of calcium chloride ($CaCl_2$) to prevent icing on roads and bridges. Predictably, this would have little use in Winnipeg, Canada, where $-40°C$ is not uncommon in winter.

The heavy water plant at Glace Bay, Nova Scotia, was designed to extract D_2O from sea water. The corrosion of the plant eventually delayed production and the redesign and use of more appropriate materials added millions to the cost of the plant.

A chemistry colleague examined his refrigerator which failed after less than 10 years of use. He noted that a compressor coil made of copper was soldered to an expansion tube made of iron. Condensing water had corroded the—guess what?—iron tube. Was this an example of designed obsolescence or sheer stupidity. One wonders, since the savings by using iron instead of copper is a few cents and the company is a well-known prominent world manufacturer of electrical appliances and equipment.

With the energy problems now facing our industry and the resulting economic problems, the engineer will be required to make judgments which can alter the cost-benefit ratio for his employer.

One must realize that perpetual motion is impossible even though the US Supreme Court has ruled that a patent should be granted for a device which the Patent Office considers to be a Perpetual Motion Machine. An example of this type of proposal appeared in a local newspaper which described an invention for a car which ran on water. This is accomplished by a battery which is initially used to electrolyze water to produce H_2 and O_2 that is then fed into a fuel cell which drives an electrical motor that propels the car. While the car is moving, an alternator driven by the automobile's motion charges the battery. Thus, the only consumable item is water. This is an excellent example of perpetual motion.

A similar invention of an automobile powered by an air engine has been described. A compressed air cylinder powers an engine which drives the automobile. A compressor which is run by the moving car recompresses the gas into a second cylinder which is used when the first cylinder is empty. Such perpetual motion systems will abound and the public must be made aware of the pitfalls.

Have you heard of the Magnatron? Using 17 oz of deuterium (from heavy water) and 1.5 oz of gallium will allow you to drive an engine 110,000 miles at a cost of $110. Are you skeptical? You should be, because it is an example of the well-known Computer GIGO Principle (meaning garbage in = garbage out).

An engineer responsible for the application of a thin film of a liquid adhesive to a plastic was experiencing problems. Bubbles were being formed which disrupted the even smooth adhesive coat. The answer was found in the dissolved gases since air at high pressure was used to force the adhesive out of the spreading nozzle. The engineer did not believe that the air was actually soluble in the hexane used to dissolve the glue. When helium was used instead of air, no bubbles formed because of the lower solubility of He compared to O_2 and N_2 in the solvent. Everything is soluble in all solvents, only the extent of solution varies from non-detectable (by present methods of measurement) to completely soluble. The same principle applies to the permeability of one substance through another.

An aluminum tank car exploded when the broken dome's door hinge was being welded. The tank car, which had been used to carry fertilizer (aqueous ammonium nitrate and urea), was washed and cleaned with water—so why had it exploded? Dilute ammonium hydroxide is more corrosive to aluminum than the concentrated solution. Hence, the reaction

$$3NH_4OH + Al \rightarrow Al(OH)_3 + 3NH_3 + 1.5H_2$$

produced hydrogen which exploded when the welding arc ignited the H_2/O_2 mixture. The broad explosive range of hydrogen in air makes it a dangerous gas when confined.

Batteries are often used as a back-up power source for relays and, hence, stand idle for long periods. To keep them ready for use they are continuously charged. However, they are known to explode occasionally when they are switched into service because of the excess hydrogen produced due to overcharging. This can be avoided by either catalyzing the recombination of the H_2 and O_2 to form water

$$2H_2 + O_2 \rightarrow 2H_2O$$

by a nickel, platinum, or palladium catalyst in the battery caps, or by keeping the charging current equal to the inherent discharge rate which is about 1% per month for the lead-acid battery.

It has recently been shown that the flaming disaster of the Hindenburg Zeppelin in 1937, in which 36 lives were lost, may have been caused by static electricity igniting the outer fabric. This was shown to contain an iron oxide pigment and reflecting powdered aluminum. Such a combination, known as a thermite mixture, results in the highly exothermic Gouldshmidt reaction (first reported in 1898):

$$Fe_2O + 2Al \rightarrow Al_2O + 2Fe \quad AH^0 = -852 \text{ kJ/mol of } Fe_2O$$

In the early days of the railway, rails were welded with the molten iron formed in this reaction. The combination of powdered aluminum and a metal oxide has been used as a rocket fuel and evidence has been obtained to indicate that after the disaster the Germans replaced the aluminum by bronze which does not react with metal oxides. Thus, the bad reputation hydrogen has had as a result of the accident is undeserved and the resulting limiting use of the airship was due to faulty chemistry and could have been avoided.

The original design and structure of the Statue of Liberty, built about 100 years ago, took into account the need to avoid using different metals in direct contact with each other. However, the salt sea spray penetrated the structure and corroded the iron frame which supported the outer copper shell. Chloride ions catalyzed the corrosion of iron. The use of brass in a steam line valve resulted in corrosion and the formation of a green solid product. The architect was apparently unaware of the standard practice to use amines such as morpholine as a corrosion inhibitor for steam lines. Amines react with copper in the brass at high temperatures in the presence of oxygen to form copper-amine complexes similar to the dark blue copper ammonium complex, $Cu(NH_3)J^+$.

Numbers are a fundamental component of measurements and of the physical properties of materials. However, numbers without units are meaningless. Few quantities do not have units, e.g., specific gravity of a substance is the ratio of the mass of a substance to the mass of an equal volume of water at 4°C. Another unitless quantity is the Reynolds Number, $R_e = \rho v l/\eta$ where ρ is the density; v is the velocity; η is the viscosity of the fluid, and l is the length or diameter of a body or internal breadth of a pipe. The ratio $\eta/\rho = \mu$ the kinematic viscosity with units of l^2/t. $R = vl/\mu$ and has no units if the units of v, l, and μ are consistent.

To ignore units is to invite disaster. Two examples will illustrate the hazards of the careless or nonuse of units. During the transition from Imperial to SI (metric) units in Canada, an Air Canada commercial jet (Boeing 767) on a trans Canada flight (No 143) from Montreal to Edmonton on July 23, 1983, ran out of fuel over Winnipeg.

Fortunately the pilot was able to glide the airplane to an abandoned airfield (Gimli, MB) used for training pilots during World War II. The cause of the near disaster was a mix-up in the two types of units involved for loading the fuel and the use of a unitless conversion factor. (See Appendix A for a detailed account of this error).

The second example of an error in units cost the USA (NASA) $94,000,000. A Mars climate probe missed its target orbit of 150 km from the Mars' surface and approached to within 60 km and burned up. The error was due to the different units used by two contractors which were not interconverted by the NASA systems engineering staff. This book uses various sets of units and the equivalences are given in Appendix A. This is designed to keep the student constantly aware of the need to watch and be aware of units.

The above examples show how what may be a simple design or system can fail due to insufficient knowledge of chemistry. This textbook is not intended to solve all the problems you might encounter during your career. It will, however, give you the vocabulary and basis on which you can build your expertise in engineering.

The exercises presented at the end of each chapter are intended to test the students' understanding of the material and to extend the topics beyond their initial levels.

The author is indebted to the office staff in the Chemistry Department of the University of Manitoba who took penciled scrawls and converted them into legible and meaningful text. These include Cheryl Armstrong, Tricia Lewis, and Debbie Dobson. I also wish to thank my colleagues and friends who contributed by critical discussions over coffee. I also wish to express my thanks to Roberta Wover who gave me many helpful comments on reading the manuscript and checking the exercises and websites. Mark Matousek having survived Chem. 224 several years ago, applied some of his acquired drawing skills to many of the illustrations shown. Nevertheless, I must accept full responsibility for any errors or omissions, and I would be very grateful if these would be brought to my attention.

Some general references are listed below:

Kirk RE, Othmer DF (1995) Encyclopedia of chemical technology, 4th edn. vol 30. Wiley, New York

(1992) Ullmann's encyclopedia of industrial chemistry, vol 26. VCH, Germany

(1987) Encyclopedia of physical science and technology, vol 15. + Year Books, Academic, Orlando

Hopp V, Hennig I (1983) Handbook of applied chemistry. Hemisphere Publishing Company, Washington

(1982) McGraw-Hill encyclopedia of science and technology, vol 15. + Year Books, New York

Steedman W, Snadden RB, Anderson IH (1980) Chemistry for the engineering and applied sciences, 2nd edn. Pergamon Press, Oxford

Muklyonov IP (ed) (1979) Chemical technology, 3rd edn. vol 2. Mir Publishing, Moscow, in English

Diamant RME (1972) Applied chemistry for engineers, 3rd edn. Pitman, London

Palin GR (1972) Chemistry for technologists. Pergamon Press, Oxford

(1972) Chemical technology: an encyclopedic treatment, vol 7. Barnes and Noble Inc., New York

Butler FG, Cowie GR (1965) A manual of applied chemistry for engineers. Oliver and Boyd, London

Munro LA (1964) Chemistry in engineering. Prentice Hall, EnglewoodCliffs

Cartweil E (1964) Chemistry for engineers—an introductory course, 2nd edn. Butterworths, London

Gyngell ES (1960) Applied chemistry for engineers, 3rd edn. Edward Arnold, London

(1957) Thorpe's dictionary of applied chemistry, 4th edn. vol 11. Longmans, Green, London

The World Wide Web is an excellent source of technical information though it is important to recognize that discretion must be exercised in selecting and using the information since the material presented is not always accurate or up to date. Some selected websites are added to the Further Readings list at the end of each chapter.

Contents

Abbreviations Used in This Text

AAGR	Average annual growth rate
AAS	Atomic absorption spectrometry
ABS	Acrylonitrile-butadiene-styrene polymer
AEM	Anion exchange membrane
AFC	Atomic fluorescence spectrometry
AFR	Air fuel ratio
AGR	Advanced gas reactor
ANFO	Ammonium nitrate fuel oil
ASTM	American Society for Testing and Materials
bbl	Barrel for oil, see Appendix A
BC	Bimetallic corrosion
BET	Brunauer-Emmett-Teller
BLEVE	Boiling liquid expanding vapor explosion
BOD	Biochemical oxygen demand
BP	Boiling point
BWR	Boiling water reactor
CANDU	Canadian deuterium uranium reactor
CASING	Cross-linking by activated species of inert gases
CC	Crevice corrosion
CEM	Cation exchange membrane
CN	Cetane number
CNG	Compressed natural gas
COD	Chemical oxygen demand
CPVC	Chlorinated polyvinylchloride
CR	Compression ratio
DC	Direct current
DNA	Deoxyribonucleic acid
DP	Degree of polymerization
DR	Drag Reducer
DR	Distribution ratio
DTA	Differential thermal analysis
ECE	Economic Commission for Europe
ECM	Electrochemical machining
EDS	Exxon Donner Solvent

EHL	Elastohydrodynamic lubrication
EIS	Electrochemical impedance spectroscopy
ENM	Electrochemical noise method
EO	Extreme pressure (lubrication)
ER	Electrorheological fluid
ETBE	Ethyl tert butyl ether
EV	Expected value
EV	Electric vehicle
FAC	Free available chlorine
FEP	Hexafluoropropylene + PTFE
FP	Flash point
GAC	Granulated activated carbon
GBC	Grain boundary corrosion
GNP	Gross National Product
GR	Gas cooled reactor
hv	photon
HAR	High aspect ratio
HDI	Hexane diisocyanate
HLW	High level waste
HMN	Heptamethylnonane
HRI	Hydrocarbon Research Inc
HWR	Heavy water reactor
IAEA	International Atomic Energy Agency
ICAPS	Inductively coupled argon plasma Spectrometry
ICE	Internal combustion engine
Is	Specific impulse
LEL	Lower explosion limit
LH_2	Liquid hydrogen
LNG	Liquified natural gas
LWR	Light water reactor
M-85	Methanol with 15% gasoline
MDF	Macro defect free
MeV	Million electron volts
MI	Machinability index
MMT	Methylcyclopentadiene Manganese II tricarbonyl
MON	Motor octane number
MPC	Maximum permissible concentration
MPN	Most probable number
MTBE	Methyl tertiary butyl ether
MW	Megawatt, 10^6 W
NMOG	Non-methane organic gases
NTP	Normal conditions of temperature and pressure, 25°C and 1 atm pressure.
OB	Oxygen balance
OECD	Organization for Economic Cooperation and Development
ON	Octane number (average of MON + RON)
OPEC	Organization of Petroleum Exporting Countries
OTEC	Ocean thermal energy conversion
PPP	Purchasing Power Parity
PAH	Polynuclear aromatic hydrocarbons

PAN	Polyacrylonitrile
PC	Pitting corrosion
PCB	Polychlorinated biphenyl
PF	Phenol-formaldehyde
pH	$pH = -\log_{10} [H^+]$, neutral water has $pH = 7$
PMMA	Polymethylmethacrylate
ppm	parts per million, $\mu g/g$ or mL/m^3
PS	Polystyrene
PTFE	Polytetrafluoroethylene
PV	Pressure volume (product)
PVAc	Polyvinylacetate
PVC	Polyvinyl chloride
PWR	Pressurized water reactor
Quad Q,	Unit of energy, see Appendix A
Rad	Radiation absorbed dose
RBE	Relative biological effectiveness
RDX	Cyclonite or hexogen
Rem	Roentgen equivalent to man
rpm	Revolutions per minute
RO	Reverse osmosis
RON	Research octane number
SAN	Styrene-acrylonitrile copolymer
SCC	Stress corrosion cracking
SCE	Saturated calomel electrode
SHE	Standard hydrogen electrode
SI	Spark ignition
SIT	Spontaneous ignition temperature
SNG	Synthetic natural gas
SOAP	Spectrographic oil analysis program
SP	Smoke point
SRC	Solvent refined coal
SSPP	Solar sea power plants
STP	Standard condition of temperature and pressure, $0°C$ and 1 atm pressure
TDC	Top dead center
TDI	Toluene diisocyanate
TEL	Tetraethyl lead
TGA	Thermal gravimetric analysis
THM	Trihalomethanes
TLV	Threshold limit value
TML	Tetramethyl lead
TNG	Trinitroglycerol
TNT	Trinitrotolluene
TOE	Tons of oil equivalent (energy)
TW	Terawatts, 10^{12} W
UC	Uniform corrosion
UEL	Upper explosion limit
UFFI	Urea-formaldehyde foam insulation
UN	United Nations
UV	Ultraviolet light, $\lambda < 380$ nm

VCI	Vapor corrosion inhibitors
VI	Viscosity index
VOC	Volatile organic compounds
VOD	Velocity of detonation
WHO	World Health Organization

Chapter 1
Energy: An Overview

1.1 Introduction

World energy pundits have long proclaimed that the fossil fuels in the Earth's crust are limited and will be exhausted some day. This argument is similar to the accepted pronouncements of cosmologists that "the entropy of the Universe is increasing towards a maximum" or that "the sun will one day burn itself out." The world energy crisis of 1973 was precipitated when OPEC curtailed oil production and fixed their own price for oil for the first time, and within a year, the price of oil went from \$2/bbl in early 1973 to more than \$11/bbl by the spring 1974. This was aggravated by the increase in oil imports to the USA, which ceased to be self-sufficient in oil in the late 1960s. By 1990, the USA imported 47% of its crude oil, compared to 73% in 1973. In 2008, the USA bought 65% of its crude oil abroad, and the cost of imported oil and refined oil products was the single largest contributor—48%—to the country's more than \$700 billion trade deficit. By 1998, the average price of crude oil had fallen to less than \$12/bbl (a mere \$16/bbl in 2008 dollars), and the oil industry's stocks were one of the worst performing investments of the 1990s. Not until the early years of the new millennium, when oil prices began to rise again, did attention return to energy supply. During the latter half of 2003, the price of crude oil reached \$25/bbl–\$30/bbl, and during 2004, it came close to, and briefly even rose above, \$40/bbl. The upward trend continued in 2005 and for the first 8 months of 2006, and the media came to comment routinely on record high prices. According to Dr. Vaclav Smil's study of myths and realities of energy supply, no records were broken once two key price corrections—adjusting for the intervening inflation and taking into account lower oil intensity of Western economies—were made. Until the early summer 2008, these doubly adjusted oil prices remained well below the records set during the early 1980s.

In August 2006, the weighted mean price of all traded oil peaked at more than \$71/bbl; it then fell by 15% within a month and closed the year at about \$56/bbl. But during 2007, it again rose steadily. By November 2006, it reached almost \$100/bbl in trading on the New York Mercantile Exchange (NYMEX; see http://www.oilnergy.com/1onymex.htm and http://www.eia.gov/emeu/international/oilprice.html), and during the first half of 2008, that price rose by half, reaching a high of \$147.27/bbl on July 11, 2008. As always, prices for the basket of OPEC oils, including mostly heavier and more sulfurous crudes, remained lower (see www.opec.org and http://en.wikipedia.org/wiki/Brent_Crude). But just after setting a record, oil prices fell by more than 20%, to about \$115/bbl. By November 12, 2008, the price had fallen below \$50/bbl, and a year later, it was around \$75/bbl, a rise largely caused by the falling value of the US dollar.

The present world price for oil is about \$100–110/bbl in 2011 and it is expected to eventually increase again. In the long term, world liquids consumption increases after 2014 and rise to more than \$130 per barrel by 2035 (IEA Outlook 2010, p. 23).

O.V. Roussak and H.D. Gesser, *Applied Chemistry: A Textbook for Engineers and Technologists*, DOI 10.1007/978-1-4614-4262-2_1, © Springer Science+Business Media New York 2013

Table 1.1 GDP (PPP) and GDP (MER) top in the world 2009

GDP (PPP) top in the world 2009			GDP (MER) top in the world 2009		
Country	PPP current US$	World (%)	Country	GDP current US$	World (%)
USA	14,256	20.42	USA	14.256	24.61
China	8,765	12.56	Japan	5,068	8.75
Japan	4,159	5.96	China	4,909	8.47
India	3,526	5.05	Germany	3,353	5.79
Germany	2,806	4.02	France	2,676	4.62
UK	2,139	3.06	UK	2,184	3.77
Russia	2,110	3.02	Italy	2,118	3.66
France	2,108	3.02	Brazil	1,574	2.72
Brazil	2,013	2.88	Spain	1,464	2.53
Italy	1,740	2.49	Canada	1,336	2.13
Mexico	1,466	2.10	India	1,236	2.13
Korea	1,364	1.95	Russia	1,229	2.12
Spain	1,361	1.95	Australia	997	1.72
Canada	1,281	1.84	Mexico	875	1.51
Indonesia	962	1.38	Korea	833	1.44
World	*69,809*		*World*	*59,937*	

The global economic recession that began in 2008 and continued into 2009 had a profound impact on world income (as measured by gross domestic product GDP) and energy use. According to Table 1.1, world energy consumption increased by 49%, or 1.4% per year, from 495 quadrillion Btu in 2007 and is expected to need 739 quadrillion Btu in 2035 (IEA Outlook 2010, p. 9). The price of oil will depend on demand as well as the financial needs of the oil producers. The successful development of alternate energy sources, e.g., fusion, could bring the price down. Since energy is an integral part of every function and product from food (which requires fertilizer) to plastics which are petroleum based, to steel or other metals which require energy for extraction, beneficiation, reduction, and fabrication, worldwide inflation can be directly attributed to the rising price of oil.

The International Energy Agency (IEA) has estimated that world increasing demand for energy will require a total investment of $20 trillion (value in 2005 dollars) by 2030 out of which about $11 trillion would be needed in the global electricity sector alone. Worldwide, the race is on to increase exploitation of existing oil fields and to find new ones. Capital expenditure in the oil industry amounts to just one-fifth of the total energy investment.

Projected oil development programs in North America required a total investment of $856 billion over 2005. In order to restore Iraqi oil production to the 1990 level, some $5 billion was needed over the following 6 years, and, in a rapid growth scenario, production could reach 5.4 million barrels per day by 2030 at a cost of $54 billion. China will need a total of $7 trillion investments, which is 18% of the total investment. The IEA states that total investment of $20 trillion is required by the global oil and gas industry to keep pace with the anticipated demand over the next 30 years, of which about $700 billion is needed to support the Middle East oil sector. Oil from the Persian Gulf region will play an increasingly important role in the world economy.

Global investments in Russia's energy sector are projected to exceed $195 billion by 2030. Peak investment was in 2010 for prospecting new oil fields and gas reserves, maintaining old ones and improving the infrastructure for transporting oil. According to the IEA, the total level of investment in oil transportation will increase sixfold by 2030. Russia is poised to become one of the leading exporters of oil and gas by 2030, gaining an important niche in many markets, including Asia (see Figs. 1.1 and 1.2).

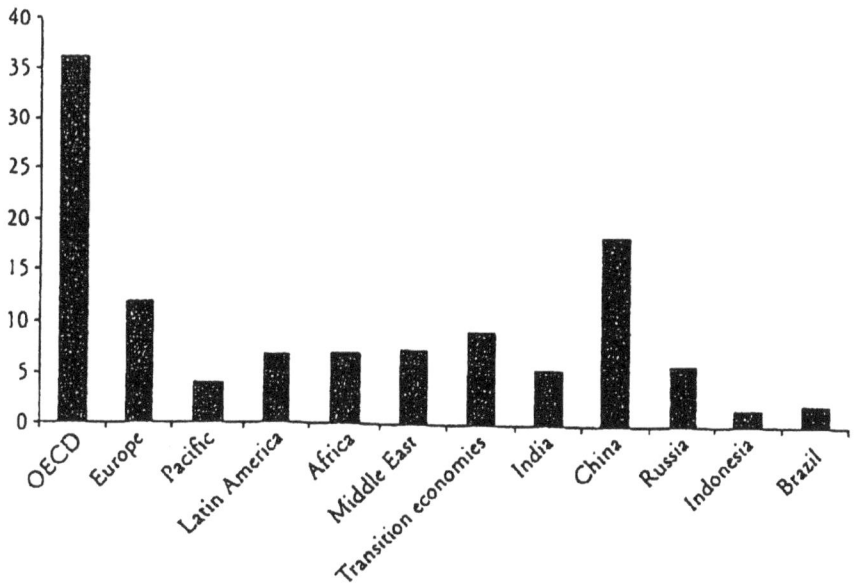

Fig. 1.1 Share of energy investment in different countries and groups, 2005–2030

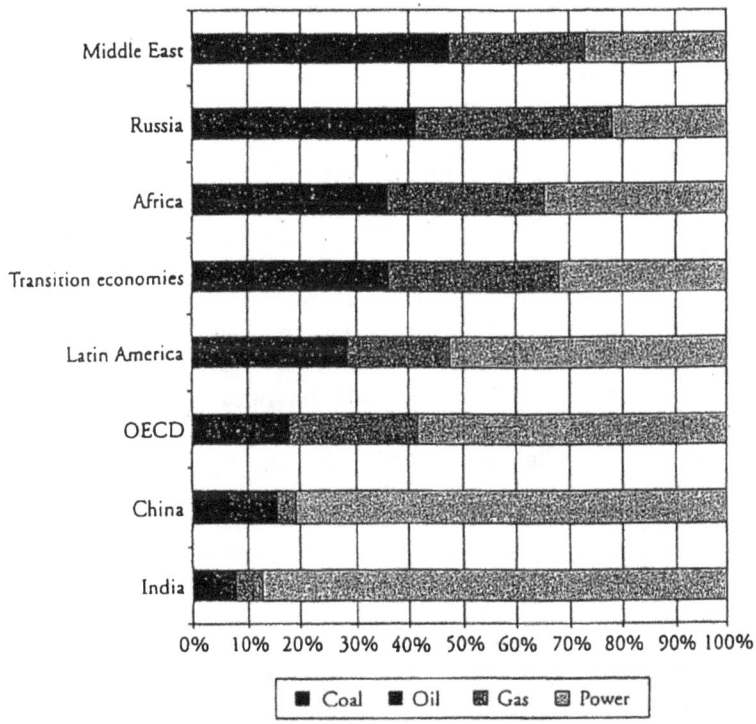

Fig. 1.2 The changing patterns of fuel share in energy investment requirements, 2005–2030 (Source: IEA 2006a)

Worldwide electricity demand will be increased to 5,100 GW electricity generating capacity by 2030 and about half of that needs to be built in Asia. Europe will need to invest about $1.7 trillion on power plants, transmission, and distribution to meet increasing demand for electricity and maintain the current capacity. Germany alone anticipates a new capacity around 40,000 MW in electricity

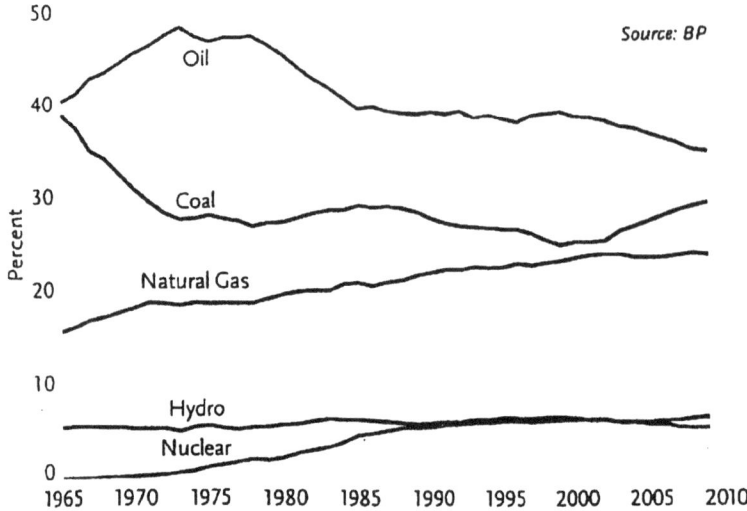

Fig. 1.3 The changing patterns of world energy consumption as a percentage of total usage, 1965–2009 (Source: BP)

production, which corresponds roughly to 60 large-scale power plants (IEA 2006). Developing countries will account for over half of the total investment over the next 20 years, or $10.5 trillion, with transition economies accounting for $1.85 trillion. Brazil's energy sector will need investments of $250 billion to meet the country's electricity demand in the next 20 years. More than $1 trillion will need to be spent on China's transmission and distribution networks—an amount equivalent to 2.1% of China's annual GDP. India will need an investment close to $100 billion in electric and oil sectors.

An illustration of the importance of energy to the world economy is shown in Fig. 1.3, where the changing patterns of world energy consumption as a percentage of total usage are plotted from 1965 to 2009. Comparison of the global per capita energy consumption and its patterns in major regions is shown in Fig. 1.4. In general, the higher the GNP[1] of a country, the larger is its per capita energy consumption. Energy is essential to progress and there is no substitute for energy. Society's use of energy has continuously increased, but sources have invariably changed with time.

It is interesting to note that the per capita use of commercial energy for the UK and the USA has been essentially constant for 100 years whereas that for Germany, Russia, and Japan showed an exponential growth (doubling time of 12 years) toward the constant US/UK values (see Tables 1.1 and 1.2). The world's population grew up fast from 2.5 billion people (1950) to 5 billion people (1987), 6 billion people (1997), and 6.8 billion people (2009). The world's population will be grown up about 9.1 billion people by 2050. The effect of the world's population growth on energy usage is obvious.

Energy can conveniently be classified into renewable and nonrenewable sources as shown in Fig. 1.5. Such a division is quite arbitrary and is based on a timescale which distinguishes hundreds of years from millions.

According to IEA Outlook 2010 in January 1, 2010, the world's total proved natural gas reserves were estimated at 6,609 trillion cubic feet. As of January 1, 2010, proved world oil reserves were estimated at 12 billion barrels (see Table A-1 in Appendix for the conversion of energy units and Tables 1.3, 1.4, and 1.5). The USA reported 22.5 billion barrels of proved reserves in 1998, proved reserves of 19.1 billion barrels were reported in 2009—a decrease of only 3.4 billion barrels despite the cumulative 24.2 billion barrels of liquids supplied from the US reserves between 1998 and 2009 (IEA Outlook 2010, p. 37).

---

[1] The gross domestic (national) product (GDP or GNP) is the sum total of the market value of goods and services produced per annum for final consumption, capital investment, or for government use.

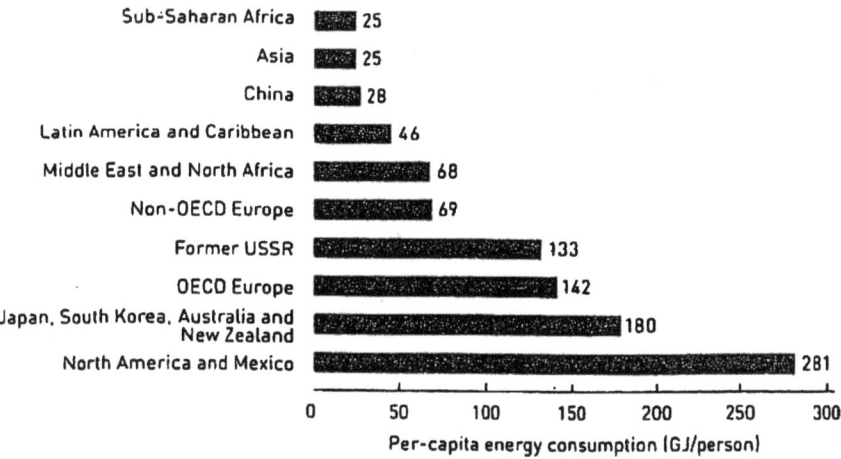

Fig. 1.4 Comparison of per capita energy consumption in major regions

Table 1.2 Comparison of the economic performance and per capita use of commercial energy in various countries

	USA	China	EU	Japan	India	Russia	Brazil
Economic size relative to the USA, current dollars, and market exchange rates 2009, %	100	33	113	35	9	9	10
GDP per capita relative to US, current exchange rates 2009, %	100	8	67	85	2	19	17
GDP per capita relative to the USA, PPP exchange rates 2009, %	100	14	66	71	6	32	23
Real GDP growth, 1999–2009 p.a. average, %	2.5	9.9	1.9	1.1	6.3	1.6	2.5
Population 2007, proportion of the USA	1.00	4.37	1.65	0.42	3.73	0.47	0.63
Share of world merchandise exports, 2008, excluding intra-EU exports, %	10.6	11.8	20.4	6.5	1.5	3.9	1.6
Trade to GDP ratio, 2004–2006 excluding intra-EU trade for EU, %	25.9	69.0	20.2	28.8	41.8	55.8	26.4
Share of world carbon dioxide emissions from energy consumption, 2007, %	20.1	21.0	15.7	4.2	4.7	5.6	1.3
Share of world military spending in US$, 2009, %	43.0	6.6	11.0	3.3	2.4	3.5	1.7
Deployed and other nuclear warheads, 2010	9,600	240	525	0	60–80	12,000	0

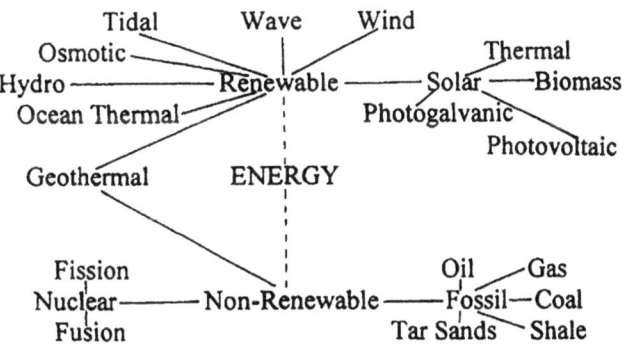

Fig. 1.5 A classification of energy sources

Table 1.3 World reserves of fossil fuels (IEA 2006)

Area	Natural gas, trillion cubic feet (2007)	Natural gas, quads (1998)	Oil, billion barrels (2007)	Oil, quads (1998)	Coal, million short tons (2005)	Coal, quads (1998)
North America	283.6	297	212.5	395	272,569	7,900
Latin America	240.7	221	102.7	430	17,941	660
Western Europe	180.3	166	15.8	110	50,781	2,755
East Europe, CIS	2,014.8	1,950	98.9	355	249,117	8,000
Africa	484.4	350	114.0	425	54,680	1,870
Middle East	2,556.0	1,723	739.2	3,770	1,528	5
Australia, Far East	419.5	430	33.4	290	283,807	8,910
Total	6,189.4	5,137	1,316.7	5,775	930,423	30,100

Table 1.4 World consumption of energy by sources in 1986, 1995, 1997, and 2006 (quads, IEA 2006)

	1986	%	1995	%	1997	%	2006	%
Oil	126.6	40.4	141.1	38.9	148.7	39.3	171.7	36.3
Coal	86.3	27.5	77.5	21.4	92.8	24.6	127.5	27
Natural gas	62.3	19.9	93.1	25.7	83.9	22.2	108.0	22.8
Hydroelectric energy	21.3	6.8	25.7	7.2	26.6	7.1	29.7	6.2
Nuclear energy	16.3	5.2	23.2	6.4	24.0	6.3	27.7	5.9
Renewable sources	0.7	0.2	1.6	0.4	1.8	0.5	4.7	1.0
Total	313.5	100	362.2	100	377.8	100	472.3	100

Table 1.5 World oil reserves by country as of January 1, 2010 (bbl—billion barrels)

Saudi Arabia	259.9 bbl/19.20%
Canada	175.2 bbl/12.94%
Iran	137.6 bbl/10.16%
Iraq	115.0 bbl/8.50%
Kuwait	101.5 bbl/7.50%
Venezuela	99.4 bbl/7.34%
United Arab Emirates	97.8 bbl/7.22%
Russia	60.0 bbl/4.43%
Libya	44.3 bbl/3.27%
Nigeria	37.2 bbl/2.75%
Kazakhstan	30.0 bbl/2.22%
Qatar	25.4 bbl/1.88%
China	20.4 bbl/1.51%
USA	19.2 bbl/1.42%
Brazil	12.8 bbl/0.95%
Algeria	12.2 bbl/0.90%
Mexico	10.4 bbl/0.77%
Angola	9.5 bbl/0.70%
Azerbaijan	7.0 bbl/0.52%
Norway	6.7 bbl/0.49%
Rest of world	72.2 bbl/5.33%
World total	1,353.7 bbl/100.00%

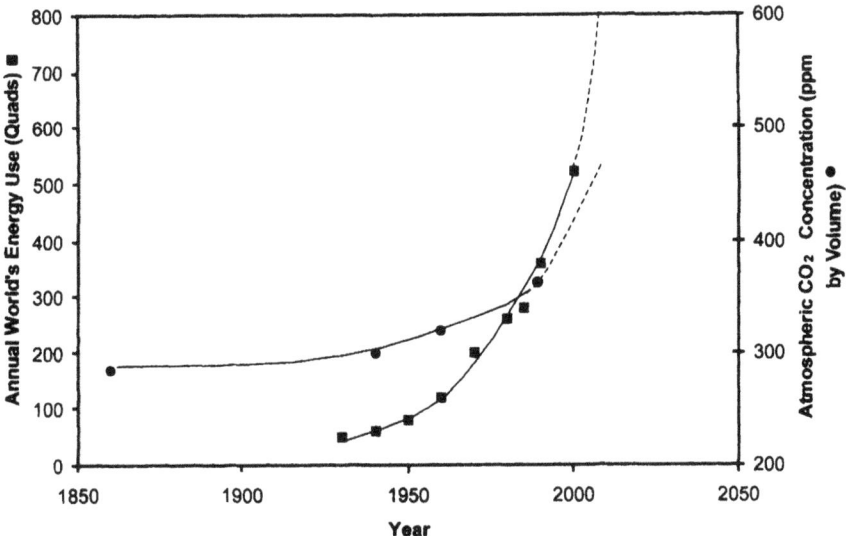

Fig. 1.6 The concentration of carbon dioxide in the Earth's atmosphere as a function of time

According to the *Oil & Gas Journal*, 56% of the world's proved oil reserves are located in the Middle East. Just below 80% of the world's proved reserves are concentrated in eight countries, of which only Canada (with oil sands included) and Russia are not OPEC members. Oil and gas production has still been slowly rising with oil production expected to peak in 2035. Gas production is expected to peak by 2050 and so will last only slightly longer, assuming that more oil and gas resources are made available. Coal is the major fossil fuel on Earth and consists of over 75% of the available energy (see Tables 1.3 and 1.4 and Fig. 1.2). Present world energy consumption is given in Table 1.4 and Fig. 1.4.

Conservative considerations of our energy consumption predict that coal will supply 1/4–1/3 of the world's energy requirements by the year 2050. Its use can be relied upon as an energy source for about another 200 years; however, other considerations (such as the greenhouse effect and acid rain) may restrict the uncontrolled use of fossil fuel in general and coal in particular.

The increase in use of fossil fuel during the past few decades has resulted in a steady increase in the CO_2 concentration in the atmosphere. This is shown in Fig. 1.6. In 1850, the concentration of CO_2 was about 200 ppm, and by the year 2025, the estimated concentration will be about double present values (350 ppm) if fossil fuels are burned at the present rate of 5 Gton of C/year. By the year 2025, the world's energy demands will have increased to over 800 Quads from 472 Quads in 2006 and 250 Quads in 1980, respectively. If a large fraction of this energy is fossil fuel, i.e., coal, then the annual increase in the concentration of CO_2 in the atmosphere is calculated to be greater than 10 ppm.

The CO_2 in the atmosphere is believed to have an adverse effect on the world's climate balance. The atmosphere allows the solar visible and near ultraviolet rays to penetrate to the Earth where they are absorbed and degraded into thermal energy, emitting infrared radiation which is partially absorbed by the CO_2, water vapor, and other gases such as CH_4 in the atmosphere (see Fig. 1.7).

There is at present a thermal balance between the constant energy reaching the Earth and the energy lost by radiation. The increase in CO_2 in the atmosphere causes an increase in the absorption of the radiated infrared from the Earth (black body radiation) and a rise in the thermal energy or temperature of the atmosphere. This is called the *greenhouse effect*. Temperature effects are difficult to calculate and estimates of temperature changes vary considerably, although most agree that a few degrees rise in the atmospheric temperature (e.g., 3°C by the year 2025) could create deserts out of the prairies and convert the temperate zones into tropics, melt the polar ice caps, and flood coastal areas. For example,

Fig. 1.7 The greenhouse effect—a schematic representation

for a 1°C rise in the Earth's temperature, the yield of wheat would be expected to drop by 20%, though rice yields might rise by 10%. If the average temperature of the oceans increased by 1°C, the expansion would cause a rise in sea level of about 60 cm (assuming no melting of glacier ice).

One uncertain factor in the modeling and predictions is that there is a lack in a material balance for CO_2, i.e., some CO_2 is unaccounted for indicating that some CO_2 sinks (i.e., systems which hold or consumed CO_2) have not been identified. The oceans and forests (biomass growth) consume most of the CO_2, and it is possible that these sinks for CO_2 may become saturated or on the other hand some new sinks may become available. With such uncertainties, it is obvious that reliable predictions cannot be made. However, the climate changes which will occur as the CO_2 concentration increases are real and a threat to world survival. Recent measurements by satellites of the temperature of the upper atmosphere over a 10-year period have indicated no overall increase in temperature. This measurement has yet to be confirmed.

Coal is considered the "ugly duckling" of fossil fuel as it contains many impurities which are released into the atmosphere when it is burned. An important impurity is sulfur which introduces SO_2 and SO_3 into the atmosphere, resulting in acid rain that can actually change the pH of lakes sufficiently to destroy the aquatic life. The acid rain is also responsible for the destruction of the forests in Europe and the eastern parts of Canada and the USA. The clean conversion of coal to other fuels may circumvent the pollution problems, but would not overcome the greenhouse effect since CO_2 ultimately enters the atmosphere.

Thus, the depletion of fossil fuels may not be soon enough and tremendous efforts are being made in the search for viable economic alternatives such as nuclear energy or renewable energy such as solar, wind, tidal, and others (Table 1.6).

1.2 Renewable Energy Sources

Ultimate sources of renewable energy are the Earth, which gives rise to geothermal energy, the Moon, which is responsible for tidal power, and the Sun, which is the final cause of all other hydro, wind, wave, thermal, and solar photodevices. A brief discussion of each source is essential for an overall appreciation of the difficulties we are facing and possible solutions to our energy requirements.

Table 1.6 World natural gas reserves by country as of January 1, 2010 (trillion cubic feet)

World	6,609 trillion cubic feet 100.0%
Top 20 countries	6,003 trillion cubic feet 90.8%
Russia	1,680 trillion cubic feet 25.4%
Iran	1,046 trillion cubic feet 15.8%
Qatar	899 trillion cubic feet 13.6%
Turkmenistan	265 trillion cubic feet 4.0%
Saudi Arabia	263 trillion cubic feet 4.0%
USA	245 trillion cubic feet 3.7%
United Arab Emirates	210 trillion cubic feet 3.2%
Nigeria	185 trillion cubic feet 2.8%
Venezuela	176 trillion cubic feet 2.7%
Algeria	159 trillion cubic feet 2.4%
Iraq	112 trillion cubic feet 1.7%
Australia	110 trillion cubic feet 1.7%
China	107 trillion cubic feet 1.6%
Indonesia	106 trillion cubic feet 1.6%
Kazakhstan	85 trillion cubic feet 1.3%
Malaysia	83 trillion cubic feet 1.3%
Norway	82 trillion cubic feet 1.2%
Uzbekistan	65 trillion cubic feet 1.0%
Kuwait	63 trillion cubic feet 1.0%
Canada	62 trillion cubic feet 0.9%
Rest of world	606 trillion cubic feet 9.2%

1.3 Geothermal

Thermal energy from within the Earth's crust is classified as geothermal energy. At depths greater than 10 km, the temperature of the magma is above 1,000°C and is a potential source not yet fully exploited. The temperature of the Earth's core is about 4,000°C. Drilling to depths of 7.5 km is presently possible and may someday reach 15–20 km. The surface source of thermal energy is due to the decay of natural radioactive elements and to the frictional dissipation of energy due to the movement of plate tectonics. The heat is usually transmitted to subsurface water which is often transformed into steam that can force water to the surface. Old Faithful at Yellowstone National Park, WY, USA, is an example of a geyser erupting 50 m every hour for 5 min (see Fig. 1.8).

The International Geothermal Association (IGA) has reported that 10,715 MW of geothermal power in 24 countries is online, which is expected to generate 67,246 GWh of electricity in 2010. This represents a 20% increase in geothermal power online capacity since 2005. IGA projects will grow to 18,500 MW by 2015, due to the large number of projects presently under consideration, often in areas previously assumed to have little exploitable resource. In 2010, the USA led the world in geothermal electricity production with 3,086 MW of installed capacity from 77 power plants; the largest group of geothermal power plants in the world is located at the Geysers a geothermal field in California. The Philippines follows the USA as the second highest producer of geothermal power in the world, with 1,904 MW of capacity online; geothermal power makes up approximately 18% of the country's electricity generation. Last January 2011, Al Gore said in The Climate Project Asia Pacific Summit that Indonesia could become a super power country in electricity production from geothermal energy. Geothermal energy is exploited near San Francisco where a 565-MW power plant is run on geothermal steam, and 15 MW of thermal energy from hot water reservoirs is used for heating and industrial heat processes. At present, there are five geothermal plants in operation in Mexico with a total output of more than 500 MW. Similar uses of geothermal energy have been developed in Italy, Japan, Iceland, USSR, and New Zealand, and it is rapidly being exploited in many other parts of the world.

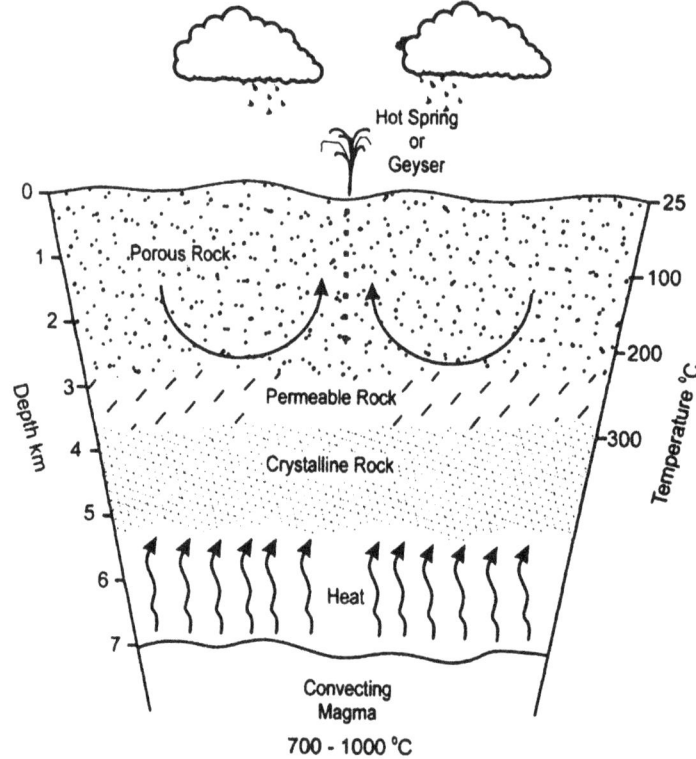

Fig. 1.8 A schematic representation of geothermal energy

It has been estimated that the geothermal energy in the outer 10 km of the Earth is approximately 10^{23} kJ or about 2,000 times the thermal energy of the total world coal resources. However, only a small fraction of this energy would be feasible for commercial utilization. Estimates of geothermal energy presently in use and converted into electrical power is about 2×10^3 MW. The greater use of geothermal energy could help save much of our energy needs and would reduce the rate of increase of CO_2 in the atmosphere.

1.4 Tidal Power

Tidal power is believed to have been used by the Anglo Saxons in about 1050. Tidal power is a remarkable source of hydroelectrical energy. The French Ranee River power plant in the Gulf of St. Malo in Brittany consists of 24 power units, each of 10 MW. A dam equipped with special reversible turbines allows the power to be generated by the tidal flow in both directions.

Several tidal projects have been in the planning stages for many years and include the Bay of Fundy (Canada-USA), the Severn Barrage (Great Britain), and San Jose Gulf (Argentina). Though tidal power is reliable, it is not continuous, and some energy storage system would make it much more practical. However, as the price of fossil fuels rises, the economics of tidal power becomes more favorable.

1.5 Solar Energy

The Sun is approximately 4.6×10^9 years old and will continue in its present state for another 5×10^9 years. The Sun produces about 4×10^{23} kJ/s of radiant energy, of which about 5×10^{21} kJ/ year reaches the outer atmosphere of the Earth. This is about 15,000 times more than man's present use of energy on Earth. We are fortunate however that only a small fraction of this energy actually reaches the Earth's surface (see Fig. 1.7). About 30% is reflected back into space from clouds, ice, and snow; about 23% is absorbed by O_2, O_3, H_2O, and upper atmosphere gases and dust; and about 47% is absorbed at or near the Earth's surface and is responsible for heating and supporting life on Earth. Of the energy absorbed by the Earth, about 56% is used to evaporate water from the sea and plants (evapotranspiration). Another 10% is dissipated as sensible heat flux. The remainder is radiated back into space, about 10% into the upper atmosphere and about 24% is absorbed by our atmosphere. A small but important fraction of the Sun's energy, about 0.2%, is consumed in producing winds and ocean waves. An even smaller fraction, 0.02%, is absorbed by plants in the process of photosynthesis of which about 0.5% of the fixed carbon is consumed as nutrient energy by the Earth's 6×10^9 people. The variation in solar intensity reaching the Earth due to its elliptical orbit about the Sun is only 3.3%. The production of fixed carbon by photosynthesis is about ten times present world consumption of energy by human society. Thus, solar energy is sufficient for man's present and future needs on Earth. The main difficulty is in the collection and storage of this energy.

Solar energy can be utilized directly in flat bed collectors for heating and hot water or concentrated by parabolic mirrors to generate temperatures over 2,000°C. The thermal storage of solar energy is best accomplished with materials of high heat capacity such as rocks, water, or salts such as Glauber's salt, which undergo phase changes, e.g.,

$$Na_2SO_4 \bullet 10H_2O_{(s)} \rightarrow Na_2SO_{4(s)} + 10H_2O_{(l)} \quad \Delta H = 81.5 \text{ kJ/mol of salt} \tag{1.1}$$

(The transition temperature for Glauber's salt is 32.383°C).

Solar energy can also be directly converted into electrical energy by photovoltaic and photogalvanic cells or transformed into gaseous fuels such as hydrogen by the photoelectrolysis or photocatalytic (solar) decomposition of water.

The Sun consists of about 80% hydrogen, 20% helium, and about 1% carbon, nitrogen, and oxygen. The fusion of hydrogen into helium, which accounts for the energy liberated, can occur several ways. Two probable mechanisms are:

The Bethe mechanism: (1939)	Q (MeV)
$^{12}C + {}^1H \rightarrow {}^{13}N + \gamma$	1.94
$^{13}N \rightarrow {}^{13}C + c^+ + v$ ($t_{1/2} = 9.9$ min)	1.20
$^{13}C + {}^1H \rightarrow {}^{14}N + \gamma$	7.55
$^{14}N + {}^1H \rightarrow {}^{15}O + \gamma$	7.29
$^{15}O \rightarrow {}^{15}N + c^+ + v$ ($t_{1/2} = 2.2$ min)	1.74
$^{15}N + {}^1H \rightarrow {}^{12}C + {}^4He$	4.96
$4{}^1H \rightarrow {}^4He + 2e^+ + 2v$	24.68
$2e^+ + 2e^- \rightarrow 2\gamma$ ($+2 \times 1.02$)	2.04
Total energy	26.72

The Salpeter mechanism: (1953)	Q (MeV)
$^1H + {}^1H \rightarrow {}^2H + e^+ + v$	0.42
$^2H + {}^1H \rightarrow {}^3He + \gamma$	5.49
$^3He + {}^3He \rightarrow {}^4He + 2{}^1H$	12.86
$4{}^1H \rightarrow {}^4He + 2e^+ + 2v$	24.68

Both reactions occur, though the Bethe mechanism requires a higher temperature and therefore predominates in the central regions of large stars.

The solar constant is 2.0 cal/cm^2 min or 1,370 W/m^2 above the Earth's atmosphere and about 1.1 kW/m^2 normal to the Sun's beam at the equator. At other latitudes, this value is reduced due to the filtering effect of the longer atmospheric path.

Ideal sites for solar energy collection are desert areas such as one in northern Chile which has low rainfall (1 mm/year) and 364 days/year of bright sunshine. The Chile site (160 × 450 km^2) receives about 5 × 10^{17} kJ/year (1 kJ/m^2/h × 60 min/h × 8 h/day × 365 day/year × 72,000 km^2 × 10^{6}m^2/km^2). This is about a third of the world's use of energy in 1995. Thus, theoretically, the desert areas or nonarid lands could be used to supply the world with all its energy requirements, and there is no doubt that before the next century has passed solar energy will probably dominate a large portion of the world's energy sources.

Figure 1.5 shows the subclassification of solar energy into thermal, biomass, photovoltaic, and photogalvanic. The most familiar aspect of solar energy is the formation of biomass or the conversion of carbon dioxide, water, and sunlight into cellulose or food, fuel, and fiber. Thus, wood was man's major fuel about 200 years ago to be displaced by coal, the modified plants of previous geological ages. Wood is a renewable energy source but it is not replenished quickly enough to be an important fuel today. A cord of wood is 128 ft^3 (8′ × 4′ × 4′) of stacked firewood. It is not recognized as a legal measure. A cord contains about 72 ft^3 of solid wood or about 4,300 lb to which must be added about 700 lb of bark or a total of about 5,000 lb, varying with the wood and its moisture content. The thermal energy of wood is from 8,000 to 9,000 Btu/lb. It has been argued that biomass used for fuel is not practicable because it displaces land which could be used for agriculture—a most essential requirement of man whose nutritional demands are continuously increasing. This objection is not valid if the desert is used, as in the case of the Jojoba bean which produces oil that has remarkable properties, including a cure for baldness, lubrication, and fuel.

Plants which produce hydrocarbons directly are well known—the best example is the rubber tree which produces an aqueous emulsion of latex—a polymer of isoprene (mol. wt. 2 × 10^6 D) (see Fig. 1.9). The annual harvest of rubber in Malaysia was 200 lb/acre/year before World War II, but by improving plant breeding and agricultural practices, the production has increased to ten times this value.

Melvin Calvin, Nobel Prize winner in Chemistry in 1961 for his work on the mechanism of photosynthesis, has been one of the principal workers in the search for plants which produce more suitable hydrocarbons, e.g., a latex with a mol. wt. of 2,000 Da which can be used as a substitute for oil. One plant he has studied, Euphorbia *(E. lathyris)* yields, on semiarid land, an emulsion which can be converted into oil at about 15 bbl/acre. Another tree, Copaiba, from the Amazon Basin, produces oil (not an aqueous emulsion) directly from a hole drilled in the trunk about l m from the ground. The yield is approximately 25 L in 2–3 h every 6 months. This oil is a C$_{15}$ terpene (tri-isoprene) which has been used in a diesel truck (directly from tree to tank) without processing.

Recent studies have shown that oils extracted from plants such as peanuts, sunflowers, maize, soya beans, olives, palm, corn, rapeseed and which are commonly classed as vegetable oils in the food industry can be used as a renewable fuel. These oils are composed primarily of triglycerides of long chain fatty acids. When used directly as a diesel fuel, they tend to be too viscous, clog the jet orifices, and deposit carbon and gum in the engine. Some improvement is obtained by diluting the oil with alcohol or regular diesel fuel or by converting the triglycerides into the methyl or ethyl esters. This is done in two steps: (1) hydrolysis and (2) esterification. The methyl or ethyl ester produced is more volatile and less viscous but is still too expensive to burn as a fuel.

The energy ratio for biomass energy, i.e., the energy yield/consumed energy for growth and processing, is variable and usually between 3 and 10. Plant breeding and genetic engineering should greatly improve this ratio.

Fig. 1.9 A tapper at Goodyear's Dolok Merangir rubber plantation on the Indonesian island of Sumatra uses an extension knife to draw latex from a rubber tree. The bark of a rubber tree is cut up part of the year and down the rest to allow the tree to replenish itself. The rubber is sold by Goodyear to other manufacturers for making such diverse products as surgical gloves, balloons, overshoes, and carpet backing (Courtesy of Goodyear, Akron, OH)

Grain, sugarcane, and other crops containing carbohydrates can be harvested for the starch and sugar which can be fermented to ethyl alcohol. The residue which is depleted in carbohydrates but richer in protein is still a valuable feed stock.

Thus, the Energy Farm, where a regular crop can be utilized as a fuel, is obviously a requirement if stored solar energy is to replace dwindling fossil fuels.

1.6 Photovoltaic Cells

The direct conversion of solar energy into electrical energy is accomplished by certain solid substances, usually semiconductors (see Chap. 18), which absorb visible and near ultraviolet (UV) light, and by means of charge separation within the solid lattice a voltage is established. This generates a current during the continuous exposure of the cell to sunlight. Typical solar cells are made of silicon, gallium arsenide, cadmium sulfide, or cadmium selenide. The main hindrance to

Fig. 1.10 Part of a typical
solar cell 100 mm in diameter

widespread use of solar cells is their high cost, which is at present about \$400/m^2 of amorphous silicon (13% efficient). A price of \$50/m^2 would make such solar cell economic and practical. The reduction in cost to some composite cells of CdS/Cu_2S has been reported. Even lower costs may be expected as a result of the major efforts being made to develop inexpensive methods of forming the polycrystalline or amorphous materials by electroplating, chemical vapor deposition, spray painting, and other processes to dispense with the expensive single crystal wafers normally used. A typical photovoltaic cell is shown in Fig. 1.10.

A 6 × 9-m2 panel of solar cells operating at 10% efficiency with a peak output capacity of 5 kW at midday would yield an average of 1 kW over the year—more than the electrical energy requirements of an average home if electrical storage was utilized to supply energy for cloudy and rainy days and during the night. More recently, a solar powered airplane crossed the English Channel using photovoltaic cells to power an electrical motor. This clearly demonstrates the potential power of solar energy.

1.7 Photogalvanic Cells

Cells in which the solar radiation initiates a photochemical reaction, which can revert to its original components via a redox reaction to generate an electrochemical voltage, are called *photogalvanic cells*. This is to be distinguished from a solar rechargeable battery where light decomposes the electrolyte which can be stored and recombined to form electrical energy via an electrochemical cell, e.g.,

$$FeBr_3 + h\nu \rightarrow FeBr_2 + 1/2Br_2$$
$$FeBr_2 + 1/2Br_2 \rightarrow FeBr_3 \ \zeta° = 0.316 \text{ V} \tag{1.2}$$

Photogalvanic cells usually consist of electrodes which are semiconductors and a solution which can undergo a redox reaction. The band gap of the semiconductor must match the energy of the redox reaction before the cell can function. Light absorbed by the electrodes promotes electrons from the valence band to the conduction band where they migrate to the surface (in n-type semiconductors) where reaction with the electrolyte can occur. This is shown in Fig. 1.11 for the system in which two photochemically active semiconductor electrodes are used, one in which the *p*-type oxidizes Fe(II) to Fe(III):

$$Fe^{2+} \rightarrow Fe^{3+} + e^- \tag{1.3}$$

The reverse reaction occurs at the other n-type electrode. Many such cells have been prepared but the efficiency is very low due to the limited surface area of the electrodes. More recently, porous transparent semiconductor electrodes have been made which can increase efficiency by some orders of magnitudes and it remains to be seen if these systems are stable over long periods. Such cells when shorted can be used for the photoelectrolysis of water or the production of hydrogen, but more will be said about this later.

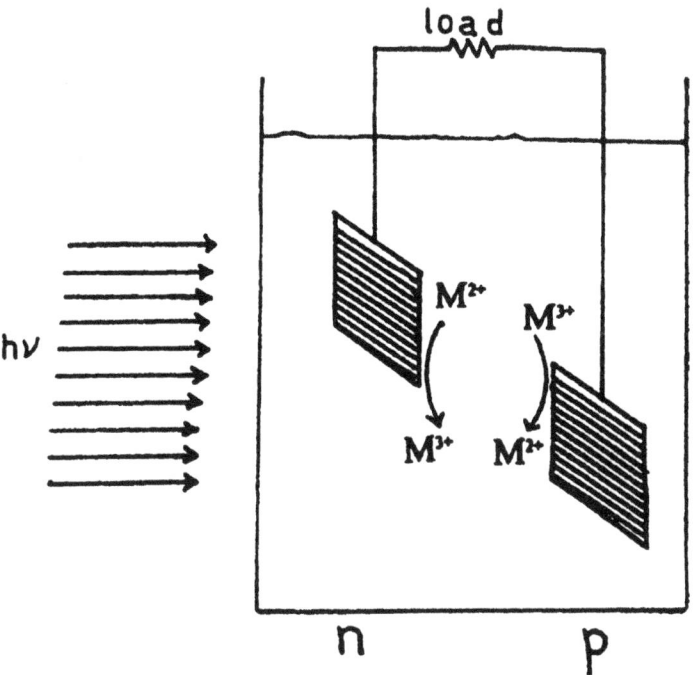

Fig. 1.11 A photogalvanic cell using n-type and p-type semiconductor electrodes and a regenerating redox system $(M^{3+} \rightleftharpoons M^{2+} + e)$ to carry the current

1.8   Wind Energy

Wind first powered sailing ships in Egypt about 2500 B.C. and windmills in Persia about 650 A.D. The use of windmills for the grinding of grain was well established in the Low Countries (Holland and Belgium) by 1430 where they are still used to this day.

The maximum power available from a horizontal axis windmill is given by

$$P_{max} = \frac{1}{2}\rho A V^3 \tag{1.4}$$

where ρ = density of air, A = cross-sectional area of the windmill disk, and V = air velocity.

However, to maintain a continuous air flow past the windmill implies that the extractable power, P_{ext}, must be less than P_{max} and it can be shown that $P_{ext} = (16/27)P_{max}$ or 59.3% is the optimum extractable power available and usually 70% of this value is realized practically. The energy which is usually extracted by an electrical generator or turbine can be stored in a bank of batteries for future use. Several very large windmills have been built to generate electrical energy, but costs are still too high to make them commonplace. The vertical axis windmill is much simpler than the horizontal axis type, but large units have not as yet been tested.

Wind farms have been successfully operating in California where many small windmills are located on exposed terrain. The continuous wind at about 17 miles/h is sufficient to make the generation of electrical energy a viable project. Many of the largest operational onshore wind farms are located in the USA. As of November 2010, the Roscoe Wind Farm is the largest onshore wind farm in the world at 781.5 MW, followed by the Horse Hollow Wind Energy Center (735.5 MW). Also, the largest wind farm under construction is the 800 MW Alta Wind Energy Center in the USA. As of November 2010, the Tranet Offshore Wind Project in the UK is the largest offshore wind farm in the world at 300 MW, followed by Horns Rev II (209 MW) in Denmark. The largest proposed project is the 20,000 MW Gansu Wind Farm in China.

1.9 Hydropower

Hydropower relies on the conversion of potential energy into kinetic energy which is used to turn an electrical generator and turbines. The development of highly efficient turbines has increased the output of power stations and allowed for their installation in many new areas. The main disadvantage in hydropower is that the transmission of electrical energy over long distances results in losses which effectively place a limit on such distances. Also, it is not convenient, because of environmental factors, to store the potential energy or the electrical energy. Several methods have been used, such as storing water behind a dam, pumping the water into another reservoir, compressing air in a large cavern, or electrolyzing water and transporting the hydrogen in a pipeline. This latter alternative has an interesting by-product, namely, heavy water which is essential for the CANDU nuclear reactor. During the electrolysis of water, the lighter isotope ^{1}H is liberated as H_2 more readily than the heavier ^{2}H or deuterium D_2 gas. Hence, the heavy water accumulates in the electrolyte and can eventually be purified by distillation.

The Yenisey River in Siberia passes north through the City of Krasnoyarsk. Two hydro dams on the Yenisey River each produce 6,000 MW which is primarily used in the electrolytic production of aluminum from imported bauxite ore. Such power source can also be used to make hydrogen and other electrochemical products.

1.10 Ocean Thermal

The oceans cover over 70% of the Earth's surface and are continually absorbing solar radiation. The penetration of the solar energy is only 3% at 100 m. This results in a temperature gradient which can be used to generate electrical energy. A variety of schemes have been proposed, and some experimental units have been tested as early as 1929 off Cuba and more recently near Hawaii. The main disadvantages to Ocean Thermal Energy Conversion (OTEC) or Solar Sea Power Plants (SSPP) are (1) the need to operate turbines with a very low temperature gradient of about 15–20°C, (2) the corrosive nature of seawater, and (3) the usually large distances from shore that the plant has to be located. The Carnot thermodynamic efficiency[2] of the heat engine is only about 3%, but by using ammonia or a low boiling organic compound as the boiler fluid, it is possible to approach maximum conversion efficiency.

It is possible to magnify the solar thermal gradient by using nonconvecting solar ponds. Much of the recent development work has been done by Tabor in Israel where Dead Sea brine is used to establish a density gradient in a pond approximately 1 m deep. The bottom of the pond is dark and absorbs the solar energy, heating the denser lower layer. With temperature differences of more than 50°C, it is possible to achieve a Carnot efficiency of over 13%. A schematic diagram of a solar pond is shown in Fig. 1.12 with an estimate of the operating parameters shown in Table 1.7. The diffusion of

---

[2] The Carnot efficiency is the theoretical maximum efficiency by which a heat engine can do work when operating between two temperatures. The greater the difference in temperatures, the greater is the efficiency by which heat can be converted into work.

Fig. 1.12 Schematic representation of the "falling pond" method of extracting heat from the bottom of a pond (From Tabor H (1980) Non-convecting solar ponds. Phil Trans R Soc Lond A295:428 with permission)

Table 1.7 Estimated annual yield of a solar pond

Area	1 km^2
1. Insulation	2,000 GWh(t)
2. Pond heat yield	400 GWh(t)
3. Equivalent fuel oil	43,000 t
Power yield	33 GWh(e)
Source temperature	87°C
Sink temperature	30°C
H-X drops	10° total
Carnot effect	13.2%
Turbine factor	0.6
Overall thermal energy	ca. 8%
Equivalent cont. power	0.4 MW
(58% load factor)	$4.7 \times 10^4 \, m^3$

salt tends to destroy the density gradient and hence will allow convection to upset the temperature gradient. To maintain the density gradient, salt-free water is added to the surface and salt-enriched water is fed into the bottom layer. The flash chamber can also yield desalinated water as a by-product as indicated in Fig. 1.12 and Table 1.7. The electrical energy is obtained from a Rankine cycle turbine which was developed for such solar energy conversion and which runs on an organic vapor.

The problems associated with solar ponds involve: (1) Surface mixing due to winds which tend to create convecting zones near the surface. This can be reduced by adding a floating netting grid to the surface. A floating plastic sheet cannot be used because it acts as a collector of dust and becomes opaque in a few days. (2) Ecological factors must be considered since brine may be slowly lost through the bottom of the pond to the aquifer. Hence, the location of the pond in flat sterile land is preferred.

Solar ponds are presently being tested in Israel where plans are in progress for a large unit located at the Dead Sea.

1.11 Wave Energy

Wave energy has been described as liquid solar energy since it originates from the Sun, which causes wind, which in turn forms waves. Wave energy has enormous potential and presents a tremendous challenge to the engineer. Its presence on the high seas is almost continuous. It has been estimated that the power available along a kilometer of shore can be over 20 MW. Several devices have been tested and usually operate air turbines which generate electricity. The power takeoff and mooring are still problems to be solved. An example of a simple wave energy device is shown in Fig. 1.13. A one-way flap allows a head of water to be stored over 5–100 waves. When the full head is reached, the water is released and allowed to drive the turbine as it empties.

As a renewable energy source, wave power is an untapped source which could help alleviate the rising cost of energy and its continued development must be encouraged.

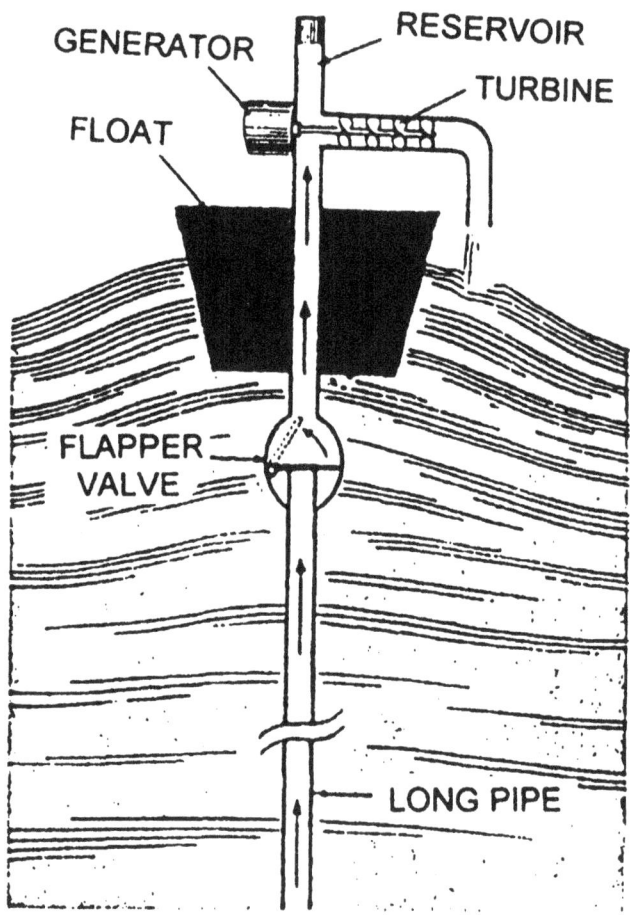

Fig. 1.13 Wave-powered generators convert the kinetic energy of waves into electricity. The Scripps wave pump amplifies the effective height of a wave by trapping water in a reservoir on each downward plunge of the buoy. Water in the long standpipe surges through a one-way flapper valve as it oscillates out of phase with the up-and-down motion of the waves. When enough water is collected after about eight to ten waves, it is released to spin the turbine-generator

1.12 Osmotic Power

As a river empties into the sea, it is possible to extract hydro power by converting gravitational potential energy into kinetic energy—mechanical energy and electrical energy. At the same time, the difference in salt concentration between the river and the sea results in a difference in chemical potential or free energy (ΔG) which can be exploited by means of the osmotic pressure.

The osmotic pressure of a solution is the pressure exerted by the solution on a semipermeable membrane which separates the solution from pure solvent. The semipermeable membrane allows only the solvent to pass freely through it while opposing the transport of solute. This is illustrated in Fig. 1.14. The solvent tends to dilute the solution and passes through the membrane until equilibrium is established, at which point a pressure differential, it, exists between solution and solvent. This pressure differential is called the *osmotic pressure*.

At equilibrium, the osmotic pressure, π, of a solution relative to the pure solvent is given by a relation which resembles the Ideal Gas Law *(PV $-$ nRT)* where

$$nV - nRT \quad \text{or} \quad n = (n/V)RT = CRT \tag{1.5}$$

where V = volume of solution, R = ideal gas constant (0.0821 L atm/°K mol), C = concentration of solute (molar mass/liter), n = mols of dissolved solute, ions, etc., and T = absolute temperature $(273° + 10°) = 283$ K.

The average molar concentration of salt in seawater is about 0.5 M and since it may be assumed that the salt is primarily NaCl, then the total molar concentration of solute is 1.0 M (i.e., 0.5 M Na^+ + 0.5 M Cl^-). Hence,

$$\pi = 1.0 \text{ mol/L} \times 0.0821 \text{ L, atm/K, mol} \times 283 \text{ K} - 23 \text{ atm}$$

This osmotic pressure represents a hydrostatic head of water of about 700 ft (30 ft/atm $\times$ 23 atm) or over 200 m. A schematic diagram of osmotic power is illustrated in Fig. 1.15 where seawater is pumped into a pressure chamber at a constant rate depending on the flow of the river. The river passes through the membrane, diluting the seawater and creating the hydrostatic head which can then turn a water wheel and generate electricity. The membrane area must be enormous to accommodate the

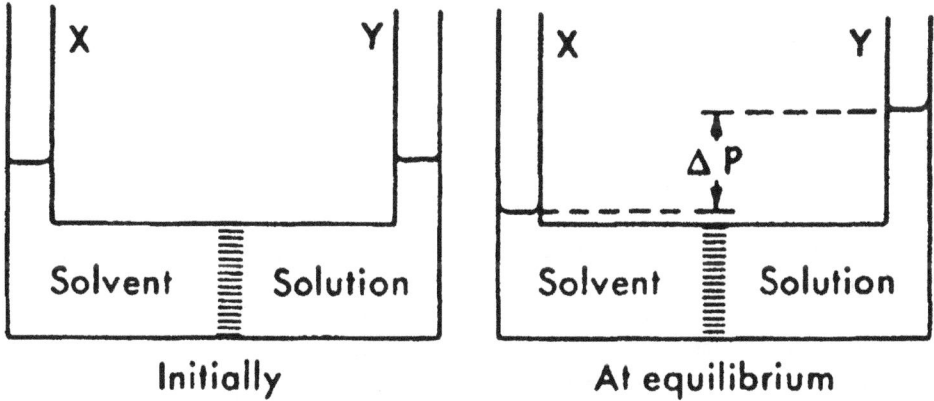

Fig. 1.14 The development of osmotic pressure is illustrated by the difference between an initial state and the final equilibrium state. Solvent, but not solute, passes through the semipermeable membrane, tending to dilute the solution and thereby allowing a differential pressure ΔP to develop. At equilibrium, the differential hydrostatic pressure is equal to the osmotic pressure

Fig. 1.15 Diagram of an osmotic salination energy converter to extract power from the natural flow of freshwater into the sea

Table 1.8 Potential power due to salinity gradients

Source	Flow rate (m³/s)	Power (W)
$\pi = 23$ atm		
Amazon River (Brazil)	2×10^5	4.4×10^{11}
La Plata-Parana River (Argentina)	8×10^4	1.7×10^{11}
Congo River (Congo/Angola)	5.7×10^4	1.2×10^{11}
Yangtze River (China)	2.2×10^4	4.8×10^{10}
Ganges River (Bangladesh)	2×10^4	4.4×10^{10}
Mississippi River (USA)	1.8×10^4	4.0×10^{10}
US waste water to oceans	500	1.1×10^9
Global runoff	1.1×10^6	2.6×10^{12}
$\kappa = 500$ atm		
Salt Lake		5.6×10^9
Dead Sea		1.8×10^9

permeability of the river. An estimate of power output is 0.5 MW/m³ of input flow resulting in an amortized cost of about 5 cents/kWh. Lower costs are expected if the salt concentration gradient is higher, e.g., seawater ($\pi = 23$ atm) emptying into the Dead Sea ($n = 500$ atm) to produce electrical power from the gravitational as well as the chemical potential. The potential power from various rivers is listed in Table 1.8.

Only with the developments in membrane technology as a result of work in reverse osmosis will osmotic power become a significant factor in world energy supply.

It has been estimated by the Scripps Institution of Oceanography that world power needs in the year 2000 will be about 33 million megawatts (33 TW). The seas can provide all this and more: wave energy 2.5 TW, tidal power 2.7 TW, current power 5 TW, osmotic power 1,400 TW, OTEC 40,000 TW. The Weitzman Institute of Science has studied this topic and has shown that with a suitable membrane, it would be a feasible source of energy.

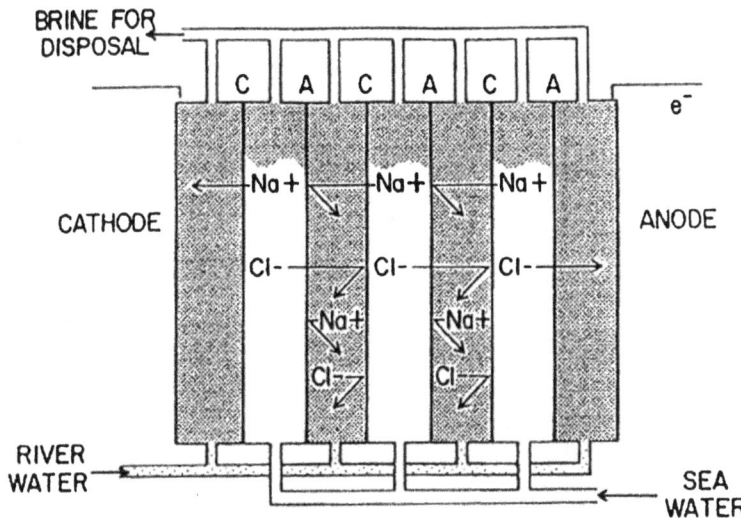

Fig. 1.16 The conversion of chemical potential into electrical energy by reverse electrodialysis

With such optimistic projections and present-day technology, it may soon be possible to rely on our renewable energy resources with confidence and assurance so long as we continue to minimize the environmental effects (Fig. 1.16).

Exercises

1. Using Internet, compare GDP per capita (PPP, 2012) for the following countries: Australia, Brazil, Canada, China, Finland, Germany, New Zeeland, Norway, Japan, Russia, South Korea, the UK, and the USA. Give the source.

2. Explain why geothermal energy is classed as both a renewable and nonrenewable energy source. Do you agree with this explanation?

3. Complete Table 1.3 with references to uranium available as a nuclear fuel, 860 Quads from the Earth and 10^5 Q from seawater.

4. It has been argued that the CO_2 absorption band in the atmosphere is almost saturated and if more CO_2 is produced there will be no additional absorption. Explain why this is incorrect.

5. From Figs. 1.7 and 1.8, it would appear that water is more important than CO_2 in governing the greenhouse effect. Comment on this.

6. Explain why burning wood and burning coal are not equivalent as far as the greenhouse effect is concerned.

7. The normal energy requirements of a house in Winnipeg in winter is 1 million Btu/day or 10^6 kJ/day. (a) What weight of Glauber's salt would be required to store solar energy for 1 month of winter use? (b) If the optimum conditions for using the Glauber's salt are a 35% by weight (Na_2SO_4) solution in water, what volume of solution would be required? (Note: Density of GS solution is 1.29 g/mL.)

8. Canada produces over 700 million pounds of vegetable oils per annum. If this were to replace all Canadian petroleum oil used, how long would it last?

9. If a typical home has 1,200 ft^2 of floor space and a sloping A-type roof at 45°–60° for optimum solar collection, the roof area facing the Sun would be approximately 850 ft^2. Calculate the energy per day available as heat and as electricity via photovoltaic cells (assume 5% efficiency) for both summer and winter.

10. What would be the rise in sea level if the average temperature of the oceans increased by 1°C at 20°C. The density of water at 20°C is 0.99823 g/mL, and at 21°C, the value is 0.99802 g/mL. Assume no ice melts and that the ocean area does not increase. The average depth of the ocean is 3,865 m.
11. Reverse electrodialysis is a method of extracting electrical energy directly from the flow of a fresh water river into the sea (salt water). Ion-exchange membranes are used to separate the flow of fresh and salt water. Draw a diagram of the system and explain how it works. See Fig.1.16 in Chap. 1 or Fig. 15.7 in Chap. 15.

Further Reading

1. Hiromichi Seya (2010) Conditions for survival. Toward a "solar energy—based society" full of vibrant life. In: Special round table conference on global environmental problems. The Asahi Glass Foundation, Tokyo, p 272
2. Smil V (2010) Energy myth and realities: Bringing science to the energy policy debate. The AEI Press, Washington, DC, p 215
3. Foot R, Walter A (2011) China, the United States and global order. Cambridge University Press, New York
4. Guel VK, Goodspeed TO, Kharas H (2010) Latin America 2040. Breaking away from complacency: an agenda for resurgence. SAGE Publications India Pvt. Ltd, New Delhi
5. Nastny L (2010) Renewable energy and energy efficiency in China: current status and prospects for 2020. World watch report 182. Worldwatch Institute, USA, p 48
6. Reddy BS, Assenza GD, Assenza D, Hasselmann F (2009) Energy power and climate change. Sage Publications India Pvt. Ltd, New Delhi, p 349
7. Lodders K, Fegley B Jr (2011) Chemistry of the solar system. RSC Publishing. Washington University, St. Louis, p 476
8. Kitasei S, Mastny L (2010) Powering the low-carbon economy: the once and future roles of renewable energy and natural gas. Worldwatch report 184. Worldwatch Institute, Washington, DC
9. Halmann MM, Steinberg M (1999) Greenhouse gas carbon dioxide mitigation. Lewis, Boca Raton
10. Halmann MH, Steinberg M (1999) Greenhouse gas carbon dioxide mitigation. CRC, Boca Raton
11. Klass DL (1998) Biomassfor renewable energy, fuels, and chemicals. Academic, New York
12. Bejan A, Vadasz P, Kroger DG (1999) Energy and the environment. Kluwer, New York
13. Bisio A, Boots S (1996) Encyclopedia of energy and the environment, vol 2. Wiley, New York
14. Campbell-Howe R (ed) (1996) In: Proceedings of 21st national passive solar conference, American Solar Energy, Boulder
15. Takahashi P, Trenka A (eds) (1996) Ocean thermal energy conversion, UNESCO energy engineering series. Wiley, New York
16. Sah SL (1995) Renewable and novel energy sources. State Mutual Book and Periodical Service Ltd., New York
17. Nansen R (1995) Sun power, the global solution for the coming energy crisis. Ocean Press, Seattle
18. Gipe P (1995) Wind energy comes of age. Wiley, New York
19. Duffield WA (1994) Trapping the earth for heat. US Geology Survey, Washington, DC
20. Avery W, Wu C (1994) Ocean thermal energy conversion. John Hopkins Applied Physics Laboratory, Baltimore
21. Heimann M (1994) The global carbon cycle. Springer, New York
22. Golob R, Brus E (1994) The almanac of renewable energy: the complete guide to emerging energy technologies. Holt, New York
23. Markvart TF (ed) (1994) Solar electricity. Wiley, New York
24. Kraushaar JJ, Ristinen RA (1993) Energy and problems of a technical society, 2nd edn. Wiley, New York
25. Potts M (1993) The independent home: living well with power from the sun, wind and water. Chelsea Green, White River Junction
26. Carless J (1993) Renewable energy: a concise guide to green alternatives. Walker, New York
27. Seymour RJ (ed) (1992) Ocean energy recovery: the state of the art. American Society of Civil Engineers, New York
28. Winteringham FPW (1992) Energy use and the environment. Lewis, London
29. Winter CJ, Sezmann RL, Vant-Hull LL (eds) (1991) Solar power plants. Springer, New York
30. (1991) Handbook of unusual energies. Gordon, New York
31. Priest J (1991) Energy, principles, problems alternatives, 4th edn. Addison-Wesley, Reading
32. (1990) Developments in tidal energy. American Society of Civil Engineers, New York

33. Scientific American Magazine, Managing Planet Earth, Sept., (1989)
34. Loftness RL (1984) Energy handbook, 2nd edn. Van Nostrand, New York
35. Shaw R (1982) Wave energy. Wiley, New York
36. Inhaber H (1982) Energy risk assessment. Gordon and Breach, New York
37. Kenneth F. Weaver, David Jeffrey, Rick Gore (1981) Energy—A special report in the public interest, National Geographic, Deluxe Edition.
38. Parker SP (ed) (1981) Encyclopedia of energy, 2nd edn. McGraw-Hill, New York
39. Bockris J-O'M (1980) Energy options—real economics and the solar-hydrogen system. Taylor and Francis Ltd, London
40. Sassin W (1980) Energy. Sci Am 243(3):118
41. Energy in Transition (1985–2010) Final report of the committee on nuclear and alternative energy systems, NRC, National Academy of Science, Washington DC (1979). W. H. Freeman, San Francisco (1980)
42. Raymond Siever (1980) Energy and environment: readings from scientific American. W. H. Freeman, San Francisco
43. Stobaugh R, Yergin D (eds) (1979) Energy future. Report of the energy project at Harvard Business School. Ballantine Books, New York
44. Messel H (ed) (1979) Energy for survival. Pergamon Press, New York
45. Sorensen B (1979) Renewable energy. Academic, New York
46. (1979) Energy—the fuel of life—by the editors of encyclopedia Britannica. Bantum Books, New York
47. RuedisiH LC, Firebaugh MW (1978) Perspectives on energy—issues, ideas and environmental dilemmas, 2nd edn. Oxford University Press, New York
48. Tillman DA (1978) Wood as an energy resource. Academic, New York
49. Considine DM (ed) (1977) Energy technology handbook. McGraw-Hill, New York
50. Lapedes DN (ed) (1976) Encyclopedia of energy. McGraw-Hill, New York
51. Morgan ND (ed) (1975) Energy and man: technical and social aspects of energy. IEEE Press, New York
52. A. Fisher (1975) Energy from the Sea, Popular Science Magazine, May p. 68, June p. 78, July, (1975)
53. Reynolds WC (1974) Energy—from nature to man. McGraw-Hill, New York
54. Anna Gyorgy (1971) No nukes. Energy and Power (full issue). Sci Am, Maple Vail, New York, PA, USA
55. Oort RH (1970) The energy cycle of the earth. Sci Am 223(3):54
56. Daniels F (1964) Direct use of the sun's energy. Ballantine Books, New York
57. Zaren AM, Erway OD (1963) Introduction to the utilization of solar energy. McGraw-Hill, New York
58. Putnam PC (1956) Energy in the future. Van Nostrand, New York
59. Daniels F (ed) (1955) Solar energy research. University of Wisconsin, Madison
60. World Energy Resource- International Energy Annual, http://www.eia.doe.gov/
61. Energy Conservation DataBook, Japan. http://www.eccj.or.jp/databook/1998e/index.html
62. Energy Sources. http://www.yahoo.com/science/energy/
63. Categories: alternate energy, biomass, fly wheels, fuel cells, geo thermal, hydropower, nuclear, renewable, solar, wind. Ocean Wave Energy Company. http://www.owec.com/
64. Ozone/Greenhouse. http://www.epa.gov./globaIwarming/index.html
65. Ozone secretariate. http://www.unep.org/ozone/
66. American Bioenergy Association. http://www.biomass.org/
67. Solar Energy Technology. http://www.sunwize.com/
68. Solar energy kits and systems. http://buelsolar.com/
69. International Solar Energy Society. http://www.ises.org/
70. Fossil Fuels Association. http://www.fossilfuels.org
71. Onta Fossil Fuels Industry. http://www.fossil-fuels.com/
72. Government views of fossil energy. http://www.fe.doe.gov/
73. Electric Power Research Institute. http://www.epri.com/
74. Hydroquebec. http://www.hydroquebec.com/hydroelectricity/index.html
75. Tidal Power. http://www.tidalelectric.com/
76. Wave Energy. http://www.waveenergy.dk/apparater/engelsk-klassificering.html
77. Wave Power Patents. http://www.waveenergy.dk/apparater/rly.htm
78. American Wind Energy Association. http://www.awea.org

Chapter 2
Solid Fuels

2.1 Introduction

Fuels are conveniently classified as solids, liquids, and gaseous fuels. Solid fuels include peat, wood, and coal and can encompass solid rocket fuels as well as metals. The earliest fuels used by man were nonfossil fuels of wood and oil from plants and fats from animals. The windmill and water wheels were other sources of energy.

Sources of power have changed with the years and will continue to change as shown in Fig. 1.2. In 1992, approximately 7% of the world's power was supplied by hydroelectric plants and the remaining 92.5% from fuels. Natural gas provides 22% of the total power; petroleum, 40%; coal, 25% (of which 7% is derived from hard coal and 18% from the soft coals); and nuclear fuels provide about 7%. But in 2005, approximately 16.6% of the world's power was supplied by hydroelectric plants; 65.9% by thermal power plants; 15.2% by nuclear power plants; and 2.1% by geothermal, solar, and wind power plants. Oil, which has displaced coal as the major fuel, will soon be replaced by natural gas which in turn will be eventually replaced by nuclear energy and environmentally friendly geothermal, solar, and wind power energy.

The origin of coal is not known with certainty. One popular theory claims that coal originated about 250 million years ago as a result of the decay of vegetation primarily from land and swamps and not of marine origin. Bacterial action undoubtedly helped with the reduction process. The first step following the exclusion of oxygen was the formation of peat—a slimy mass of rotting organic matter and debris. Under the pressure of sediments, the peat became dehydrated and hard, forming low-grade coal, called *lignite*. Under further pressure and time, the reactions of condensation and consolidation (50-fold decrease in volume) converted the lignite into a higher grade coal—bituminous coal. A highest grade coal—anthracite—has the highest percentage of carbon. A simplified flow description for coal formation is shown in Fig. 2.1. The three-component ternary phase diagram for the C, H, and O content of the various grades of coal is shown in Fig. 2.2 where comparison is made with cellulose and lignin,[1] the general precursor to coal. The simple weight percentage point of cellulose is indicated by a filled triangle (▲) and the atom percentage is shown as a filled square (■). The bond-equivalent points are meant to account for the bonding (valency) of the elements, namely, 4 for C, 2 for O, and 1 for H. The atom percentage multiplied by the bond factor and normalized (to 100% for all the elements) is called the *bond-equivalent percentage* and is shown as filled circles (●). The values for CH_4, CO_2, and H_2O are also shown in Fig. 2.2 as open circles (○) The direction from cellulose to anthracite shows clearly that the loss of water and oxygen must occur during the coalification process.

---

[1] Wood consists of about 25% lignin which acts to bind the cellulose fibers together.

O.V. Roussak and H.D. Gesser, *Applied Chemistry: A Textbook for Engineers and Technologists*, DOI 10.1007/978-1-4614-4262-2_2, © Springer Science+Business Media New York 2013

Fig. 2.1 Possible route for the formation of coal from plant matter

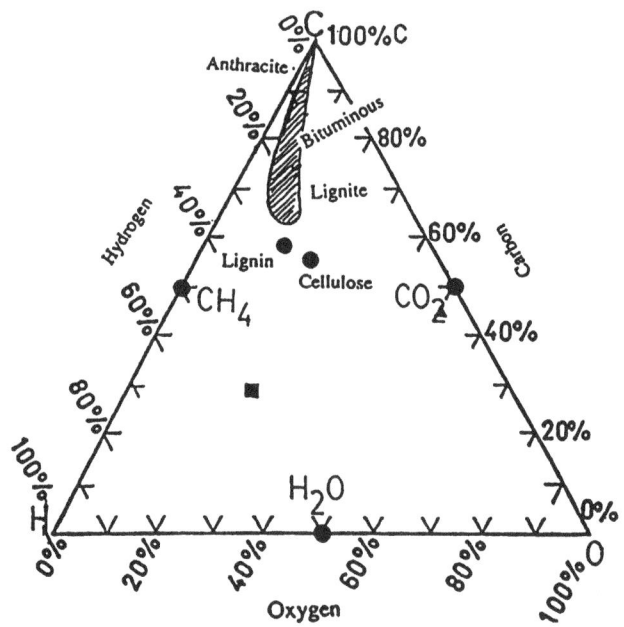

Fig. 2.2 The three-phase ternary bond-equivalent diagram for C-O-H showing the transition from cellulose to anthracite. Cellulose ($C_6H_{10}O_5$) is also shown as ▲ for weight % and as ■ for atom % in the diagram. The other points are bond-equivalent %

Wood was obviously man's first fuel, followed by animal fats and vegetable oil. There is evidence that candles were used during the first Minoan civilization about 3000 B.C. Coal was used by the Chinese about 100 B.C. and the "black stone" was reported to be used by Greek smiths about 250 B.C. The Romans in Great Britain also used coal. Marco Polo describes the mining of "black stone" in his travels 1271–1298 A.D.

Coal is primarily known as a fuel but it is also a valuable chemical. It can be reacted with lime, CaO, at high temperatures (electric arc) to form calcium carbide,

Table 2.1 World coal resources and use in 2006/1996[a]

| | Recoverable reserves 10^9 t | | | Uses 10^6 t | |
	Hard 2005/1996	Soft 2005/1996	Total 2005/1996	Production 2005/1996	Consumption 2005/1996
Russia	54.1/54	173.1/173	173.1/173	322/331	246.4/284
USA	122.0/122	263/275	263/275	1,162/1,087	1,112.3/1,028
China	68.6/68.6	126.2/126.2	126.2/126.2	2,620.5/1,690	2.584.2/1,532
Germany	0.2/26.4	7.4/73.8	7.4/73.8	220.5/286	269.5/277
Australia	40.9/52.1	84.4/99.6	84.4/99.6	0/276	156.4/135
UK	0.1/1.1	0.2/1.6	0.2/1.6	19.9/62	74.2/78
Canada	3.8/5.0	7.2/9.5	7.2/9.5	72.7/87	65.4/62
Poland	6.6/13.4	8.2/15.8	8.2/15.8	171.1/232	154.8/195
South Africa	54.5/61.0	56.5/61.0	56.5/61.0	275.3/230	207.8/169
India	57.6/80.2	62.3/82.2	62.3/82.2	498.9/352	550.7/342
Rest of the world	65.1/77.5	143.5/170.2	143.5/170.2	1,443.1/922	1,322.1/1,167
Total	473.2/565.3	930.4/1,088.1	930.4/1,088.1	6,806.9/5,625.0	6,743.8/5,269

[a]From International Energy Annual (2006, 1996)
Hard = anthracite and bituminous; soft = lignite and subbituminous

CaC_2, which when hydrolyzed, forms acetylene (C_2H_2):

$$3C + CaO \rightarrow CaC_2 + CO \tag{2.1}$$

$$CaC_2 + 2H_2O \rightarrow Ca(OH)_2 + C_2H_2 \tag{2.2}$$

Coal is also used to prepare active carbon which is used for the purification of air and water and in numerous industrial processes.

The distribution of coal in the world is given in Table 2.1. The recoverable reserves consist of the coal which can be economically mined with presently known technology and conditions. The proven reserves are the coal known to be present, within +20%, by extensive drilling and experience and are about twice the recoverable reserves. The estimated total world coal resources are about eight times larger and are based on geologically favorable formation within the earth's crust and on previous experience within the countries concerned. It excludes under sea or under ice sites.

Of the world's reserves, the distribution between hard and soft coal is about equal on a global basis. There is enough coal to last the world for 200–1,000 years depending on the rate of usage and the rate of exploration.

Coal in 1947 was at a record high price of $4.16 per ton at the mine. More than 600 Mt were mined that year in the USA, but already, oil was + beginning to displace it. The old steam engines, which used a ton of coal for every four miles hauling a heavy freight, consumed a quarter of the coal production (125 Mt). By 1960 when the railroad had almost completely converted from coal to diesel, the rail industry used only 2 Mt of coal.

A drop in coal-fired generation in the Organization for Economic Cooperation and Development (OECD) countries is offset by big increases elsewhere, especially China, where 600 gigawatts of new capacity exceeds current capacity of the USA, European Union (EU), and Japan (IEA Outlook 2010). According to the American Worldwide Institute's report (No. 182, Oct. 2010), China is the first world's largest coal user today. China consumed 2.7 billion tons of coal in 2008 or 43% of the world total and in 2.5 times higher than the second largest consumer, the USA.

2.2   Wood and Charcoal

Wood, a renewable source, is not an important industrial fuel today. However, its use continues in some rural areas where it is often supplemented with liquid propane. In some underdeveloped countries, wood is still the principal source of energy. Dry wood contains from 1% to 12% moisture whereas green wood contains from 26% to 50% water. The resinous woods, like pine or cedar, yield about 18.5 MJ/kg of air-dried wood or, allowing for the moisture content, about 21 MJ/kg on a dry weight basis. Hardwoods have a heating value of about 19.4 MJ/kg. The energy available in present forest stocks is estimated to be equivalent to about 270×10^9 t of coal or about 2/3 of the equivalent oil reserves.

Wood-burning fireplaces have become a popular form of heating in the past few years. The normal open hearth fireplace is not an efficient producer of heat since its draft sends most of the hot air up the chimney, creating a partial vacuum pulling cold air into the home. The recent introduction of glass doors to close in the fire and the introduction of outside air to the fire for the combustion process has improved the efficiency of the fireplace, especially with forced air circulation around the fire chamber.

Open fires, however, are a source of pollution since the smoke produced contains large quantities of polynuclear aromatic hydrocarbons such as benzo(a)pyrene, a carcinogen also found in cigarette smoke. In some communities, it has been necessary to restrict the burning of wood because of the resulting air pollution.

2.3   Peat

Peat is formed when dead vegetation is saturated with water which prevents the action of aerobic bacteria. Thus, most of the carbon of the cellulosic matter is retained, and with ageing, peat is formed. It accumulates at an average rate of 0.7 mm/year or worldwide at 210 Mt of carbon. Canada (40%) and Russia (36%) have more than 3/4 of the world's peat land (320 Mha or 150×10^3 Mt of carbon). In Russia, peat deposits occupy about 1/10 of the total country's terrain. It is a spongy watery mass when first obtained from the peat bog. Six tonnes of dry peat yield about 1 t of fuel. A commercial grade of peat contains about 25% water. Air-dried peat has a heating value of about 16.3 MJ/kg. Peat is rich in bitumens, carbohydrates, and humic acids, and as a chemical source, it can yield waxes, paraffins, resins, and oils. Peat also is a source of pharmaceutical and curative preparations as well as a livestock-feed supplement.

Peat is not used in North America as a fuel to any great extent, but in Europe, it has been employed in domestic heating for centuries. Peat is often harvested by massive machinery for industrial use (Fig. 2.3). The machine deposits the cut and macerated peat in long furrows, which are then crosscut into blocks and conveyed to dry storage or dryers. Ireland harvests 4 Mt annually as fuel for generating 20% of the nation's electricity. In Finland, about 3 Mt of peat, with an energy content of 30×10^9 MJ, is used annually. This is about 6% of the total fossil fuel energy used. Peat as a primary fuel was used in 15 power plants in 1983 to generate 950 MW. Peat is also used to generate power for electricity in Russia.

The pyrolysis of peat in the absence of oxygen is being studied in New Zealand, where the oil-wax product is then hydrogenated to form the equivalent to a common crude oil. New Zealand has enough peat to produce 400 Mbbl of oil.

The growth of peatlands consumes CO_2, and the harvesting of peat and its combustion as a fuel has a twofold adverse effect on the CO_2 balance in our atmosphere, first by releasing CO_2 into the atmosphere and second by removing it as a sink for CO_2.

Fig. 2.3 This machine cuts and macerates the peat and then sends it along the extended arm which deposits it in rows. The peat is cut into blocks by the discs as the equipment moves forward

2.4 Coal

Coal, the generic term applied to solid fossil fuels, ranges from lignite, which is basically a matured or modified peat, to meta-anthracite, which is more than 98% carbon. The qualities of different coals are classified in different ways either depending on the chemical composition, the heating value, or even the ash content and its fusion temperature. The International Classification of Hard Coals by Type (1956) has been widely accepted. The criteria used include volatile matter and calorific value as well as the swelling and caking properties of the coal. The Economic Commission for Europe has proposed its own classification and codification system. In North America, the common classification of coals is by the ASTM method. The rank classification system[2] is based on the application and commercial use of coal, namely, in combustion for electric power generation and in the preparation of coke for the metallurgical industry.

2.5 Analysis of Coal

The elemental analysis of coal, i.e., its C, H, O, N, S, and ash residue, may be important to a chemist who wishes to use coal as a chemical or source of carbon, but to an engineer who wants to burn the coal in a heat- or power-generating plant or a coking oven, other parameters are more important—most notably its heat of combustion, moisture level, volatile matter, carbon and sulfur content, as well as the ash. This is called the *proximate analysis* and it is determined as follows:

Moisture content. A sample of coal is ground to pass 20/60 mesh and weighed. It is dried in an oven at 110°C and reweighed. The loss in weight represents the moisture content of the coal.

---

[2] The ASTM (American Society for Testing and Materials) has provided detailed definitions (D121-85) and tests (D-388-88) for rank classification.

Table 2.2 Classification of coals by rank[a]

Class/group	Fixed carbon, C (%)	Volatile matter, V (%)	Calorific value, ΔH (MJ/kg)
Anthracite			
Meta-anthracite	$98 \leq C$	$V < 2$	
Anthracite	$92 < C < 98$	$2 < V < 8$	
Semianthracite	$86 < C < 92$	$8 < V < 14$	
Bituminous			
Low-volatile bituminous	$78 < C < 86$	$14 < V < 22$	$32.6 < \Delta H$
Medium-volatile bituminous	$69 < C < 78$	$22 < V < 31$	$30.2 < \Delta H < 32.6$
High-volatile A bituminous	$C < 69$	$31 < V$	$26.7 < \Delta H < 30.2$
High-volatile B bituminous			$24.4 < \Delta H < 26.7$
High-volatile C bituminous[b]			
Subbituminous			
Subbituminous A coal			$24.4 < \Delta H < 26.7$
Subbituminous B coal			$22.1 < \Delta H < 24.4$
Subbituminous C coal			$19.3 < \Delta H < 22.1$
Lignitic			
Lignitic A			$14.7 < \Delta H < 19.3$
Lignite B			$\Delta H < 14.7$

[a]Based on dry, moisture and mineral free (ash free basis)
[b]Variable values depending on the agglomerating properties of the coal when freed of volatile matter

Volatile content. The coal sample is heated in an inert atmosphere up to 900°C. The loss in weight varies with the temperature since some of the coal is decomposed into oils and tars which volatilize at various temperatures.

Ash content. A coal sample is heated in a muffle furnace at 900°C in the presence of air to combust the coal leaving the ash residue which is heated to constant weight. If the sample of coal is first freed of volatile matter, then the loss in weight represents the fixed carbon in the coal.

Heat content. This is determined in a bomb calorimeter where a dry sample of coal is burned in an excess of oxygen and the heat evolved is measured. This is often referred to as the caloric value or the heat content of the coal. This includes the combustion of the volatile and tar components as well as the fixed carbon and is therefore related to the actual heat-generating value of the coal. The heat of combustion of pure carbon is 32.8 MJ/kg.

2.6 ASTM Classification

An abbreviated version of the ASTM classification of coal by rank is shown in Table 2.2. The highest rank—meta-anthracite—contains the highest percentage of carbon and the lowest amount of volatile matter. The four classes are divided into 13 groups according to carbon content and heating value. Thus, in the bituminous class, there may be gas coals, coking coals, and steam coals. The proximate and ultimate analysis of typical coals is given in Table 2.3. It is possible to calculate the heat content of coal from its ultimate analysis. From the standard heats of formation of CO_2 (-393.5 kJ/mol), H_2O (-285.8 kJ/mol), and SO_2 (-296.8 kJ/mol), it is possible to derive the formula

$$Q = (1/100)[32.76C + 142.9(H - 0/8) + 9.3S] \qquad (2.3)$$

Table 2.3 Proximate and ultimate analysis of five Canadian coals (moisture free)

	BC	NS	NS	Sask	Sask
Proximate analysis					
Volatile carbon	26.1	34.6	35.4	43.5	41.4
Fixed carbon	58.6	49.8	61.7	43.1	46.1
Ash	15.3	15.6	2.9	13.4	12.5
Ultimate analysis					
Carbon	74.4	66.0	84.7	61.1	66.1
Hydrogen	4.3	4.5	5.6	3.6	2.2
Sulfur	0.8	4.9	1.3	1.1	0.6
Nitrogen	1.2	1.4	1.3	1.0	1.3
Ash	15.3	15.6	2.8	13.4	12.5
Oxygen[a]	4.0	7.6	4.3	19.8	17.3
Calorific value[b] (MJ/kg)	28.6	23.8	33.9	17.3	18.3

[a]Determined by difference
[b]Coal sample as received

where Q is the exothermic heat evolved in the combustion of the coal in units of kJ/g and where C, H, O, and S represent the weight percentage of the element in the coal. Oxygen, which is assumed to be present as H_2O, is normally determined by difference. The assumption concerning oxygen is not too unreasonable since more than half of the oxygen in coal is normally present as OH. If the ash content is significant, then a correction must be made for it. The nitrogen content of coal is usually small and often ignored. Since $\Delta H_s^0(NO_2) = 33.2$ kJ/mol, the added term to the equation would be 2.4 N. Several other formulas have been proposed which are correct for some of the structural aspects of coal.

2.7 Ash

The combustion of coal results in the formation of an ash—the noncombustible component of coal—part of which is carried off with the combustion products as very fine particulate powder called *fly ash*. The heavy ash remaining in the combustion chamber is called *bottom ash* or *boiler slag*. The continuous mining process produces a coal containing about 28% ash. Coal used to generate steam usually has about 15% ash.

The ash is an undesirable component of the coal and is usually reduced during the cleaning process that coal is normally subjected to after being shipped from the mine. This cleaning process removes the rocks, clay, and minerals which invariably mix with the coal.

Some applications such as chain-grate stokers require a minimum of 7–10% ash to protect the metal parts of the furnace. One of the most important characteristics of the bottom ash is its melting point (or fusion temperature) which determines the ease with which it is removed from the furnace. The melting point of these inherent impurities in a coal can be represented by a three-component phase diagram of oxides such as Fe_2O_3, Al_2O_3, and SiO_2 or by minerals such as $Al_2O_3 \cdot 2SiO_2$ (clay), SiO_2 (quartz), and MgO. $Al_2O_3 \cdot 2.5SiO_2$ (feldspar) where lines join the common melting points of the various mixtures. The choice of the components depends on the coal and the impurities in the ash. A high fusion ash melts above 1,316°C whereas a low fusion ash melts below 1,093°C. In some instances, iron oxide or sand (SiO_2) is added to the coal to increase the melting point of the ash so as to favor clinker formation which results in easy removal. Ash with a low fusion temperature can form an undesirable glassy coat of the furnace and grill. In the case where steam is generated, the deposition of this adherent glassy deposit on the fireside of the steel boiler tubes causes the ash to insulate the steel tubes from the heat and thus reduce the efficiency of steam generation.

Table 2.4 Trace element analysis of coal and its various combustion (ash) fractions[a] (µg/g)

Element	Symbol	Coal	Bottom ash	Precipitated[b] ash	Bag[c] ash	Stack[d] ash
A. Barium	Ba	150	1,200	1,600	1,540	1,700
Calcium	Ca	4,000	49,000	34,000	21,000	23,000
Cesium	Cs	0.8	10	7	6	8
Magnesium	Mg	420	4,700	5,400	440	4,800
Potassium	K	1,220	16,000	13,000	11,000	13,300
Rubidium	Rb	23	75	160	90	90
Sodium	Na	500	7,100	4,600	3,400	4,000
Strontium	Sr.	120	1,200	1,200	1,000	1,100
B. Aluminum	Al	11,000	100,000	112,000	103,000	103,000
Beryllium	Be	0.9	–	8	10	10
Dysprosium	Dy	1.0	6	10	5	6
Europium	Eu	0.3	3	3	3	3
Hafnium	Hf	0.7	–	9	3	7
Lanthanum	La	5.0	70	60	50	50
Lutetium	Lu	0.2	1	2	0.6	2
Samarium	Sm	1.0	10	12	9	12
Tantalum	Ta	0.3	2	3	1	3
Terbium	Tb	0.3	4	4	0.8	3
Thorium	Th	1.4	20	15	10	12
Uranium	U	0.7	–	9	8	9
Ytterbium	Yb	0.4	4	5	3	4
C. Chromium	Cr	12	160	140	220	370
Cobalt	Co	3	20	40	35	60
Copper	Cu	6	20	84	142	170
Iron	Fe	6,300	140,000	90,000	70,000	70,000
Manganese	Mn	24	360	270	250	510
Nickel	Ni	4	70	35	200	340
Scandium	Sc	3	–	40	20	20
Silver	Ag	2	50	40	7	40
Titanium	Ti	560	5,000	6,500	6,000	6,000
Vanadium	V	20	130	180	240	240
D. Bromine	Br	14	Nil	6	30	12,000
Chlorine	Cl	1,200	Nil	Nil	800	900,000
Fluorine	F	80	–	100	400	33,000
Iodine	I	1	Nil	Nil	Nil	1,300
E. Antimony	Sb	0.5	3	6	11	30
Arsenic	As	12	35	90	230	300
Boron	B	16	–	220	2,000	2,600
Cadmium	Cd	0.3	3	2	4	20
Gallium	Ga	10	30	60	140	140
Lead	Pb	6	70	50	160	200
Mercury	Hg	0.4	0.2	0.5	0.5	500
Selenium	Sc	3	2	14	14	300
Zinc	Zn	27	–	330	380	1,100

[a]Values rounded off—average of several coals
[b]Electrostatic precipitators 99 % efficient
[c]Filter bags

Table 2.5 Chemical analysis of fly ash (Saskatchewan)

Component	Content (%)	Component	Content (μg/g)
Na (as Na$_2$O)	3.02	As	<1
K (as K$_2$O)	0.48	B	291
Ca (as CaO)	10.87	Cd	2
Mg (as MgO)	1.09	Cr	20
Al (as Al$_2$O$_3$)	21.02	Cu	26
Si (as SiO$_2$)	57.64	Co	13
Fe (as Fe$_2$O$_3$)	2.81	Ga	107
Ti (as TiO$_2$)	0.76	Pb	36
P(as P$_2$O$_5$)	0.09	Mn	292
S (as SO$_3$)	<0.1	Hg	<0.5
		Mo	<50
		Ni	32
		Ag	2.9
		V	72
		Zn	48
		Se	<0.5

Fly ash is usually precipitated by a Cottrell electrostatic precipitator and/or collected in filter bags. The composition of the fly ash usually differs from the bottom ash due to thermal fractionation of the oxides. This is illustrated in Table 2.4 where the composition of the bottom ash is compared with various fractions of the fly ash which is classified into five groups:

(a) The alkali and alkaline earth metals
(b) Refractory metals
(c) Transition metals
(d) Halogens
(e) The volatile elements.

As of 2006, about 125 million tons of coal—combustion by-products, including fly ash—were produced in the USA each year, with about 43% of that amount used in commercial applications, according to the American Coal Ash Association website. As of 2005, US coal-fired power plants reported producing 71.1 million tons of fly ash, of which 29.1 million tons were reused in various applications.

In 1995, the US utilities produced 65 Mt/y of fly ash. The 21 coal-fired, thermal-generating plants in Canada produced 16.6 GW consuming 37.8 Mt of coal of which 10% was ash. The fly ash produced was 2.5 Mt. The composition of the fly ashes varies considerably. One particular fly ash from Saskatchewan (see Table 2.5) was recently tested for the extraction and recovery of gallium which can be converted to GaAs. In the near future, GaAs will be replacing silicon as the semiconductor of the electronic industry.

Another potential use of some fly ash is as a source of alumina (Al$_2$O$_3$), replacing bauxite which is being exhausted at an ever increasing rate.

2.8   Coal and Its Environment

Though coal is cheap and plentiful, it is a "dirty" fuel which contaminates our environment and contributes to the CO_2 imbalance in our atmosphere as well as oxides of nitrogen from the combustion process. The principal contaminant in coal is sulfur which burns to form sulfur dioxide (SO$_2$)

which is oxidized to sulfur trioxide (SO_3) in the atmosphere. In the commercial production of sulfuric acid, sulfur (S) is burned to form SO_2:

$$S + O_2 \rightarrow SO_2 \qquad (2.4)$$

The SO_2 can dissolve in water to form sulfurous acid (H_2SO_3). The SO_2 is also catalytically converted to SO_3 using V_2O_5 as a catalyst:

$$2SO_2 + O_2 \xrightarrow{V_2O_5} O_3 \qquad (2.5)$$

The SO_3 is then treated with water or sulfuric acid solution to form H_2SO_4.

Similar reactions can occur in the atmosphere with the resulting formation of acid (H_2SO_4) rain which can fall considerable distances from the source. Thus, the acid rain falling in Norway and Sweden primarily originated from coal burning in the Ruhr valley and the UK. Similarly, the acid rain reaching Ontario comes primarily from the US iron and steel centers and industrial Ohio Valley though it is claimed that a large part of the SO_2 is also coming from the nickel smelters in Sudbury, Ontario. A joint US–Canadian study has determined the sources and recommended appropriate solutions to prevent the lakes from becoming too acidic to support aquatic life (fish).

Acid rain is not the only environmental contaminant from coal burning. The average concentration of mercury in coal is about 0.3 µg/g (i.e., less than 1 ppm). A 755-MW steam turbine-driven power station burns approximately 7,100 t of coal per day. This corresponds to about 2.5 kg/day of mercury being sent up the stack. With present North America estimates of coal consumption at about 10^{10} t/year, about 3,000 t of mercury is put into the environment. This is about four times the natural source.[3]

Coal also contains uranium and thorium and their radioactive decay products. Though as much as 98% of the fly ash is precipitated, the 2% remaining escapes up the stack. For a power plant burning 2×10^4 t/day if there is approximately 10% of fly ash, then approximately 40 t/day fly ash escapes up the stack. Measurements have shown that this contains about 500 µCi ^{226}Ra per day. The fly ash behaves like ordinary smoke and its dispersion follows standard equations. If we assume a wind speed of 1 m/s, a stack of 120 m high will result in a maximum concentration at ground level at 400 m from the stack of 9×10^{-14} µCi ^{226}Ra per cm^3 air. The maximum permissible concentration (MPC) for ^{226}Ra in air is 10^{-11} µCi/cm^3. Thus, the concentration of ^{226}Ra is two orders of magnitude lower than the MPC, but since fly ash contains other radionuclides (^{230}U, ^{210}Pb, etc.), the long-term effects on those living close to or downwind from coal burning power plants must be carefully monitored.

The use of fly ash in concrete does not solve the disposal problem but only shifts it to another locale. If cement is composed of 30% fly ash, it has been estimated that the radon diffusing out of the concrete (porosity—5%) into a room ($10 \times 10 \times 4$ m) would be about 10^{-9} µCi/cm^3—100 times lower than the MPC. However, with higher porosity, concrete and lower ventilation rates the margin of safety decreases, and it means that concrete containing fly ash should not be used in structures for habitation though it would be permissible for use in foundations, bridges, and roads.

A recent coal-fired power station located between Los Angeles and Las Vegas has shown that it is possible to burn coal with as much as 3.5% sulfur without contaminating the environment with SO_2 and NO_x. This is done using two existing technologies, coal gasification, and combined cycle generation, i.e., generating electricity simultaneously from turbines running on gas and steam. The plant pulverizes 1,000 t of coal per day, which is converted to synthesis gas by reaction with water and

---

[3] The average concentration of mercury in rock is about 0.1 µg/g and since about 10^{10} tone of rock is weathered annually, it is estimated that about 800 tone of Hg is put into the atmosphere each year. This mercury is distributed uniformly around the earth and presents little environmental danger. It is the non-uniform or concentrated dumping of mercury that is dangerous and must be avoided.

oxygen. The ash and minerals fuse and are removed, whereas the H_2S is removed after the gas is cooled by generating steam. Water is then added to the clean gas to reduce NO_x formation upon combustion and more steam is generated to drive the turbines. The cost of power for a 600 MW plant is estimated to be 4–5 cents per kWh.

2.9 Fluidized Bed Combustion

The fluidized bed reactor is about 60 years old, but only in recent years has its application to coal combustion taken on commercial significance. The fluidized bed is the dispersion of a solid, usually in powder form, by a gas, under flow conditions such that the solid takes on the properties of a gas. Such reactors can be designed to operate continuously. Thermal conduction (heat transfer) in such systems can be high, and, as a result, in the case of coal and air, the combustion can occur at much lower temperatures than in the fixed bed system. Thus, the addition of limestone ($CaCO_3$) or dolomite ($CaCO_3 \cdot MgCO_3$) to the fluidized bed system can result in the reduction of oxides of sulfur and oxides of nitrogen.

This is accomplished by the reactions which can occur in the combustion bed:

$$CaCO_3 \rightarrow CaO + CO_2 \tag{2.6}$$

$$CaO + SO_x \rightarrow CaSO_{(x+1)} \tag{2.7}$$

The lower combustion temperature keeps the equilibrium for the NO_x reaction toward the $O_2 + N_2$. See Exercise 18:

$$O_2 + N_2 \rightleftharpoons 2NO \tag{2.8}$$

This would mean that high sulfur coal could eventually be used in combustion processes without contaminating the environment.

Unfortunately, the effect of increased CO_2 levels in the atmosphere will still contribute to the greenhouse effect. Only the combustion of biomass, H_2, or renewable energy sources does not enhance the greenhouse effect.

Recent work by the Coal Utilization Research Laboratory in Leatherhead, England, has shown that a slurry of 68% coal in water burned very well in a pressurized fluidized bed reactor. However, much more work remains to be done before such mixtures can be pumped through pipelines and burned directly in specially designed reactors.

2.10 Coke

Coke is produced when a bituminous coal is devolatilized by heating (in the absence of air) to temperatures ranging from 900°C to 1,200°C. As the temperature is slowly raised, physical and chemical changes take place. Adsorbed water is lost at temperatures up to 250°C. Some CO and CO_2 are liberated up to 300°C with some pyrogenic water. Above 350°C, the coal becomes plastic and begins to decompose between 500°C and 550°C to form gases and tars. The product at this point is called *semicoke* which still off-gases up to 700°C where the hot coke acts as a catalyst for the decomposition of the volatile products. Above 700°C, the coke is hardened and agglomerated.

The by-product consists of coke-oven gas (55–60% H_2, 20–30% CH_4, 5–8% CO, 2–3% heavy hydrocarbons, 3–5% N_2, and 1–3% CO_2, as well as traces of O_2) having a calorific value of about 17 MJ/m^3.

Coal tar is also produced during the coking process. It contains over 300 substances which usually include 5–10% naphthalene, 4–6% phenanthrene, 1–2% carbazole, 0.5–1.5% anthracene, <0.5% phenol, ~1% cresol, and ~1% pyridene compounds and other aromatic hydrocarbons. About 50–60% of the tar consists of high-molecular-weight hydrocarbons.

The principal use of coke is in the blast furnace for steel making, the manufacture of calcium carbide (CaC_2), and other metallurgical processes. The annual world production of coke was about 5.28×10^8 t in 2009 and 4×10^8 t in 1992. The US production of coke has decreased in two times from 22 Mt (2000) to 11.2 Mt (2010) as steel production moved to Asia and the Far East. The coals used for coke production are usually a blend of two or more coals consisting of a high-volatile coal blended with low-volatile coals in a ratio of from 90:10 to 60:40. Such blends increase the rate of coking and produce a better product. The large-sized fused coke (≈ 75 mm) is used in blast furnaces whereas the smaller sized (≈ 20 mm), called *breeze*, is used in boiler-firing, iron ore sintering, electric smelting, and other applications where a purer grade of carbon is required.

The ash content of a metallurgical grade of coke must be less than 10% with a sulfur content <1.5% and a volatile component < 1%. The calorific value of such coke is between 31.4 and 33.5 MJ/kg. One tonne of coal normally produces from 650 to 750 kg of coke.

The development of the coal gasification process allows for the removal of contaminants which make the coal a "dirty fuel." If a new sink for CO_2 can be invented, the need to go fully nuclear could be delayed.

Exercises

1. Discuss the statement "Coal is a dirty fuel."
2. What is the difference between a primary and secondary fuel?
3. What does charcoal and coke have in common? Compare their properties.
4. Describe a laboratory experiment by which wood could be converted into coal.
5. Explain why the proximate analysis of a coal is useful.
6. Explain why chloride salts should be washed from coal before it is used in combustion processes.
7. Wet coal can be air-dried at room temperature. How does this differ from the moisture content of a coal?
8. Explain why the components in an ash from a coal has a different relative concentration compared to the initial coal.
9. Two recent formulas have been proposed to evaluate the heat of combustion of coals:

$$- \Delta H_c (kJ/g) = 357.8C + 1135.7H + 59.4N + 111.9S - 84.5O \qquad (2.9)$$

Lloyd and Bavenport

$$- \Delta H_c (kJ/g) = 351C + 116.1H + 62N + 104S - 110S \qquad (2.10)$$

Boie's formula

Compare the accuracy of these formulas with (2.3) using the data given in Table 2.6.

10. The three-component ternary phase diagram of a bottom ash is shown in Fig. 2.4. (a) How much sand (SiO_2) must be added to each ton of coal (type A) (point x) if the melting temperature of the ash is to be raised from 1,000°C (point x) to 1,200°C (point y)? The ash content of the coal is 10% and the oxides in the ash are graphed as weight %. (b) A second coal (type B) has 15% bottom ash

Table 2.6 Elemental analysis of some solid fuels

	wt.%					
	C	H	N	S	O	$-\Delta H_c$ (kJ/g) measured
Oil shales						
(A) Australia	84.6	11.5	0.56	0.30	3.0	43.7
(B) France	85.1	11.4	1.12	0.32	2.06	43.19
(C) Sweden	85.0	9.0	0.71	1.72	3.6	40.58
Coal						
(D) South Africa	84.3	5.9	2.5	0.70	6.6	35.60
(E) Athabasca tar	82.5	10.0	0.47	4.86	1.7	41.14
Sand						
(F) Syncrude oil						
(a)	87.8	10.1	0.34	0.12	1.6	42.54
(b)	79.2	5.1	1.30	1.30	13.1	32.36

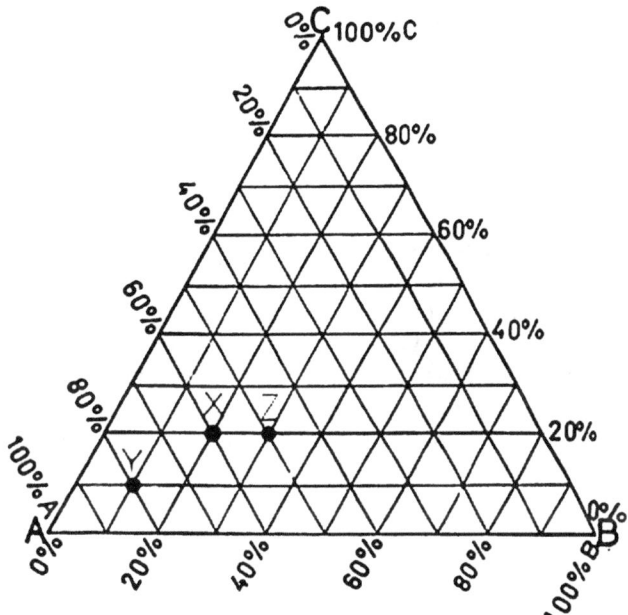

Fig. 2.4 A = SiO_2, B = Al_2O_3, C = Fe_2O_3. Ash at point X has a melting point of 1,000°C and at point Y the melting point is 1,200°C

with a composition as shown in Fig. 2.4 (point z). What would be the composition of the ash if a blend of the coals was burned A:B of 2:1?

11. The proximate analysis of a coal can be determined automatically by thermal gravimetric analysis (TGA). This method involves the automatic recording of the mass of a coal sample while the temperature of the sample is programmed and the atmosphere of the sample controlled. This is illustrated in Fig. 2.5 where the mass of the sample is recorded at 25, 110, and 900°C under the inert atmosphere of nitrogen. Air or oxygen is then introduced while the sample is still at 900°C. (a) Describe the changes which occur at each temperature stage. (b) Determine the proximate analysis of the sample from the results in Fig. 2.5.

Fig. 2.5 Proximate analysis of a coal by TGA

12. Explain why the elements listed in Table 2.4 are classified into the five specific groups.
13. Explain why the calorific value of the volatile fraction of a coal (per unit mass) is normally greater than that of the fixed carbon of the coal.
14. What is coal gas?
15. Explain why coal is often used as a source of active carbon.
16. Why do some coals have a higher calorific value than pure carbon?
17. Explain why using peat for fuel has an adverse effect on the CO_2 balance of our atmosphere.
18. The equilibrium constant, Kp, for the reaction $N_2 + O_2 = 2NO$ as a function of temperature was determined experimentally to be log Kp $= a - b/T$ where $a = 1.63$ and $b = 9.452$ (K). Calculate the equilibrium pressure of NO when air is heated to 800, 1,000, and 1,200 K.

Further Reading

1. http://eia.doe.gov/cneaf/coal/quarterly/qcr.pdf
2. Smil V (2010) Energy myth and realities. Bringing science to the energy policy debate. The AEI Press, Washington, DC
3. Nastny L (2010) Renewable energy and energy efficiency in China: current status and prospects for 2020, vol 182, World watch report. Worldwatch Institute, Washington, DC
4. Reddy BS, Assenza GD, Assenza D, Hasselmann F (2009) Energy power and climate change. Sage Publications, India Pvt. Ltd, New Delhi
5. Pajares JA, Tascon JMD (1995) Coal science, vol 2. Elsevier, Amsterdam
6. Rogers S et al (1995) In: Proceedings of coal liquefaction and gas conversion: contractor's review conference, 2 vol. DIANE, Upland
7. National Research Council Staff (1995) Coal: energy for the future. National Academies Press, Washington, DC
8. Tomeczek J (1994) Coal combustion. Krieger, Philadelphia
9. Speight JG (1994) The chemistry and technology of coal, 2nd edn. Marcel Dekker, New York
10. The IEA International conference, OECD Publications and Information Centre (1994) Clean and efficient use of coal: the new era for low-rank coal. OECD, Washington DC
11. Smith KL et al (1994) The structure and reaction processes of coal. Plenum, New York

12. Berkowitz N (1993) An introduction to coal technology, 2nd edn. Academic, New York
13. Keefer RF, Sajwan KS (eds) (1993) Trace elements in coal and coal combustion residues. Lewis, El Cajon
14. Law BE, Rice DD (eds) (1993) Hydrocarbons from coal. American Association of Petroleum Geologists, Tulsa
15. Yurum Y (ed) (1992) Clean utilization of coal. Kluwer, Norwall
16. Schobert HH et al (eds) (1991) Coal science II, vol 461, ACS symposium series. ACS, Washington, DC
17. Schobert HH (1989) The geochemistry of coal: (I) the classification and origin of coal (II). The components of coal. J Chem Educ 66:242–290
18. Van Krevelen DW et al (eds) (1981/1989) Coal science technology, vol 16. Elsevier, New York
19. Ward CR (ed) (1984) Coal geology and coal technology. Blackwell Science Publ, Maiden
20. Gorbaty ML, Larson JW, Wender I (eds) (1983) Coal science, vol 2. Academic, New York
21. Valkovic V (1983) Trace elements in coal, vol 2. CRC Press, Boca Raton
22. Meyers RA (ed) (1982) Coal structure. Academic, New York
23. Meyers RA (ed) (1981) Coal handbook. Marcel Dekker, New York
24. Greene RP, Gallagher JM (eds) (1980) World coal study-future coal prospects. Ballinger Publ Co, Cambridge
25. Ezra D (1978) Coal and energy. Ernest Benn Ltd, London
26. Tillman DA (1976) Wood as an energy resource. Academic, New York
27. Wood Heat Organization. http://www.woodheat.org
28. International Peat Society. http://www.peatsociety.fi/
29. International Energy Agency Clean Coal Center. http://www.iea-coal.org.uk/
30. World Coal Institute. http://www.wci-coal.com/
31. American Coal and Coke Chemical Institute. http://www.accci.org/
32. Coal Association of Canada. http://www.coal.ca
33. Retexo-RISP. http://www.flyashrecycling.com
34. ISG Resources. http://www.flyash.com
35. International Ash Utilization Symposium Series. http://www.flyash.org/
36. ASTM-Coal and Coke Term. http://www.ASTM.org7DATABASE.CART/PAGES/0121.htm

Chapter 3
Crude Oil

3.1 Introduction

Oil is a major liquid fuel, and it is also the basis of most other liquid fuels. It is formed by refining petroleum or crude oil, which is a very complex mixture of components, composed of many different types of hydrocarbons of various molecular weights. Crude oils are usually classified by the major type of hydrocarbons in the oil. The three major classes are (1) paraffinic (alkanes), (2) aromatic with naphthenic and asphaltic components, and (3) mixed oils containing significant quantities of both aliphatic and aromatic compounds. Typical gas chromatograms of two crude oils are shown in Fig. 3.1. Such chromatograms are often used as a "fingerprint" of an oil in identifying the origin of oil spills. The origin of petroleum is not known with certainty. It is believed to have formed from the accumulation of various marine organic deposits which, by partial bacterial decay, heat, and pressure, eventually formed the crude oil which migrated through the pores, cracks, and fissures in the rocks, forming oil fields in underground structures illustrated in Fig. 3.2.

Crude oil varies in its properties and composite from one country to another and from field to field. The differences are due to viscosity, volatility, and composition (e.g., sulfur content). A list of 14 different crude oils and their primary distillation products is given in Table 3.1.

3.2 Early History

Petroleum and its products, notably asphalt, have been used to waterproof boats since 6000 BC. The Bible refers to many applications from Noah's Ark to Moses' basket. Job refers to "the rock poured me out rivers of oil" (Job 29:2), and even today, one can find asphalt floating up to the surface of the Dead Sea. The Egyptians used asphalt in the construction of the pyramids. It is believed that the Chinese were drilling for oil in the third century AD. In North America, the native Cree Indians used bitumen from the Alberta oil sands to seal their canoes as well as for medicinal purposes. Oil seepage on the surface was the earliest source of crude oil. The first to use petroleum products commercially was Dr. Abraham Gesner of Nova Scotia. He distilled a kerosene fraction (about 1840) to use as an illuminating fuel. This coal oil replaced the smoky smelly fuel from whale oil and vegetable oils used for lighting. Seventy plants were producing coal oil for lighting by 1861. However, by 1855, a Canadian company was producing a better grade of kerosene from oil seepages near Sarnia, Ontario, where the first oil-producing well was drilled (15 m) in 1858. By 1863, there were 30 refineries in Ontario producing oil for $4/bbl which rose to $11/bbl in 1865. However, the oil found in

O.V. Roussak and H.D. Gesser, *Applied Chemistry: A Textbook for Engineers and Technologists*, DOI 10.1007/978-1-4614-4262-2_3, © Springer Science+Business Media New York 2013

Arabian Light Crude Oil

1. heptane
2. octane
3. nonane
4. decane
5. undecane
6. dodecane
7. tridecane
8. tetradecane
9. pentadecane
10. hexadecane
11. heptadecane
12. octadecane
13. nonadecane
14. eicosane
15. heneicosane
16. docosane
17. tricosane
18. tetracosane
19. pentacosane
20. hexacosane
21. heptacosane
22. octacosane

min. 4 8 12 16 20 24 28 32 36 40 44 48 52 56 60 64

Prudhoe Bay Crude Oil

1. benzene
2. heptane
3. toluene
4. octane
5. m-xylene & p-xylene
6. o-xylene
7. nonane
8. decane
9. undecane
10. dodecane
11. tridecane
12. tetradecane
13. pentadecane
14. hexadecane
15. heptadecane
16. pristane
17. octadecane
18. phytane
19. nonadecane
20. eicosane
21. heneicosane
22. docosane
23. tricosane
24. tetracosane
25. pentacosane
26. hexacosane
27. heptacosane
28. octacosane

4 8 12 16 20 24 28 32 36 40 44 48 52 56 60 64 68 72
min.

Fig. 3.1 Typical gas chromatograms of two crude oils where each peak represents one (or more) individual compounds. The height (or area) of the peak is proportional to its concentration in the injected sample. The time taken for the substance to elute from the column, which is temperature programmed to speed up the analysis, is used to identify the substance. The presence and the ratio of the two branched hydrocarbons, pristane (2,6,10,14-tetramethylpentadecane) and phytane (2,6,10,14-tetramethylhexadecane), can be used to identify the source of the oil

ANTICLINE

GAS

OIL

WATER

Fig. 3.2 Oil formation in sedimentary basins where natural gas is usually trapped with the oil

Table 3.1 Various crude oils and yields of main products from primary distillation (per cent by volume)

Country (crude oil name)	Gases	Gasoline	Kerosene	Gasoil/diesel	Residue (fuel oil component)	Sulfur content
North America						
USA (Alaska)	0.8	13.4	11.6	21.5	53.3	Low 1.0% on crude, 1.5% on residue
Western Europe						
UK (Forties)	4.3	22.5	12.2	21.9	39.5	Very low 0.3% on crude, 0.6% on residue
Norway (Ekofisk)	3.3	31.2	13.6	21.6	30.8	Very low 0.1% on crude, 0.3% on residue
Middle East Saudi						
Arabia (Arabia Light)	1.7	20.5	12	21.1	45.1	Very high 3% on crude, 4.5% on residue
Qatar (Qatar)	4.4	29.1	15.9	20.6	30.7	Medium 1.1% on crude, 2.6% on residue
South and Central America						
Venezuela (Tia Juana Pesado)	–	1.4	3.6	14.7	80.8	High 2.7% on crude, 3.0% on residue
Mexico (Isthmus)	1.8	22.9	13.1	22	40.4	High 1.6% on crude, 3.0% on residue
Africa Nigeria (Nigerian Light)	2.9	25.8	14.4	27.7	29.4	Very low 0.09% on crude, 0.02% on residue
Libya (Libyan Light)	2.8	21.6	12.9	22.1	40.9	Low 0.5% on crude, less than 1.0% on residue
Far East and Australasia						
Indonesia (Sumatran Light)	0.5	11.5	9.5	20.6	58.3	Very low 0.08% on crude, 0.1% on residue
Malaysia (Miri Light)	1.9	28.1	16.7	32.1	21.3	Very low 0.04% on crude, 0.1% on residue
Australia (Gippsland)	2.3	36	13.8	24.7	23.5	Very low 0.08% on crude, 0.2% on residue
CIS China						
(Ural) Russia	2.2	20.9	14.7	19.5	43.1	High 1.5% on crude, 2.6% on residue
China (Daqing)	0.5	9.8	6.9	16.5	66.4	Very low 0.1% on crude, 0.15% on residue

Pennsylvania proved to contain less sulfur and thus produced a better quality kerosene. The Canadian oil price then dropped to $0.50/bbl.

3.3 World Production of Crude Oil

In 1950, world oil production was 519 Mt or 10 million bbl/day. This doubled by 1960 (1.05 Gt) and doubled again by 1970 (2.3 Gt).

The production continued to increase (3.0 Gt or about 22 billion bbl in 1979) and seemed to level off at about 20 billion barrels during the 1980s reflecting the changing attitude toward energy. However, the world's annual production of crude oil has gone to over 22 billion bbl in 1992 in spite of an economic slump. The percentage contributed by different geographical areas in 1998 is shown in Fig. 3.3. The 1992 production and consumption levels for the major producers and consumers is given in Table 3.2.

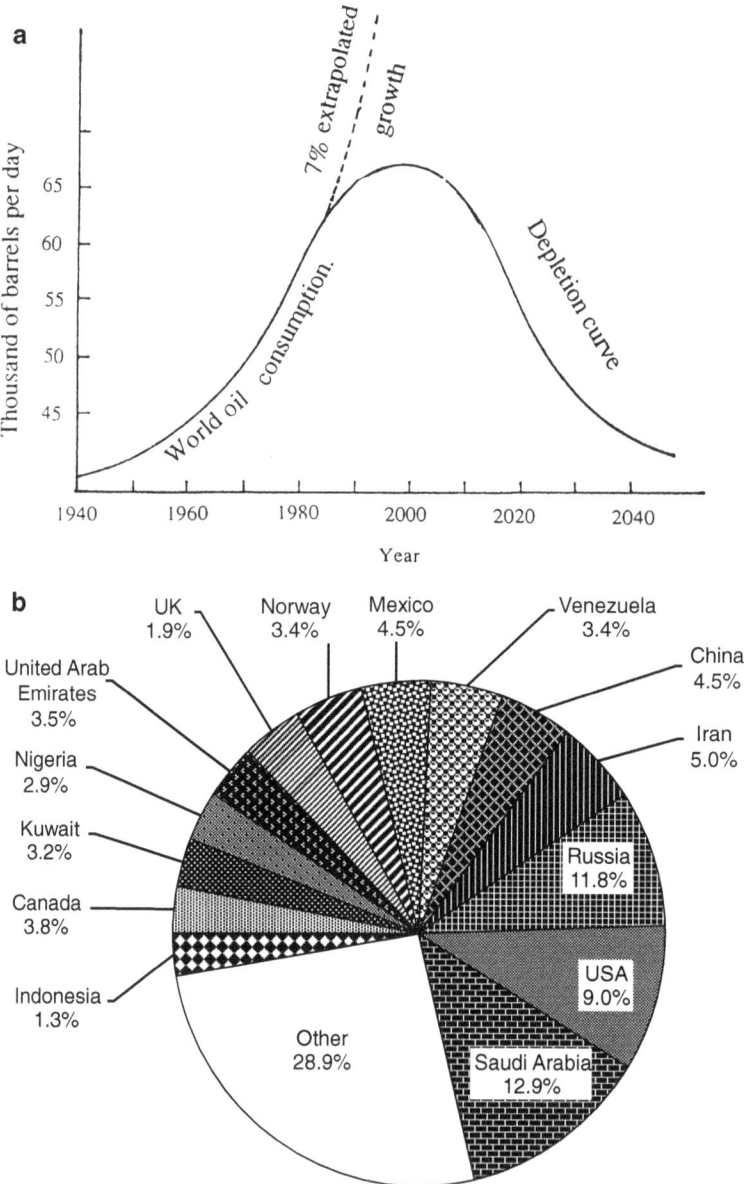

Fig. 3.3 Percentage of world crude oil production for various countries in 2005 (World total = 82.4 M bbl/day) (see also Microsoft Excel file)

The USA used to be the world's leading producer of crude oil (7.75 M bbl/day in 1965, approximately 8.7 M bbl/day in 1980, and 7.2 M bbl/day in 1992). The decrease in production was more than balanced by increased imports of 7.0 M bbl/day in 1992. Thus, the percentage contributed by the USA to the world supply decreased from 45% in 1956 to 13.7% in 1979 to less than 12% in 1992. Even though Canada increased its production during this same time period, 3% in 1956 went to 2.6% of the total world production in 1992. In the same interval, new producing areas were developed, notably in the Sahara, the Arctic, Australia, and the North Sea. The Middle East also showed a marked increase in oil production from 11.6 M bbl/day in 1983 to 17.6 M bbl/day in 1992. Soviet Union (USSR) increased production of crude oil from 9.3% of the world's oil in 1956 to about 20% of

Table 3.2 World crude oil reserves and marketing (IEA 2006)

Country	Reserves 10^9 bbl (2007)	Production capacity 10^6 bbl/ day (2006)	Consumption 10^6 bbl/ day (2006)
Algeria	11.9	2.1	0.3
Australia	4.2	0.6	0.9
Brazil	12.5	2.1	2.2
Canada	25.6	3.2	2.3
China	16.2	3.7	7.2
Egypt	3.6	0.7	0.6
India	3.8	0.8	2.7
Indonesia	4.8	1.1	1.7
Iran	133.0	4.1	1.2
Iraq	125.1	2.0	0.6
Japan	–	–	5.2
Kuwait	100.1	2.7	0.3
Libya	34.9	1.8	0.3
Mexico	11.7	3.7	2.1
Nigeria	37.2	2.4	0.3
Norway	7.1	2.8[a]	0.2
Russia	74.4	9.7	2.1
Saudi Arabia	262.2	10.6	2.8
South Korea	–	–	2.2
UK	3.8	1.6	1.8
United Arab Emirates	70.6	2.9	0.4
USA	20.9	7.4	20.7
Venezuela	52.9	2.8	0.6
Other	126,8	13.6	26.3
Total	1,143.4	82.4	85.0

[a]North Sea

Uncertainty. $\pm$: $a = 1, b = 2, c = 3, d = 4, e = 5, f = 6, g = 10$

the world's oil in 1992. Russian Federation produced about 12% of the world's oil in 2006 (about 9.7 million barrels/day in 2006) after Soviet Union crashed in 1990.

The Russian oil reserves are as vast as the country. The major producing areas are West Siberia and the Volga-Urals region which produce about 400 Mt and 200 Mt/year, respectively. Many of the oil fields of northern Siberia are not connected by pipelines and thus await further development. This is similar to Canada where arctic gas and oil fields are not as yet being tapped.

World oil production and consumption was projected to reach 75.9, 82.4, 86.5, and 111 M bbl/day in the years 2000, 2005, 2010, and 2035, respectively. Total world reserves have been estimated to be about from 1.354 billion barrels to 12 billion barrels. Based on projected uses and available reserves at present prices, it would appear that oil will be available for only 30–50 years or until the year 2040. It must be pointed out that all these predictions are based on present technology, and in the USA alone, there is an additional 3×10^{11} bbl of oil which can only be economically recovered when the price of oil has risen to about $30/bbl.

Another source of oil is oil shales and tar sands which require special processing to extract the oil. Tar sands are found throughout the world, and deposits are known to exist in 49 countries. Major deposits are found in Venezuela and Canada. Alberta's tar sands contain an estimated 9×10^{11} bbl of crude bitumen from which about 3×10^{11} bbl of synthetic crude oil can be obtained. The tar sands consist of particles of sand held together by a water and oil coating as shown in Fig. 3.4. The process by which the oil is separated from the water and sand involves subjecting the mixture to hot water and steam and skimming off the oil. Since over 90% of the oil sands cannot be surface mined

Fig. 3.4 Typical structure of tar sand consisting of an oil-sand-water mixture

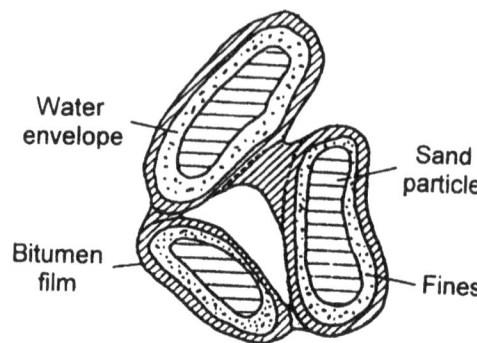

economically, more elaborate techniques are being developed. For in situ mining, the steam can be injected into the drill holes to heat up the sands and fluidize the oil. Other processes involve controlled ignition of the tar sands and air pressure to drive the freed oil toward the production holes.

3.4   Crude Oil Processing

Crude oil is treated by physical and chemical processes to produce the various petroleum products. The early use of oil was in the preparation of kerosene. This was accomplished by batch distillation which separated the mixture of hydrocarbons by boiling points (vapor pressure). The modern distillation process (see Fig. 3.5) is designed to operate continuously. The temperature gradient of the column separates the crude oil into fractions according to specific boiling point ranges. These are shown in Table 3.3.

The demand for the various fractions in a crude oil seldom coincides with the distillation yields. Hence, it is necessary to chemically alter the various proportions of the natural oil fractions. This is called *reforming* and is accomplished by the use of various catalysts. The main reactions are:

(a) The dehydrogenation of cyclic paraffins to aromatic hydrocarbons and H_2.
(b) Isomerization of normal paraffins to branched paraffins (alkylation).
(c) Hydrocracking of paraffins and cycloparaffins to short chain alkanes.
(d) Visbreaking reduces the viscosity of the oil by cracking the heavy components.
(e) Polymerization is the formation of large molecules from the catalytic recombination of two or more smaller ones.

These reactions are designed to provide the fuels required and the chemicals needed for the petrochemical industry. Another reaction which is important in reformulation is the removal of sulfur from the oil as H_2S by catalytic reactions of Co and Mo on Al_2O_3. Nitrogen forms NH_3 on Ni/Mo catalysts.

3.5   Petroleum Products

The principal consumer of petroleum products is the transportation industry with the internal combustion engine (ICE) as the major application. This is shown in Table 3.4. The ICE includes the spark ignition and diesel engines, common for the automotive vehicle, and the gas turbine used for aviation and industrial applications. Each class of engine requires a special type of fuel. The heats of combustion of the fuels may not be significantly different, but the composition of the components and their rates of reactions with oxygen could determine its application. The characteristics of gasoline

Fig. 3.5 Distillation, the first step in oil refining, separates crude oil into a number of products. First, the crude oil is heated by being pumped through pipes in a furnace. The resulting mixture of vapors and liquid goes to a tower where the vapors rise, condense on trays, and are drawn off through pipes as products. The part of the crude that does not boil in the first distillation step is reheated and distilled in a vacuum. Again, vapors rise, condense, and are drawn off as products. Some of the very heavy oil, which does not boil even under the reduced pressure, is used in factories and ships or made into asphalt

Table 3.3 Primary distillation fractions from crude petroleum

Fraction	Boiling point range (°C)	Composition	Uses
Gas	up to 25	C_1–C_4	Petrochemical fuel
Petroleum ether	25–60	C_5–C_7	Solvent
Gasoline	0–17	C_6–C_{11}	Fuel for ICE
Kerosene	17–220	C_{11}–C_{13}	Jet fuel
Heating oil	220–250	C_{13}–C_{16}	Fuel oil
Lubricating oils	350–400	C_{16}–C_{20}	Lubricants and grease
Paraffins	400–500	C_{20+}	Candles
Asphalts	>500	C_{30+}	Road surfaces

Table 3.4 Consumption of petroleum products in Canada (in thousands m^3/day, IEA 2006)

Name	1983	1990	1994	2005
Motor gasoline	91.0	103.5	109.1	111.5
Aviation turbo fuel	10.7	15.7	13.0	19.0
Diesel fuel	38.4	46.2	49.8	9.3
Light fuel oil	23.5	17.6	15.5	86.4
Heavy fuel oil	25.2	29.1	19.8	–
Lubricating oil	2.6	2.5	2.4	–
Lubricating grease	0.10	0.083	.062	–
Total domestic demand	236.1	267.3	267.3	372.3

and diesel fuels are discussed in Chap. 4, and lubrication oils and greases are discussed in Chap. 8. Aviation fuel has not changed much since its early use in the jet engine. The stability of the fuel has improved, and specific corrosion inhibitors, antioxidants, and antistatic additives have been included in formulations. Combustion in a gas turbine is continuous in contrast to the intermittent combustion of the ICE, and thus, the "antiknock" quality of the fuel is not important. The thrust of the engine is generated by the energy released during the combustion process which takes place at constant pressure.

The heat generated thus causes an increase in volume and gas flow to the turbine blades. An important restriction is the material of construction which must withstand the high temperatures developed.

The jet fuel must have a low tendency to deposit carbon which can lead to high thermal radiation causing damage to the engine components. Likewise, sulfur in fuel must be very low since it enhances carbon deposition and corrosion. These factors and the physical properties require the fuel to have a low freezing point and be free of wax which, at low temperatures, can precipitate out and clog fuel lines. At high altitudes, temperatures as low as $-50^\circ C$ are normal, and $-60^\circ C$ is the specified maximum freezing point for aviation fuel. Similarly water must be absent. Anti-icing additives are present in the fuel to ensure that at low temperatures, any water which does come out of the fuel will not form large ice crystals which can clog the fuel lines.

Aviation fuel (avogas), which is designed for use in piston engines, still contains lead [0.53 mL of tetraethyl lead (TEL) per liter of fuel]. One hazard, tolerable on the ground but deadly in the air, is the vapor lock, and special care is taken to ensure that this does not happen. Hence, automotive fuel, though often equivalent to avogas but half the cost, cannot be used as a substitute.

Oil seepages on the surface were the early sources of crude oil. Kerosene was initially used primarily for lamps. Today, it is generally used for heating as space heaters, where a catalytic surface maintains a hot radiating source of heat. In countries where electricity is expensive, e.g., Japan and Israel, it is not uncommon to see such appliance heaters in homes and offices. Such open flames are sufficiently hot to produce NO_x as well as some hydrocarbon products which are readily detected by their odor. The main characteristics of a domestic grade of kerosene, which burns on a catalytic surface, are low sulfur and controlled smoke point and volatility.

Coal for heating was replaced by oil during the 1940s, and oil was replaced by natural gas soon after. The natural gas pipeline, however, cannot reach every home, and as a result, fuel oil as well as bottled gas (propane) is still used in many locations. The fuel oil differs only slightly from diesel fuel, and because its ignition property is not important, it usually contains more aromatics and olefins than diesel fuel. The most important property of fuel oil is its storage stability, and additives prevent sludge formation. Viscosity is another controlled characteristic of the fuel because the size of the droplets in the burning spray is critical for efficient combustion.

World production of petroleum wax is about 2 Mt/year. The major product is candles (about 25%), replacing beeswax and tallow used a century ago. The present applications include coating on cartons for juice, milk, and other food products and for waterproofing and preserving freshness. The major concern with the use of wax or any petroleum product for food and edibles is the need to have it completely free of the polynuclear aromatic hydrocarbons (PAH). Trace amounts have been detected in liquid petroleum used for oral consumption. However, our daily diet usually contains much more PAH than what is in a spoonful of liquid petroleum taken occasionally. Bitumen or asphalt is the black or dark brown solid residues from the distillation of crude oil. Naturally occurring bitumen is found in the USA, Europe, and other parts of the world. Commercial bitumen comes in various degrees of hardness and softens at a temperature of $25–135^\circ C$. When air is blown into molten bitumen, oxidation or dehydrogenation occurs, and the material becomes rubbery and more penetrating than the regular grades. Bitumens can be applied as a melt, emulsion, or from a solution. The application of bitumen to roads usually involves the inclusion of sand and gravel, and in North America, this is normally referred to as asphalt.

The inclusion of other aggregates which have been tested and added for various road systems includes sulfur, shredded discarded tires, limestone (<200 mesh), and crushed concrete. Other applications of bitumen are in roofing, flooring, and as an anticorrosion coating on surfaces exposed to corrosive atmosphere, aggressive soils, and chemicals. The main advantage of bitumen, in many of its applications, is that it is cheaper than alternative materials. The world production of bitumen in 1993 was about 17 Mt.

The reserves of oil are limited, and its uses as a fuel for its heat value may eventually become a luxury which few will be able to afford. The petrochemical industry supplies the plastics and resins we use daily, the synthetic fiber for our clothes, and the detergents for our soaps and washings, as well as the chemicals and solvents for industrial use. World petrochemicals amount to about 1/7 of total steel production and about seven times the aluminum produced by weight.

3.6 Synthetic Oil

It is possible to produce oil from coal either by direct hydrogenation at high temperatures and high pressures or by the syngas route followed by the Fischer–Tropsch process. Though these processes can make an oil which is more expensive than wellhead crude, in exceptional strategic circumstances (war, oil embargo, etc.), it has been produced and used as a viable substitute.

The first process was studied by Berthelot in 1867 and was further developed in Germany by Bergius in 1910. The early Bergius process involved the reaction of H_2 under atmospheric pressure with pulverized coal suspended in an oil heated to about 450°C in the presence of a catalyst such as stannous formate or Mo. The liquid oil product is separated from the solid residue and processed as ordinary crude oil. Modern developments in this coal liquefaction approach include (1) Exxon Donner Solvent (EDS) process, (2) the HRI H-Coal process, and (3) the Gulf Solvent Refined Coal SRC-II process. The major improvement of these processes over the Bergius process is in the catalyst used, allowing for milder reaction conditions.

The EDS system is a direct liquefaction procedure in which coal is chemically reacted at about 540°C for 30–100 min with a recycle solvent that is rehydrogenated between passes to the liquefaction reactor. Each tonne of dry coal can give up to 3 bbl of synthetic crude oil. Some examples of donor solvents are given in Table 3.5. A schematic block diagram of the overall process is shown in Fig. 3.6. A pilot plant processing 225 t/day has been in operation since June 1980, and the product yields obtained are given in Table 3.6. The advantage of the EDS process is that it makes maximum use of petroleum refining technology, uses a wide range of coals, and has no catalyst to be poisoned or recycled.

The H-Coal process was developed by Hydrocarbon Research Inc. and uses a catalytic ebullated reactor. The pulverized coal-oil slurry is mixed with hydrogen (150 atm) fed into the reactor (450°C) where the Co/Mo catalyst converts the coal and hydrogen to oil.

The SRC-II process, like the EDS process, is claimed to be free of catalysts and the problems associated with their use. A schematic diagram of the process is shown in Fig. 3.7. The coal is dissolved in a process-derived solvent at elevated temperatures and reacts with hydrogen under pressure to form the oil. The solvent is recovered by vacuum distillation, and the residue (ash plus insoluble carbon) is separated.

The second method by which synthetic oil can be made is by the Fischer–Tropsch process. This can use either coal or natural gas as the starting material since the process uses synthesis gas (a mixture of CO and H_2) which can be made from coal by the water gas reaction:

$$C + H_2O \rightarrow CO + H_2 \quad \Delta H^0 = 131.28 \text{ kJ/mol} \tag{3.1}$$

Table 3.5 Some potential donor components in solvent

Compound	Structure	Compound	Structure
Tetralin		Tetrahydrophenanthrene	
Tetrahydroaccnaphthene		Octahydrophenanthrene	
Hexahydrofluorene		Hexahydrophyrene	

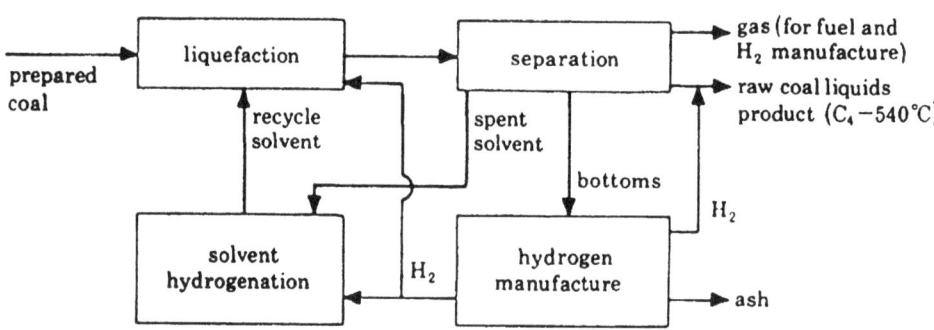

Fig. 3.6 Simplified block diagram of the EDS process

Table 3.6 EDS products and their disposition

Stream	Disposition	Yield (% of liquids)
C_4-177°C naphtha	Motor gasoline	40–80
	Chemicals	
177–204°C distillate	Jet fuel	10–30
	Turbine fuel	
204–399°C solvent	Home heating oil	
	Fuel oil blendstock	
399–538°C vacuum gas oil, coker gas oil	Fuel oil blendstock	10–30
Scrubber liquids	Fuel oil blendstock	

or natural gas or other hydrocarbon blends by steam reforming

$$CH_4 + H_2O \rightarrow CO + 3H_2 \quad \Delta H^0 = 206.1 \text{ kJ/mol} \tag{3.2}$$

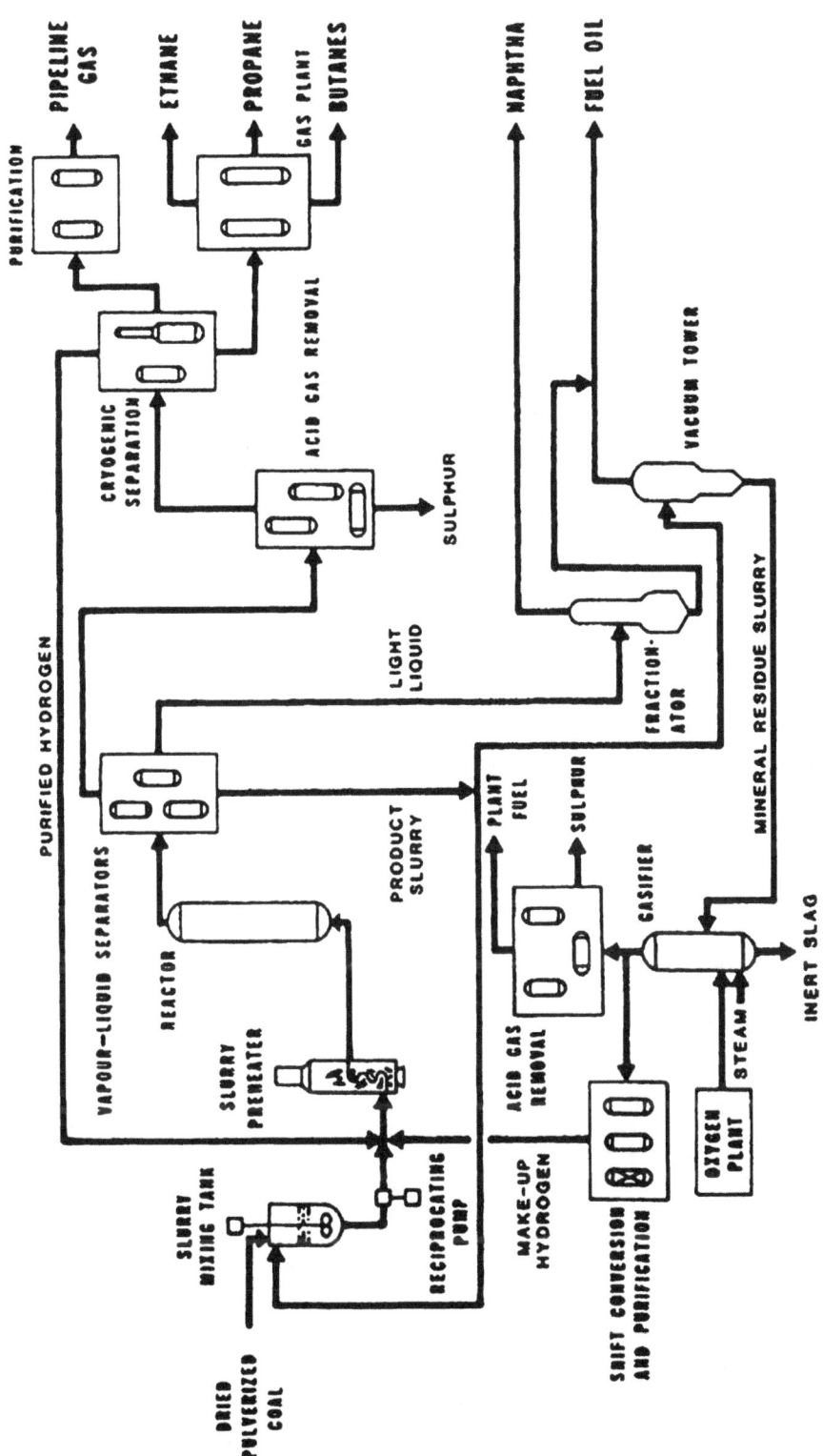

Fig. 3.7 The SRC-II process

Fig. 3.8 A Lurgi dry-ash gasifier

The syngas is then balanced for a preferred H_2/CO ratio of 2 since the reaction is

$$2H_2 + CO \rightarrow (CH_2)_x + H_2O \tag{3.3}$$

The reaction occurs on various catalysts such as iron oxide which is the most commonly used material. Some oxygenated compounds such as alcohols and acids (1%) are also formed.

The Fischer–Tropsch process involves the following steps:

1. Manufacture of syngas
2. Gas purification—removal of S, H_2O, etc.,—and H/C balance
3. Synthesis of oil on fresh catalyst
4. Condensation of liquids and removal of light fractions (gasoline)
5. Distillation of remaining products: 35% gas–oil and 30% paraffins

The gasoline is primarily straight chain hydrocarbons of low octane rating and must therefore be reformulated for use.

The only commercial plant converting coal into oil is in South Africa and known as the Sasol I and II. Sasol I was built in 1956 and produced 10,000 bbl/day of synthetic crude oil using fixed bed (with precipitated iron catalyst) and fluidized bed (with powdered iron catalyst) reactors. A typical gasifier is shown in Fig. 3.8.

The construction of Sasol II plant was started in 1976 and completed in 1980; it costs 3×10^9 and produced 40,000 bbl/day oil from 36 gasifiers, each of which processed 1,200 t/day of coal. The third plant, Sasol III, was designed to be the most advanced conversion plant in the world; it costs 5.5×10^9 and is set to produce slightly more than Sasol II. Total synthetic crude oil production

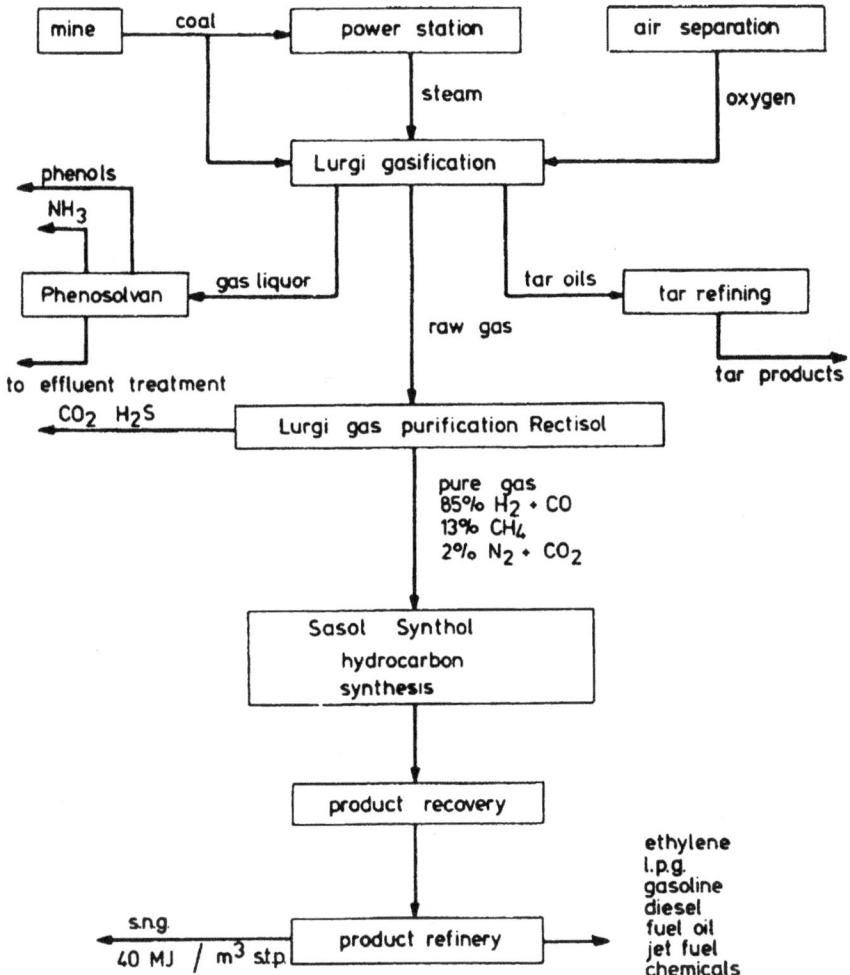

Fig. 3.9 Sasol Synthol process: simplified flowsheet for Sasol II

was expected to be 120,000 bbl/day and represented over half of the country's needs. A simplified flow sheet of Sasol II and III is shown in Fig. 3.9. A photograph of Sasol II is shown in Fig. 3.10.

It should be noted that the production cost of the synthetic crude oil is higher than the world price for wellhead crude, but the value of independence cannot be assigned a simple price.

Exercises

1. The populations of the USA, Canada, and Japan in 2006 were approximately 313, 34, and 126 M, respectively. Calculate the annual per capita oil consumption for each country (see Table 3.2) and comment on the differences.
2. Calculate the time required to exhaust the world's oil reserves if the 6 billion people on earth all consume oil at the rate of 28 bbl each per year.
3. Explain the shape of the theoretical depletion curve of world oil reserves (Fig. 3.11).
4. Discuss the problems associated with the use of waste glass as an aggregate in asphalt for roads.
5. Write the reaction for the syngas production from ethane.

Fig. 3.10 General view of Sasol II at Secunda in the Eastern Transvaal

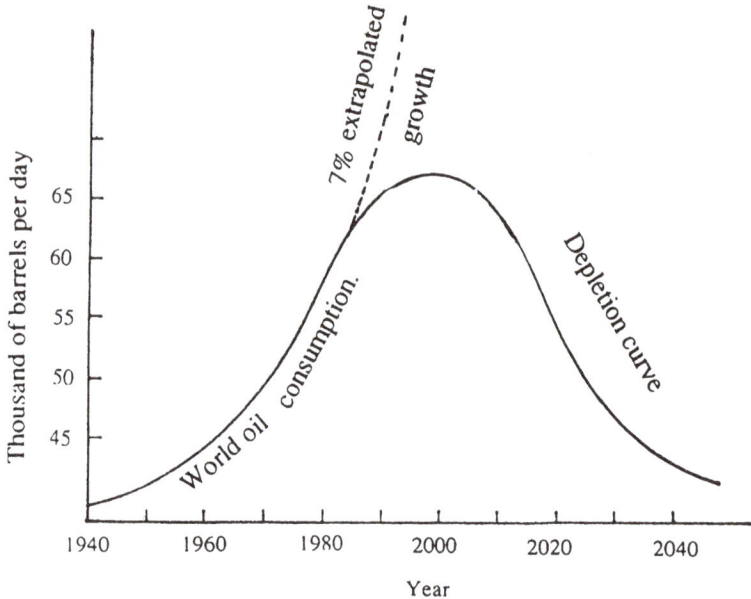

Fig. 3.11 World consumption of oil and theoretical depletion curve for world oil reserves

6. The reaction $CO + H_2O \rightarrow H_2 + CO_2$ is called the water-shift reaction and is often used in the Fischer–Tropsch process. Explain why this reaction is used in this process.
7. Why is the direct liquefaction of coal with hydrogen to form oil a more sensible process than the production of crude oil via the Fischer–Tropsch process for the production of synthetic fuel?
8. Estimate the standard heat of reaction for reaction 3.3.

Further Reading

1. Smil Vaclav (2010) Energy myth and realities. Bringing science to the energy policy debate. The AEI Press, Washington, DC, p 215
2. Nastny L (2010) Renewable energy and energy efficiency in China: current status and prospects for 2020. World watch report 182. Worldwatch Institute, USA, p 48
3. Reddy BS, Assenza GD, Assenza D, Hasselmann F (2009) Energy power and climate change. Sage Publications, New Delhi, p 349
4. Berkowitz N (1997) Fossil hydrocarbons. Academic, San Diego
5. Kopsch H (1996) Thermal methods in petroleum analysis. VCH Pub, New York
6. Ogden PH (ed) (1995) Recent advances in oilfield chemistry. CRC Press, Boca Raton
7. Matar S, Hatch LF (1994) Chemistry of petrochemical processes. Gulf Publ Co., Houston
8. Liebsen D (1993) Petroleum pipeline encyclopedia. OPRI, Boulder
9. Kennedy JL (1993) Oil and gas pipeline fundamentals, 2nd edn. Penn Well Bks, Tulsa
10. Speight JG (1991) The chemistry and technology of petroleum, 2nd edn. M. Dekker Inc., New York
11. Campbell CJ (1991) The golden century of oil, 1950–2050: the depletion of a resource. Kluwer, Norwall, MA
12. (1983) Petroleum handbook (Shell), vol 6. Elsevier, New York
13. Hocking MB (1977) The chemistry of oil recovery from bituminous sands. J Chem Educ 54:725
14. Berkowitz N, Speight JG (1975) The oil sands of Alberta. Fuel 54:138
15. Oil and Gas News and Prices. http://www.petroleumplace.com/
16. Canadian Association of Petroleum Producers. http://www.capp.ca/
17. Alberta Syncrude. http://www.syncrude.ca
18. American Petroleum Institute. http://www.api.org/
19. Institute of Petroleum. http://www.petroleum.co.uk/
20. World Petroleum Congress. http://www.world-petroleum.org/
21. Organization of Petroleum Exporting Countries (OPEC). http://www.opec.org/
22. National Petroleum Council. http://www.npc.org/

Chapter 4
Liquid Fuels

4.1 Introduction

Liquid fuels are a major energy factor which determines the course of transportation, and the automobile is one of the most important consumers of such fuel. These liquid fuels include diesel oil, gasoline, liquid propane, alcohol (both methyl and ethyl), as well as the less common liquids such as ammonia and hydrazine. Though the boiling point of propane ($-42°C$) and ammonia ($-33°C$) are below ambient temperature (in most places), these substances are still classed as liquids because they can be stored as liquids at room temperature ($25°C$) at the modest pressure of about 10 atm.

Whale oil was used for lighting and heating long before the extensive use of petroleum oil. Similarly, vegetable oils (or biomass liquids and saps) are also potential fuels (see Chap. 1), but their modest production at present precludes their wide spread use. Perhaps when the energy farm has developed, it will be possible to consider biomass fuels as an alternative to fossil-based fuels.

Of the various liquid fuels, diesel oil requires the minimum of preparation from petroleum crude oil and therefore is usually the cheapest of the fuels.

4.2 Diesel Engine

The diesel engine was described in a patent in 1892 by Dr. Rudolf Diesel who originally designed it to operate using coal dust as the fuel. However, the characteristics of such fuel were not very reproducible, and it was quickly replaced by oil.

In the diesel engine, air is pulled into the cylinder and compressed to approx. 35 atm. This compression is effectively adiabatic and causes the temperature of the air to increase to about 550°C. At the end of the compression stroke, when the piston is at the top of the cylinder [top dead center (TDC)], an oil spray is injected into the hot air where it ignites on vaporization. The heat of combustion raises the temperature of the gas mixture which now expands at constant pressure as the piston moves down, increasing the volume of the gases. The heat generated and the larger volume of product gases thus further increase the volume as the pressure drops.

An ideal PV diagram for a diesel engine is shown in Fig. 4.1 where the four strokes are:

1. Compression $1 \rightarrow 2$
2. Expansion/combustion $2 \rightarrow 3 \rightarrow 4$
3. Exhaust $4 \rightarrow 1 \rightarrow 5$
4. Air intake $5 \rightarrow 1$

O.V. Roussak and H.D. Gesser, *Applied Chemistry: A Textbook for Engineers and Technologists*, DOI 10.1007/978-1-4614-4262-2_4, © Springer Science+Business Media New York 2013

Fig. 4.1 Ideal PV diagram of
a four-stroke diesel engine.
(1) Compression 1 → 2. (2)
Expansion/combustion
2 → 3 → 4. (3) Exhaust
4 → 1 → 5. (4) Air intake
5 → 1

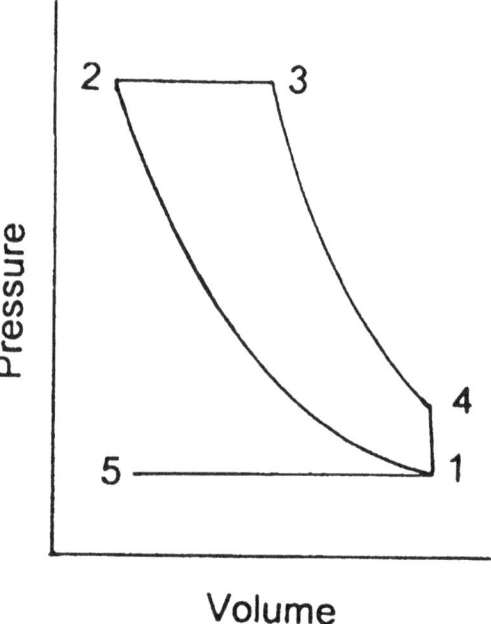

In the compression stage, air is compressed to about 1/20 of its initial volume, i.e., the engine has a compression ratio (CR) of at least 20, and this high ratio accounts for the high efficiency of the engine.

Diesel engines are classified into indirect and direct injection engines. The former have a precombustion chamber where the fuel is initially injected and where combustion starts after an induction delay. The fuel-rich flame then expands into the main chamber. Such engines are common in high-speed diesel passenger cars.

Direct injection engines are normally used in larger engines with lower speed and with cylinder bores greater than 12 cm in diameter. They are more efficient and easier to start than indirect injection engines of comparable size.

Diesel engines are also classified according to their speed (rpm). Low-speed engines run at 100–500 rpm and are usually large stationary installations or marine engines. Medium-speed engines operate from 500 to 1,200 rpm and are usually installed in power generators, power shovels, tractors, and locomotives. The diesel-electric locomotive runs at about 20–28% efficiency, compared to the coal-fired steam engine which was only 5–8% efficient, or the electric powered engine which is about 23% efficient. The high-speed diesel engines run at 1,200–2,000 rpm and are found in the automobile, light trucks and buses, and aircraft. The type of fuel used in a diesel engine is determined, to a great extent, by the speed of the engine.

4.3 Diesel Fuel

The diesel fuel injected into the hot compressed air ignites only after a short delay. This ignition delay depends on the composition of the fuel. It is longer for aromatic hydrocarbons and cycloparaffins than for olefinic and paraffinic fuels. Thus, the best grade of diesel fuel consists of long straight chain hydrocarbons which can spontaneously ignite in the hot compressed air. The absence of an induction period is important since it results in a loss of efficiency. Long ignition delay times result in rapid combustion and sharp pressure rise, causing the engine to run roughly. The grade of diesel fuel is determined by an empirical scale and is based on the ignition characteristics of the fuel, which is compared with a blend of hexadecane (cetane) rated as 100 and 1-methylnaphthalene

Table 4.1 Comparison of components in diesel fuel and performance characteristics

	Conventional diesel fuel (Wt. %)	Cracked gas oil (Wt. %)	Synthetic diesel fuel (Wt. %)	Ignition quality	Cold flow character	ΔH comb by volume	Density	Smoking tendency
n-Paraffins	39	19	17	Good	Poor	Low	Low	Low
iso-Paraffins							Low	Low
				Low	Good	Low		
Naphthenes	34	16	37	Moderate	Good	Moderate	Moderate	Moderate
Olefins				Low	Good	Low	Low	Moderate
Alkyl benzenes	18	34	36	Poor	Moderate	High	High	High
2-Ring aromatics	8	28	8	Poor	Moderate	High	High	High
3-Ring aromatics	1	3	2	Poor	Moderate	High	High	High

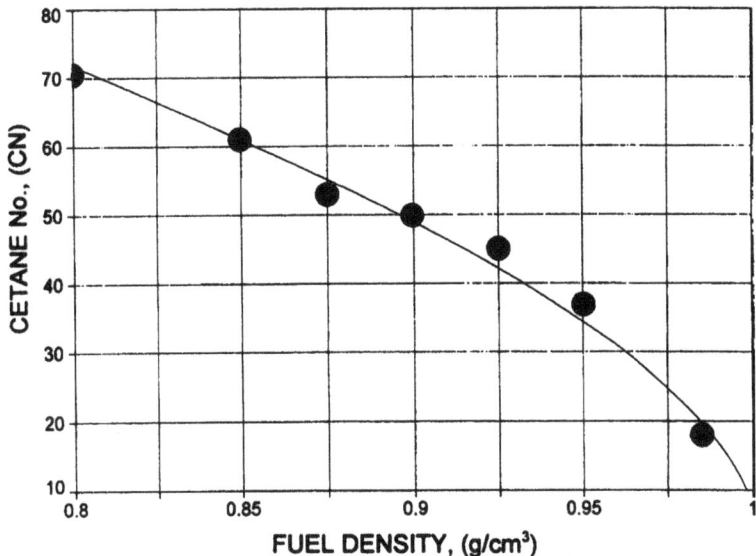

Fig. 4.2 General relationship between cetane number (CN) and density of the fuel

(α-methylnaphthalene) which is rated at 0. The recent use of 2,2,4,4,6,8,8-heptamethylnonane (HMN), CN = 15, has been introduced as a low cetane standard because it can be obtained easily in high purity. Low-grade fuels have a cetane number (CN) equal to about 20 which is suitable for low-speed engines, whereas a high-grade fuel will have CN = 70. High-speed engines require a fuel with CN = 50 or more. Table 4.1 shows some of the performance characteristics of diesel fuel.

A crude indication of the CN of a fuel can be obtained from its density. A plot of CN against density for various grades of fuel is shown in Fig. 4.2. In general, the CN is reduced by the presence of aromatics in the fuel.

Corrosion inhibitors and gum reducers are other additives designed to improve the quality of a fuel. In cold weather, it is customary to add ethanol to the fuel (not more than 1.24 mL/L) to prevent ice from clogging fuel lines.

The addition of barium compounds to diesel fuel has been shown to reduce the emission of black smoke. However, its use is limited, and proper engine maintenance can do much to reduce smoke in the exhaust.

Table 4.2 North American specification for diesel fuel (average of US and Canadian)

	Type AA	Type A	Type B	ASTM 1-D	2-D	4-D
Flash point °C min.	40	40	40	38	52	54
Cloud point °C max.	−48	−34	–	–	–	–
Pour point °C max.	−51	−39	–	–	–	–
Kinematic viscosity 40 °C cSt	min. 1.2 to	1.3	1.4	1.3	1.9	5.8
	max.–	4.1	4.1	2.4	4.1	26.4
Distillation						
90 % recovered °C max.	290	315	360	288	282″ min. to 338 max.	
Water and sediment, % vol. max.	0.05	0.05	0.05	0.05	0.05	0.5
Total acid number, max.	0.10	0.10	0.10	–	–	–
Sulfur, % mass max.	0.2	0.5	0.7	0.5″	0.5°	2
Corrosion, 3 h @ 100°C max.	No. 1	No. 1	No. 1	No. 3	No. 3	–
Carbon residue (Ramsbottom)	0.15	0.15	0.20	0.15	0.35	–
On 10 %″ bottoms, % mass max.						
Ash, % wt., max.	0.01	0.01	0.01	0.01	0.01	0.1
Ignition quality, CN, min.	40	40	40	40°	40″	30

Additives can be incorporated to improve the quality of diesel fuel. For example, ethyl nitrite, ethyl nitrate, and isoamyl nitrite when added to an oil will increase its CN value. The addition of 2.5% by volume of amyl nitrite to a diesel oil (CN = 26) increased the CN to 44. Additives such as amyl nitrate ($C_5H_{11}ONO_2$) when added at about 0.1% by volume will increase the CN by 4, and 0.25% will add 7 to the CN of the fuel. Other nitrates such as heptyl and octyl nitrates are also ignition improvers. A similar effect is obtained when ammonium nitrate is added to oil. A 2% by weight addition of a solution of 5 M NH_4NO_3 in water can increase the CN of a diesel fuel from 39 to 42. Other additives will prevent gum formation, decrease surface tension permitting a finer spray, or reduce the change in fuel properties due to changes in temperature.

Diesel fuel is a complex mixture of hydrocarbons which includes paraffins, naphthenes, and aromatics, and at low temperatures, phase separation can occur causing engine failure. The temperature, Tps, at which phase separation occurs is called the *cloud point* and is determined under standard conditions of cooling, e.g., 1°C/min. Long chain polymer additives at as low as 0.1% can lower the cloud point, Tps, by several degrees.

The recommended cloud point of a fuel is 6°C above the pour point, which is the temperature at which the fuel ceases to flow readily. A fuel is normally blended so as to make the pour point at least 6°C below normal driving temperatures. The ASTM (D-975-78) specification of diesel fuel oils is given in Table 4.2.

Vegetable oils are composed primarily of glycol esters of fatty acids with the general formula where R′, R″, and R are the same or different alkane or alkene radicals with carbon chains from 16 to about 20. The triglycerides are too viscous and of too high a molecular mass and too low a vapor pressure to be a useful diesel fuel. However, it is possible to convert the triglycerides to monomethyl esters by a transesterification process which can be represented by the equation

$$
\begin{array}{l}
\quad\quad\quad\quad\quad\quad\; O \\
\quad\quad\quad\quad\quad\quad\; \| \\
H_2C\!-\!O\!-\!C\!-\!R' \\
\;| \\
HC\!-\!O\!-\!C\!-\!R'' \\
\;| \quad\quad\quad\; \| \\
\quad\quad\quad\quad\; O \\
H_2C\!-\!O\!-\!C\!-\!R \\
\quad\quad\quad\quad\; \| \\
\quad\quad\quad\quad\; O
\end{array}
$$

Oil	CN
Babassu	63
Palm	62
Peanut	54
Soybean	45
Sunflower	49

Table 4.3 Experimental CNs for selected transesterified vegetable oils (methyl esters)

The methyl esters are usually straight chain hydrocarbons and have relatively high cetane values shown in Table 4.3. The ethyl esters are between 2 and 5 units higher than the corresponding methyl esters. Such diesel fuel, though effective, is still too expensive and not competitive with petroleum-based fuels.

$$
\begin{array}{l}
H_2C-O-\overset{\overset{O}{\parallel}}{C}-R' \\[4pt]
HC-O-\underset{\underset{O}{\parallel}}{C}-R'' + 3CH_3OH \overset{H^+}{\longrightarrow} \\[4pt]
H_2C-O-\underset{\underset{O}{\parallel}}{C}-R
\end{array}
\qquad
\begin{array}{l}
H_2C-OH \\[4pt]
HC-OH + 3R-\overset{\overset{O}{\parallel}}{C}-O-CH_3 \\[4pt]
H_2C-OH
\end{array}
\qquad (4.1)
$$

4.4 Ignition Temperature, Flash Point, Fire Point, and Smoke Point

Four important tests which are used to characterize an engine fuel are the spontaneous ignition temperature (SIT), flash point, fire point, and smoke point. These tests are standardized, and specialized fuels have specific requirements as defined by these tests. The SIT is dependent on the composition of the fuel and the conditions of the walls of the cylinder. Diesel fuels require low SIT with short delay times of the order of 1–2 ms. The SIT of heptane (CN = 60) is 330°C, whereas benzene with CN = −10 has a SIT of 420°C.

The flash point of a fuel is obtained by slowly increasing the temperature (5.6°C/min) of the liquid fuel in a standard container (flash cup) until sufficient vapor is given off to produce a flash as a flame is passed over the mouth of the cup every 30 s. The temperature of the oil at which this occurs is the flash point. This is an index of the volatility of the oil or liquid. It is used as an indication of the fire hazard of combustible liquids. For example, the Canadian specification for heating fuel oil stipulates a minimum flash point of 43°C, whereas the flash point of diesel fuel varies from 38°C to 52°C. The flash point and boiling point of various substances are compared in Table 4.4.

The fire point is the temperature to which the oil must be heated so that the vapor pressure is sufficient to maintain the flame after the flame source is removed.

The smoke point of a fuel is an arbitrary scale related to the height of a flame of the fuel burning in a standard lamp without smoking. Some values are given in Table 4.4 and show that aromatic hydrocarbons have low smoke point values, whereas saturated normal alkanes have the highest smoke points. Additives such as ferrocene [Fe(C$_5$H$_5$)$_2$] increase the smoke point. Fuels that have low smoke points tend to deposit carbon during the combustion process.

Water emulsified in diesel fuel can reduce smoke in the exhaust and carbon deposits in the engine. It was shown that, while 5% or less water increased the fuel consumption, more water showed an increase in the thermal efficiency, the maximum increase being 9.7% when the fuel contained 22.5% water.

Table 4.4 Boiling point (BP), flash point (FP), spontaneous ignition temperature (SIT), and smoke point (SP) of selected substances

Substance	BP (°C)	FP (°C)	SIT (°C)	SP (mm)
Acetone	56.5	−18		
Benzene	80	−11	560	8
n-Pentane	36	−49	308	150
n-Hexane	69	−23	247	149
n-Heptane	98	−4	447	147
Isopentane	28	−51	420	−135
Neopentane	9.5	−	440	−140
1-Pentane	30	−28	298	84
Toluene	110	4	536	6
p-Xylene	138	27	464	5
PCB	380 + 10	222	−	−
Kerosene	260 ± 10	54	−	27
Diesel 2-D	−	52	−	−
10-W-30	−	226	−	−

"In air." Specifications provide for modification of these requirements appropriate for individual situations

The internal combustion engine is a notorious polluter, but the diesel engine has the additional emission of particulate matter ($10–25$ μg/m^3) which is rich in polynuclear aromatic hydrocarbons such as phenanthrene, fluoranthene, benzo(a)pyrene, and benzoperylene—all of which are carcinogenic.

4.5 The Spark Ignition Internal Combustion Engine

The four-stroke cycle spark ignition (SI) internal combustion engine (ICE) was initially proposed by Beau de Rockas in 1862 and first built by N. A. Otto in 1876. This engine has become the major piston engine in use today. The PV cycle of the engine is shown in Fig. 4.3 where the four strokes are indicated.

1a Compression of the air/fuel mixture $1 \rightarrow 2$
1b Spark ignitions, constant volume combustion $2 \rightarrow 3$
2a Expansion of product gases $3 \rightarrow 4$
2b Exhaust at constant volume $4 \rightarrow 1$
3 Exhaust of cylinder $1 \rightarrow 5$
4 Induction stroke air/fuel intake $5 \rightarrow 1$

The efficiency of the engine is a maximum of about 25% at a CR of about 15. Fuel injection, as in the diesel engine, means better control of the air/fuel ratio under variable temperature conditions.

The use of fuel injection with the two-stroke SI engine has added a new dimension to small and efficient engines. The major advantages over the four-stroke engine are lower cost and low weight/power ratio. These engines have been primarily used in motorcycles, outboard motors, lawn mowers, chain saws, and similar lightweight engines. Similarly the rotary two-stroke Wankel engine has reached the production stage in the Mazda car, but further extension of its use has not materialized. Two-cycle diesel engines have also been produced.

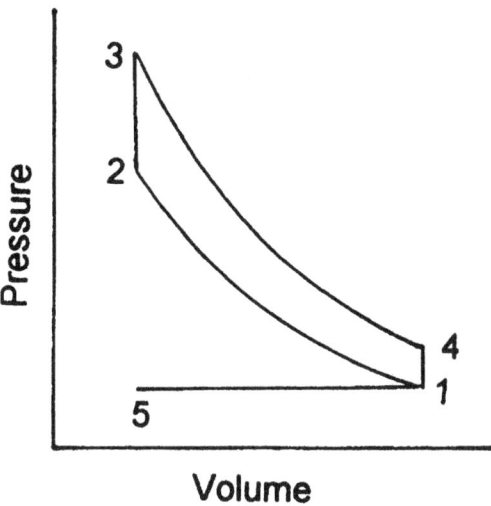

Fig. 4.3 Ideal PV diagram of the four-stroke spark ignition ICE. (**1a**) Compression 1 → 2. (**1b**) Spark ignition, constant volume combustion 2 → 3. (**2a**) Expansion of product gases 3 → 4. (**2b**) Exhaust at constant value 4 → 1. (**3**) Exhaust of cylinder 1 → 5. (**4**) Induction stroke air/fuel intake 5 → 1

4.6 Gasoline Fuel

The fuel for the SI–ICE was in the early years centered on alcohol. In 1895, Nikolaus Otto recommended that alcohol be used in his engine. One of Henry Ford's first models, the quadricycle, was meant to run on alcohol. His later model "T" was designed to run on either alcohol or gasoline requiring only a simple adjustment of the carburetor. Efforts in 1906 to extend the use of alcohol in the USA as a fuel for the automobile by rescinding the 40 cent/gal liquor tax failed. A 50/50 mixture of alcohol gasoline was used in France after World War I due to the oil shortage. The Great Depression of the 1930s made farm-produced alcohol a cheap fuel which was available in 2000 midwestern service stations.

However, the cheaper oil eventually predominated, and gasoline has become the common fuel for the automobile. The quest for greater power and efficiency has resulted in an increase in the CR from about 5 in 1928 to over 13 in 1995. This has resulted in the need to reformulate the fuel to meet the stringent demands of the newer engines.

The problem in the pre-1920 years was the tendency of the engine to "knock"—a violent explosion in the cylinder which, at times, cracked the cylinder head and pistons. This knock was shown to occur after ignition by the spark, and it was believed that the delayed vaporization of the fuel droplets caused the post-ignition explosion. Thomas Midgley Jr., a mechanical engineer working for Dayton Engineering Laboratory Co. (Delco), decided to add a colored component to the fuel and selected iodine. This had a beneficial effect and eliminated the engine knock. When several red dyes were tested, they showed no improvement, whereas ethyl iodide reduced the knock, proving that the iodine was the effective element. However, iodine and its compounds were too expensive and corroded the engine, and so a search was on for a simple alternative. By 1919, the research team found that 2 mL/L of aniline was better than 1 g of iodine. On December 9, 1921, after testing about 33,000 compounds, they found that 0.025% tetraethyl lead was a superior antiknock substance when compared to the 1.3% aniline which was the comparison standard. Thus, a smooth running fuel became a reality.

One difficulty caused by the added lead was the buildup of the yellow lead oxide (PbO) in the engine which coated the spark plugs and valves. This problem was solved by adding ethylene dibromide to the fuel, which converted the lead to the more volatile lead bromide ($PbBr_2$). The first sale of "ethyl" gas was on February 1, 1923. A list of antiknock additives is shown in Table 4.5. The present interpretation of knock has recently been reached by photographing the combustion process through a glass-top cylinder and observing the smooth propagation of the flame front from the spark to the farthest part of the combustion chamber. The gas to be burned last is called the end gas and is

Table 4.5 Relative effectiveness of antiknock compounds and some antiknock fuels (based on aniline = 1)

Benzene	0.085
Isooctane (2,2,4-trimethylpentane)	0.085
Triphenylamine	0.090
Ethyl alcohol	0.101
Xylene	0.142
Dimethyl aniline	0.21
Diethylamine	0.495
Aniline	1.00
Ethyl iodide	1.09
Toluidine	1.22
Cadmium dimethyl	1.24
m-Xylidine	1.40
Triphenylarsine	1.60
Titanium tetrachloride	3.2
Tin tetraethyl	4.0
Stannic chloride	4.1
Diethyl selenide	6.9
Bismuth triethyl	23.8
Diethyl telluride	26.6
Nickel carbonyl	35
Iron carbonyl	50
Lead tetraethyl	118

usually located in the combustion chamber furthest from the spark plug. This end gas is heated by compression, by the combustion taking place, and by the approaching flame front. The spontaneous ignition of this end gas results in an explosion and knock unless the normal flame reacts and consumes the end gas before it ignites. This is shown in Fig. 4.4.

Thus, antiknock additives are inhibitors of autoignition. The lead oxide aerosol which forms in the combustion is a free radical trap and prevents the chain reaction from branching and progressing to explosive rates.

4.7 Grading Gasoline

As the automobile engine developed, it became necessary to grade the gasoline on a realistic scale. This was established by determining the behavior of various organic substances as a fuel in a standard engine and observing the onset of knock as the CR is increased. The effect of structure on the critical CR was discovered for the various isomers of heptane.

The best isomer—trimethylbutane (called *triptane*)—proved to be too difficult to make in the large quantities required for testing, though it is superior to isooctane. The octane was selected as the rating of 100 on the octane scale. The compound *n*-heptane was selected as the 0 rating.

Other octane enhancers such as methyl tertiary butyl ether (MTBE) can act in a different manner by suppressing cool-flame reactions by consuming OH free radicals. MTBE is sufficiently soluble in water to contaminate the aquifer when MTBE/fuels have been spilled. As a result, efforts are under way to ban MTBE as a fuel additive. Ethanol is considered to replace it.

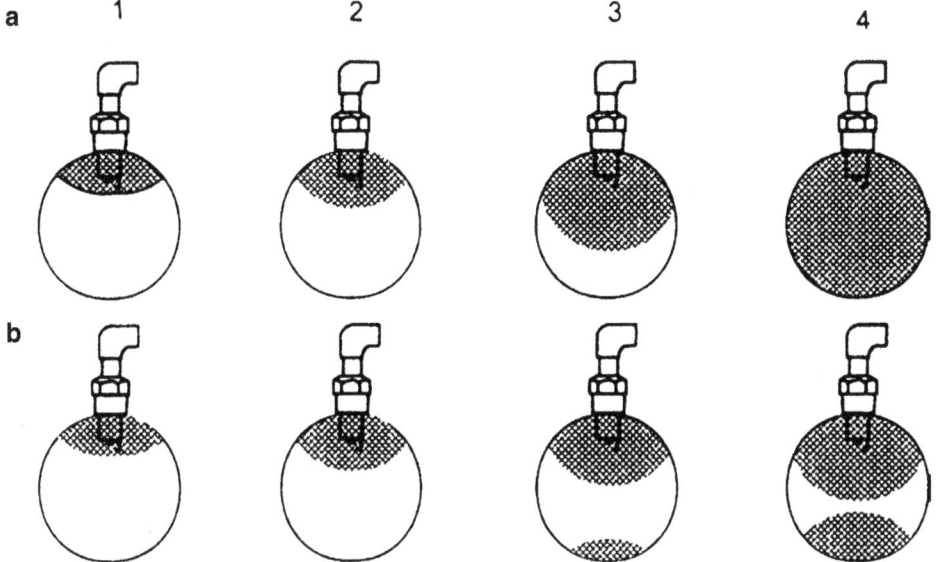

Fig. 4.4 Flame propagation from a spark in a SI–ICE. (Time increases from left to right.) (**a**) Normal process—the flame moves uniformly from the spark to the end gas. (**b**) Knock condition where the end gas ignites at B3 before the flame reaches the end gas

Table 4.6 Mercury and lead concentrations in the glacial samples[a]

Time of deposition	Mercury content[b]	Lead content[b]
800 BC	62	1
1724	75	10
1815	75	30
1881	30	–
1892	66	–
1946	53	–
1952	153	200
1960	89	–
1964 (fall)	87	–
1964 (winter)	125	–
1965 (winter)	94	–
1964 (spring)	230 ± 18	
1965 (summer)	98	–

[a]The sample deposited in 1724 was recovered from Antarctica, the others from Greenland
[b]Nanogram (Hg/Pb) content (per kilogram of water)

A good illustration of the global contamination by lead is a comparison of the lead and mercury content of the Greenland ice field at various depths. This is shown in Table 4.6 where the level of Hg in the ice has not changed significantly in 3000 years, whereas the level of Pb has increased 200-fold, showing the direct effect of the automobile. This implies that the contamination by mercury is of regional interest and not of global importance.

Because of the poisonous properties of lead and its compounds, some countries have replaced lead by other octane enhancers. One such additive is methylcyclopentadiene manganese (III) tricarbonyl (MMT).[1] This substance is probably as toxic as the lead compounds, but since much smaller quantities are used (1/20 of lead), it is argued that it is safer than lead. Actually, manganese will be contaminating the environment just as lead has done if its use is continued. Aromatic hydrocarbons such as benzene and toluene also act as octane enhancers even though they are known carcinogens.

The addition of water to gasoline is reported to improve the quality of the fuel. The water is sucked up into the carburetor and vaporized as it passes around the hot exhaust manifold and enters the intake manifold. As with diesel fuel, the mechanism by which the water acts is not fully understood.

The mechanism of oxidation of hydrocarbon (which can be represented by RH and $RCH_2CH_2CH_3$) involves the formation of peroxides (ROO) and hydroperoxides (ROOH) by the following series of reactions:

1. Initiation by spark

$$RCH_2 \text{---} CH_2 \text{---} CH_3 \rightarrow RCH_2 \text{---} CH_2 \rightarrow RCH_2 \text{---} CH_2$$
$$+CH_3$$
$$O_2 \rightarrow 2O$$

2. Chain branching

$$RCH_2 \text{---} CH_2 \text{---} CH_3 + O \rightarrow RCH_2 \text{---} CH \text{---} CH_3 + OH$$

3. Peroxide formation

$$RCH_2 \text{---} CH_2 + O_2 \rightarrow RCH_2CH_2O_2$$
$$RCH_2CH \text{---} CH_3 + O_2 \rightarrow RCH_2CHCH_3$$
$$I$$
$$O_2$$
$$CH_3 + O_2 \rightarrow CH_3O_2$$

4. Hydroperoxide formation

$$RCH_2\text{--}CH_2O_2^{\cdot} + RH \rightarrow RCH_2CH_2O_2H + R^{\cdot}$$
$$RCH_2\text{--}CH\text{--}CH_3 + RH \rightarrow RCH_2\text{--}CH\text{--}CH_3 + R^{\cdot}$$
$$\qquad\quad | \qquad\qquad\qquad\qquad\qquad\qquad | $$
$$\qquad\; O\text{--}O^{\cdot} \qquad\qquad\qquad\qquad\qquad O\text{--}O\text{--}H$$
$$CH_3O_2^{\cdot} + RH \rightarrow CH_3OOH + R^{\cdot}$$

---

[1] MMT, as well as lead, is no longer used as fuel additives in Canada.

5. Hydroperoxide decomposition

$$RCH_2CH_2O_2H \rightarrow RCH_2CH_2O^. + OH$$

$$RCH_2\text{---}CH\text{---}CH_3 \rightarrow RCH_2\text{---}CH\text{---}CH_3 + OH$$
$$\quad\quad | \quad\quad\quad\quad\quad\quad | $$
$$\quad\quad O_2H \quad\quad\quad\quad\quad O^.$$

$$CH_2OOH \rightarrow CH_3O^. + OH$$

$$RCH_2CH_2O \rightarrow RCH_2^. + CH_2O$$

$$RO^. + RH \rightarrow ROH + R^.$$

Recent kinetic modeling of the combustion mechanism involving about 2000 reactions has made it possible to calculate RON values. A list of comparable values is given in Table 4.7.

The price of gasoline in various parts of the world is given in Table 4.8. The major single component of the cost to the motorist is a government tax which varies from about 33% to 75%. The cost of gasoline should be compared to other liquid products such as bottled spring water which can sell for $0.50 – $0.75 per liter or distilled water at about $2.00 per gallon (4.5 L) in Canada. Considering the processing involved in marketing gasoline compared to water, one must conclude that gasoline is still a bargain in most countries.

The problem of environmental contamination by vehicle exhaust is further illustrated by the presence of lead compounds in French wines from a vineyard at the intersection of two major autoroutes. It was recently shown that triethyl lead and trimethyl lead were present in older wines.

Table 4.7 Comparison of calculated and measured research octane number for various hydrocarbons

Fuel	Formula	Measured RON	Predicted RON	Ignition time (ms) After TDC
n-Heptane(PRFO)	C_7H_{16}	0	0	55.0
n-Hexane	$C \ll H_{14}$	25	20	55.2
PRF25	25 % C_8H_{18} + 75 % C_7H_{16}	25	25	55.3
2-Methylhexane	C_7H_{16}	42	40	55.5
PRF50	50 % C_8H_{18} + 50 % C_7H_{16}	50	50	55.8
n-Pentane	C_5H_{12}	62	55	55.9
3-Ethylpentane	C_7H_{16}	65	80	57.4
2-Methylpentane	C_6H_{14}	73	80	57.3
3-Methylpentane	C_6H_{14}	74	80	57.2
PRF75	75 % C_8H_{18} + 25 % C_7H_{16}	75	75	56.7
3,3-Dimethylpentane	C_7H_{16}	81	50	55.7
2,4-Dimethylpentane	C_5H_{12}	86	75	56.4
2,2-Dimethylpropane	C_7H_{15}	83	70	56.2
PRF90	90 % C_8H_8 + 10 % C_7H_{16}	90	90	58.0
2-Methylbutane	C_5H_{12}	92	100	59.1
2,2-Dimethylbutane	C_6H_{14}	92	100	59.9
2,2-Dimethylpentane	C_7H_{16}	93	>90	No ignition
n-Butane	C_4H_{10}	94	85	57.6
2,3-Dimethylbutane	C_6H_{14}	100	90	58.0
Isooctane (PRF 100)	C_8H_{18}	100	100	59.5
Isobutane	C_4H_{10}	102	>90	No ignition
Propane	C_3H_8	112	>90	No ignition
2,2,3-Trimethylbutane	C_7H_{16}	112	>90	No ignition
Ethane	C_2H_6	115	>90	No ignition

Table 4.8 Gasoline prices and tax component in the OECD, 1992 and 2000[a]

Country	Gasoline prices (SUS/L)	Tax component (% of total)
Australia	0.499	46.2
Austria	0.970	64.8
Belgium	0.987	70.0
Canada	0.455 (0.48)	46.2
Denmark	0.961	67.2
Finland	1.013	68.0
France	0.992 (1.04)	77.2
Germany	0.981 (0.97)	72.4
Greece	0.820	69.1
Ireland	1.001	66.6
Italy	1.236(1.00)	75.8
Japan	0.977 (0.94)	46.1
Luxembourg	0.746	62.0
Netherlands	1.141	72.4
New Zealand	0.5411	46.6
Norway	1.284	71.4
Portugal	1.083	75.4
Spain	0.943 (0.76)	69.8
Sweden	1.137	69.2
Switzerland	0.759	62.5
Turkey	0.745	63.7
United Kingdom	0.882(1.00)	69.5
United States	0.298 (0.36)	33.9

OECD = Organization for Economic Cooperation and Development Values in ( ) refer to year 2000.

The trimethyl lead and triethyl lead originate from the tetramethyl lead (TML) and tetraethyl lead (TEL) which were added to gasoline as antiknock agents. The levels of lead in the wines are still at least 10–100 times less than the $0.5/.tg/L$ limit of lead in drinking water.

Some general properties of gasoline and diesel fuel compared to some alternate fuels are given in Table 4.9. A major parameter which is missing from the table is the unit cost, which, to a great extent, determines the choice to be made.

Exercises

1. Smoke emission from a diesel engine is environmentally undesirable. Explain why and discuss methods which can eliminate this aspect of diesel fuel use.
2. How can the ignition temperature (SIT) of a fuel be altered?
3. Calculate the air/fuel ratio for gasoline. (Note: Assume gasoline can be represented by nonane $[C_9H_{20}]$.)
4. The TLV (threshold limit value) of mercury in Canada and USA is 0.1 mg/m^3. If a droplet of mercury (density = 13.6 g/mL) 3 mm in diameter is spilled and allowed to evaporate completely in a room 3 m × 3 m × 2.5 m. What will its concentration be, relative to the TLV?
5. Show how dimethyl ether can be prepared from methanol.
6. Compare the energy density of dimethyl ether with diesel fuel.
7. What advantages are there in using dimethyl ether in a diesel engine?
8. Discuss the special characteristics and requirements of aviation fuel in comparison to automotive fuels.

Table 4.9 Fuel comparison chart

	Gasoline $(CH_2)_x$	Methanol CH_3OH	Ethanol C_2H_5OH	Methane CH_4	Propane C_3H_8	Hydrogen H_2	Ammonia NH_3	Gasohol 10 % C_2H_5OH	Diesel $(CH_2)_x$
RON/MON	90–96 / 82–87	106 / 92	132 / 103	105	101 / 95			128 / 99	CN / 45
Heat of combustion MJ/kg	43.9	21.0	26.8	50.1	46.5	120.0	18.6	43.1	42.5
Density (gm/mL)	0.73	0.791	0.789	0.466	0.50	0.071 @	0.771		0.85
Temp. (°C)	25°	25°	25°	−161°	20°	−252°	−33°		25°
Heat of Combustion MJ/L	32.3	16.6	21.1	23.3 liq	23.3 liq	8.5 liq	14.3		35.3
Latent Heat of Vaporization MJ/kg	0.35	1.18	0.92	0.51	0.34	0.45	1.37		
VP @ 20 °C (Torr) (atm)	450–680	99	38	45 atm @ −82 °C[a]	8 (atm)	12.8 atm @[a] −239 °C[a]	8 atm		1.1 @ 100 °C
F.P. °C	−80	−94	−117	−182	−187	−259	−77	−80	−50
B.P. °C	30–150	65	78	−161	−42	−253	−33	30–150	150–300
Ignition temp. °C	371	446 385	422 370	540	432	500	650	300	250
Flash point °C	−43	11	13					−40	>40
Explosion limits % vol. in air	1.4–7.5	6.0–36	3.3–19	5.0–1.5	2.1–9.5	4.0–75	16–25	1.4–7.5	
Stoichiometric air/fuel ratio (kg air/kg fuel)	14.7	6.4	9.0						

[a]Critical Temp. and Critical Pressure

Further Reading

1. Ralbovsky E (1997) Introduction to compact and automotive diesels. Delmar, Albany, New York
2. Thiessen FJ, Dales DN (1996) Diesel fundamentals, 3rd edn. Prentice Hall, New York
3. Halberstadt H (1996) Modern diesel locomotive. Motorbooks Intl, Osceola, Wisconsin
4. (1995) Motor gasoline industry—past, present, and future. Gordon Pr., New York
5. Dagel JF (1993) Diesel engines, 3rd edn. Prentice Hall, New York
6. (1993) Diesel fuels for the nineties. Soc. Auto. Engineers, Warrendale, Pennsylvania
7. Taylor CF (1985) The internal-combustion engine in theory and practice, 1 and 2, 2nd edn. Rev. MIT Press, Cambridge, MA
8. Moseley CG (1980) Chemistry and the first great gasoline shortage. J Chem Educ 57:288
9. Pierre JR (1975) The fuel consumption of automobiles. Sci Am
10. Universal Diesel Liquifier (fuel atomizer). http://www.universaldiesel.com/
11. Caterpillar. http://www.cat.com/
12. Detroit Diesel. http://www.detroitdiesel.com/
13. Fuel Technologies Inc. http://www.fueltechnic.com/
14. Biodiesel Fuel. http://www.veggievan.org/
15. Canadian Renewable Fuels Association. http://www.greenfuels.org/ Fuel prices, http://www.cheap-fuel.com/
16. Gasoline prices. http://www.chartoftheday.com/20000913b.htm

Chapter 5
Alternate Fuels

5.1 Introduction

The search for alternate fuels has been stimulated by the environmental concerns about the emissions in exhaust from gasoline and diesel fuel engines. Urban air pollution due to the automobile is due to excessive production of CO, NOx, and hydrocarbons which in sunlight can give rise to ozone and smog. NOx is formed during all high-temperature combustions using air. Thus, low-temperature combustion processes result in lower NOx emission because of the equilibrium

$$N_2 + O_2 \rightleftarrows 2\ NO \tag{5.1}$$

Similarly the presence of oxygen in the fuel, for example, methanol or ethanol, reduces the emission of smog-forming hydrocarbons.

The relative emission characteristics of fuels are shown in Table 5.1 and Fig. 5.1. It indicates that the electric vehicle is the least polluting energy source. However, this implies that the origin of the electrical energy used in recharging the batteries is environmentally benign.

The present alternate fuels in use today are propane, methanol, and ethanol.

5.2 Propane

The fuel properties of propane, also referred to as bottled gas, are shown in Table 4.9 and show that the boiling point of the liquid is −42°C. At 20°C, the vapor pressure is 8 atm, and thus, the fuel must be stored in a pressurized cylinder meant to withstand at least 21 atm, 62°C if mounted in an enclosed space.

About 80% of the propane is obtained from natural gas where it occurs in concentration which varies from 1% to about 5%. The other 20% is formed during the catalytic cracking of petroleum oil. The world production of propane is tabulated in Table 5.2 and shows an average annual growth rate (AAGR) of about 1.5%. In 2006, Canada's consumption of about 1.25×10^{10} L/year (IEA 2006) is less than half its production. In 2006, Canada exported 57% of its propane (7.2×10^9 L/year—IEA 2006) to USA. For comparison, the annual Canadian consumption of gasoline and diesel fuel was 40.7 GL (2005) and 9.3 GL (2005), respectively (IEA 2006). Hence, all the propane available can only satisfy about one-fifth of the total gasoline usage in Canada. The amounts produced are small in comparison to the gasoline used, and therefore, as an alternate fuel, it has limited use. This applies to the other alternate fuels where production is still at a fraction of the full requirements. Propane is, like

O.V. Roussak and H.D. Gesser, *Applied Chemistry: A Textbook for Engineers and Technologists*,
DOI 10.1007/978-1-4614-4262-2_5, © Springer Science+Business Media New York 2013

Table 5.1 Qualitative summary of emissions from alternative-fuel vehicles versus gasoline-fuel vehicles

Environmental issues	Gasoline[a]	Reformulated gasoline	Diesel	Methanol[b]	M-85[b]	Ethanol	Liquid petroleum gas	Compressed natural gas	Hydrogen[b]	Electric[c]
Tailpipe, evaporative emission levels[d]										
NMOG	Base	Lower	Lower	Lower	Lower	Lower	Lower	Lower	None	None
CO	Base	Lower	Lower	Equal	Equal	Equal	Lower	Lower	None	None
Nox	Base	Equal	Higher	Equal	Equal	Equal	Higher	Higher	Higher	Lower
Impact on urban ozone relative to gasoline										
Ozone-forming potential	1.00	.80 ± 0.2	.80 ± 0.2	0.65 ± 0.2	.70 ± 0.2	.80 ± 0.2	.65 ± 0.2	.40 ± 0.2	.00 ± 0.2	.00 ± 0.2
Reduction in peak ozone[e] in Los Angeles	0%	1.4%	1.4%	2.4%	2.0%	1.4%	2.4%	4.1%	6.8%	6.8%
Level of concern over airborne toxic emissions										
Particles	Low	Low	High	Low	Low	Low	Low	Low	–	–
1,3-Butadiene	Medium	medium	Unknown	Low	Low	Low	Low	Low	–	–
Benzene	High	Medium	Low	Low	Medium	Low	Low	Low	–	–
Formaldehyde	Medium	Medium	Medium	High	High	Medium	Low	Low	–	–
Acetaldehyde	Low	Low	Medium Low	Low	Low	High	Low	Low	–	–
Global warming factors										
g CO$_2$/BTU	0.094	0.094	0.096	0.093	0.094	0.083	0.081	0.071/0.09[f]	0.15	0.18
Fuel economy (miles/gal)	34	34	39/48[g]	22	22	28	31.2	8.0	11.3	–
Energy economy	3353	3353	3257/2614	2923	2923	2923	2829	3353	2591	1705
(BTU/mile) g CO$_2$/mile	315	315	313/252[g]	272	275	243	229	238/301[f]	389	307

Reprinted with permission from Chang et al. (1991) and Ford Motor Co. Copyright 1991 American Chemical Society

[a] Current industry average

[b] Natural gas as the feedstock source of methanol and hydrogen

[c] No direct emissions from vehicles, but associated power plant Nox emissions may be substantial

[d] Relative to gasoline vehicles with Tier 1 (Clean Air Act Amendments) standards, using 9 lb/in.[2] Reid vapor pressure gasoline

[e] Year 2015 with the pre-1987 emissions control regulations, except for LDV. These are semiquantitative estimates

[f] There is a debate over global warming impact of methane as compared to carbon dioxide. If the atmospheric lifetime of methane is fully accounted, as we believe it should, then CNG vehicles have a substantial global warming benefit. However, if the lifetime is discounted, CNG vehicles have approximately the same greenhouse impact as gasoline vehicles

[g] The fuel economy improvement of diesel compared to gasoline vehicles is uncertain; estimates range from 15% to 40% better

Fig. 5.1 US methane emissions by source, 2008 (EPA)

Table 5.2 World annual production of propane—G.L. (IEA 2006)

	2005	1990	1995	AAGR[a] %
USA	33.2	70.0	66.0	(1.6)
Western Europe	40.3	34.8	34.9	0.2
Japan	8.9	5.0	4.9	–
Canada	3.2	7.8	8.0	0.9
Latin America	14.4	26.9	29.5	2.9
Africa	5.1	11.2	14.5	8.7
Asia/Australia	69.6	18.7	19.8	2.9
Middle East	26.4	33.1	41.0	7.2
Eastern Block	24.1	0.8	0.8	–
World total	227.5	208.3	219.4	1.4

[a]Average annual growth rate (%)

natural gas, a clean burning automotive fuel with a low ozone-forming potential, good fuel economy, and low emission of toxic organic hydrocarbons such as benzene and formaldehyde. However, acetaldehyde and olefins are some of the adverse components in the exhaust.

The cost of propane is usually less than gasoline on a volume basis, and major users are the fleet operators such as taxis, delivery trucks, and vans (see Fig. 5.2). Liquid propane is readily available in most locations. Its use is, however, restricted to ICE-SI engines and not for the diesel engines. Propane use in Canada and the USA has been failing because of the cost of conversion requirement for large storage tanks, fewer refueling station than for gasoline, and the restrictions in underground parking. The most recent incidents of propane explosions happened at the Falk Corporation facility in Wisconsin, USA (2006), and at the Sunrise Propane Industrial Cases facility in Toronto, Ontario, Canada (2008), causing several employee deaths, evacuation of thousands of people, and significant damage of buildings and houses (http://en.wikipedia.org/wiki/2008_Toronto_explosions).

Fig. 5.2 US natural gas prices, 1976–2010 (Source: EIA, BLS)

5.3   Methanol

Methanol was produced by the destructive distillation of wood until about 1923 when BASF in Germany showed that methanol can be formed by the catalytic reduction of CO with H_2.

$$CO + 2H_2 \xrightarrow{\text{catalyst}} CH_3OH \ \Delta H^\circ = -90.77 \ \text{kJ/mol} \tag{5.2}$$

This is the method used commercially today where the syngas is usually prepared from natural gas by the steam-reforming reaction.

$$CH_4 + H_2O_{(g)} \xrightarrow{\text{catalyst}} CO + 3H_2 \ \Delta H^\circ = +206.2 \ \text{kJ/mol} \tag{5.3}$$

The energy-intensive step is the formation of the syngas, and as a result, the economics of the overall process is favorable only for large-scale plants producing over 1,500 t/day methanol. It seems wasteful to break 4C–H bonds to form CO + H_2 and then reform three of the C–H bonds. This can be avoided by the direct partial oxidation of CH_4 to CH_3OH at high pressures. It was shown by Bone in 1930 that at high pressures (50–100 atm), methanol is a major intermediate product in the oxidation of natural gas. This has been extensively studied at the University of Manitoba since 1982 by the author. A small pilot plant has operated in Kharkov, Ukraine, from 1985 to 1990 producing 100 t/year methanol. A larger plant designed to produce 10,000/year is in the planning process. The bench top optimum efficiency is achieved at about 50 atm, 10% O_2, 400°C where 80% selectively and 12% conversion result in a 10% yield of methanol. The process is represented by the reaction

$$3 \ CH_4 + 2\frac{1}{2}O_2 \rightarrow 2 \ CH_3OH + CO + 2H_2O. \tag{5.4}$$

The present uses of methanol are shown in Fig. 5.3 and given in Table 5.3 which shows that the formation of formaldehyde is still the major application of methanol. The direct use of methanol as a fuel is expected to surpass this in the near future. The octane enhancer methyl tert-butyl ether is

Fig. 5.3 Uses of methanol

capable of satisfying the oxygen requirements in gasoline and to replace lead (TEL) as an additive. It is made by reacting methanol with isobutene

$$CH_3OH + CH_2 = C - CH_3 \rightarrow CH_3OCH_2 - CH - CH_3$$
$$\qquad\qquad\quad | \qquad\qquad\qquad\qquad | \qquad\qquad\qquad\qquad (5.5)$$
$$\qquad\qquad\quad CH_3 \qquad\qquad\qquad\quad CH_3$$

The present optimistic predictions of methanol production and demand are shown in Fig. 5.4 where it can be seen that the demand will soon exceed the production. The construction of a full-sized conventional methanol plant (2,000 t/d) takes at least 3 years from start to full time on stream and requires a suitable source of natural gas. Small gas fields which usually accompany crude oil

Table 5.3 Conversion of methanol into various products (%)

	1982	1994
Formaldehyde	31	39
Methylamine	4	
Chloromethanes	9	
Acetic acid	12	7
Methyl ester	8	6
Solvent	11	7
Fuel-antifreeze	25	2
MTBE		13
Miscellaneous		26

Fig. 5.4 World production of natural gas and the proportion that is wasted by flaring and venting

production are normally vented or flared. The extent of such wasted energy is shown in Fig. 5.5, and it is this lost gas which can be readily converted to methanol by the direct partial oxidation process.

The main apparent disadvantage of methanol is its low energy density by volume, 16.6 kJ/L, about half that of gasoline. This does not imply that the fuel tank must be twice as large because if the engine was designed for the fuel, the methanol fuel economy would be close to that of gasoline.

The use of gasohol or methanol (or ethanol) with gasoline (M-85) has become the intermediate stage in the extensive use of alcohol fuels. Because of the possibility of phase separation when water is present in 10% methanol in gasoline, a cosolvent such as tertbutyl alcohol (TBA) must be added. M-85 does not require a cosolvent, though the emissions are worse than M-100.

Cold starting difficulties with methanol fuel are due to its high heat of vaporization and low volatility. Proper engine design can correct for such deficiencies. Another problem is corrosion due to the acid (formic acid) formed by the oxidation of the methanol, and all fuel and engine components must be carefully selected. The methanol flame is not readily visible, whereas M-85 (15% gasoline) has the advantage of appearing more like a yellow gasoline flame.

Fig. 5.5 World production capacity for methanol and projected world demand

The aldehyde emission from alcohol fuel must be destroyed by a catalytic after-burner if environmental contamination is to be avoided. Methanol has been used in diesel engines, though it has a CN between 0 and 5. However, because it is not readily self-ignited, special engine design must correct for such deficiency. A glow plug or fuel additives can aid self-ignition. Partial conversion of the methanol to dimethyl ether is another approach to the problem which has recently been studied.

A major problem in the use of gasohol is in the materials of construction of the fuel system. Alcohol tends to dissolve the oxidation products and gums from gasoline, and special precautions must be taken when first switching to gasohol in an old vehicle. This is even more important when pure methanol is used as the fuel. Special gaskets and O-rings are required. However, these present no problem if design and construction are started with such fuel in mind. Besides, being independent of petroleum, alcohol offers other inherent advantages, such as cooler, cleaner combustion; improved power; reduced carcinogens and NOx emission; reduced hydrocarbon emission which can be photochemically activated and smog forming; and smaller and lighter engine designs are possible.

Ethanol and methanol have cetane values from 0 to 5 making them poor compression ignition fuels for a diesel engine. However, alcohol-diesel fuel blends have been used successfully, and emulsifiers have been added to help blend the two components which have limited miscibility.

When methanol is catalytically converted to CO and H_2

$$CH_3OH(g) \rightarrow CO(g) + 2H_2(g) \Delta H^\circ = 90.7 \text{ kJ/mol} \tag{5.6}$$

the resulting synthesis gas has a higher heat of combustion (-766 kJ) than the methanol (-676 kJ). If the exhaust heat is used to decompose the methanol, then a higher energy fuel is obtained; that is, approximately 20% increase in fuel economy has been proposed. This is valid if only thermal energy is considered. In an internal combustion spark engine, the air and fuel are compressed (which require work), and the combustion then expands the gas, and work is done. Hence, the ratio, R = moles of gaseous products/moles of gaseous reactant, must be used to normalize differences in fuels.

Assuming $N_2/O_2 = 4$ and no excess O_2:

$$CH_3OH + 1.5O_2 + 6N_2 \rightarrow CO_2 + 2H_2O + 6N_2 \tag{5.7}$$

8.5 moles $\rightarrow$ 9 moles $R_1 = 1.058$

$$CO + 2H_2 + 1.5O_2 + 6N_2 \rightarrow CO_2 + 2H_2O + 6N_2 \tag{5.8}$$

10.5 moles $\rightarrow$ 9 moles $R_2 = .857$, $R_1/R_2 = 1.235$

The product gases have the same volume in each reaction:

Hence, the additional work of compression (W_c) is essentially due to the $P\Delta V$ or $\Delta n\, RT$ where $\Delta n = 2$ moles. The remaining question is what is the value of T? We can select 400°C as an estimated value of the temperature. This value can be calculated from an adiabatic compression of the gas (mostly air) from $T_1 = 350$ K and assuming that CR = 10.

It is worth considering an extension to this argument by considering the increase fuel economy if we remove the nitrogen from the combustion process. Firstly, a higher temperature is obtained, and secondly, less work is done in compressing the gas. If we assume that gasoline is represented by octane, C_8H_{18}, then in air, the reaction is

$$C_8H_{18} + 12.5O_2 + 50N_2 \rightarrow 8\ CO_2 + 9H_2O + 50N_2 \tag{5.9}$$

63.5 moles $\rightarrow$ 67 moles $R_3 = 1.055$

$$C_8H_{18} + 12.5O_2 \rightarrow 8\ CO_2 + 9H_2O \tag{5.10}$$

13.5 moles $\rightarrow$ 17 moles $R_4 = 1.259$, $R_4/R_3 = 1.19$

The difference in moles of compression of reaction 5–9 and 5–10 Δ n is 63.5–13.5 or Δ n = 50.

Hence, $\Delta n\, RT = 50 \times 8.314 \times 673 = 280$ kJ. The heat of combustion of octane is 5,470 kJ/mol. This extra work of compression is only about 5% of the combustion energy. However, the calculated adiabatic combustion flame temperature for octane in air is about 1,400 K, whereas in oxygen, the flame temperature is over 9,000 K. This represents a substantial increase in energy of combustion due to the removal of nitrogen.

Thus, besides obtaining more heat from the reaction (since the heat capacity of nitrogen absorbs some of the energy to reach the high temperature), the removal of nitrogen can give rise to about 20% higher fuel economy. This is now being done for stationary furnaces where only the thermal improvement is obtained. It would be interesting to run an automobile on enriched oxygen which can be obtained by using suitable membranes. More will be said about this later.

One aspect which is important to note is that methanol is toxic and its TLV (threshold limit value) is 200 ppm requiring its dispensing in well-ventilated areas.

The ultimately efficient fuel system would be an electric vehicle running on a methanol fuel cell. Such a system could meet the most stringent environmental emission requirements as well as high energy efficiency. This will be discussed more fully in Chap. 9.

5.4 Ethanol

Ethanol is one of the few possible nonfossil fuels which can be made from a variety of renewable sources such as grapes, corn, straw, and sugarcane. The fermentation of sugar and starch from various agricultural sources has been known for centuries.

The industrial production of ethanol from the acid catalyzed hydration of ethylene was developed by Union Carbide Corp. in 1930:

$$CH_2 = CH_2 + H_2SO_4 \rightarrow CH_3 - CH_2 - OSO_3H$$
$$\downarrow H_2O \tag{5.11}$$
$$H_2SO_4 + CH_3 - CH_2OH$$

The sulfuric acid, initially 96–98%, is diluted to about 50% during the reaction and has to be reconcentrated. This is one of the more costly steps in the process. Other solid phase acidic catalysts have been used to replace the H_2SO_4 in the hydration process.

Fig. 5.6 US fuel ethanol production, million US liquid gallons (2000–2010)

The ethylene is formed by the thermal cracking of ethane by the reaction

$$C_2H_{6(g)} \rightarrow C_2H_{4(g)} + H_{2(g)} \tag{5.12}$$

Ethanol from ethane is a fossil-based fuel and must be distinguished from grain alcohol. Ethanol from syngas has been extensively studied, but catalysts have not been found, as yet, to be efficient enough to make the process cost-effective. Similarly, the conversion of syngas to ethanol by alternate routes has been considered:

$$CH_3OH_{(g)} + CO_{(g)} + 2H_{2(g)} \rightarrow CH_3CH_2OH_{(g)} + H_2O_{(g)}$$
$$2\ CO_{(g)} + 4H_{2(g)} \rightarrow C_2H_5OH_{(g)} + H_2O_{(g)} \tag{5.13}$$

or

$$2\ CH_3O_{(g)}H \rightarrow CH_3OCH_{3(g)} + H_2O_{(g)} \tag{5.14}$$

$$CH_3OCH_{3(g)} \rightarrow C_2H_5OH_{(g)} \tag{5.15}$$

These catalytic processes are attractive because the relative value of ethanol/methanol is about 4, whereas the C_2H_6/CH_4 value is about 2, though in the last year ending in June 1995, the price of methanol doubled. Hence, the direct conversion of ethane to ethanol by the reaction analogous to reaction 5.4

$$3C_2H_6 + 3^1/_2O_2 \rightarrow 2C_2H_5OH + 2CO + 3H_2O \tag{5.16}$$

is a convenient approach to the utilization of surplus ethane.

According to Vaclav Smil's analysis of ethanol production in the USA, commercial production of fuel ethanol began in 1980, and it took nearly 15 years to surpass 5 billion liters, accelerated expansion began only in 2002; more than 15 billion liters were shipped in 2005; 25 billion liters in 2007; and 35 billion liters (9 billion gallons) in 2008 (see Fig. 5.6). Much higher targets were established in June 2007 when the American Senate passed an energy bill that mandated no less than 36 billion gallons of ethanol by the year 2022. The USA produced about 45 GL of ethanol (2010) or 52% of the word's production from over 110 plants in comparison to 4 GL in 1992 from over 15 plants. The estimated annual US market is about 100 GL today. The total annual world ethanol production was 85.9 GL in 2010 and over 31 GL in 1998, respectively. The production of ethanol in the European Union was slightly reduced from 22 GL (1998) to 4.3 GL (2009). Russia increased ethanol production in three times from 2.5 GL (1998) to 7.6 GL (2010).

The present use of ethanol as a fuel additive is made possible by the subsidy which varies from one location to another but is approximately 60¢/US gal. of ethanol or approximately 16¢/L. This subsidy applies only to ethanol formed from grain, that is, nonfossil fuel sources. The present processes of production include fermentation using yeasts as the enzyme catalyst which converts sucrose to CO_2 and ethanol

$$C_6H_{12}O_6(aq) \xrightarrow{\text{catalyst}} 2C_2H_5OH \ (aq) + 2 \ CO_2(g) \tag{5.17}$$

This is then followed by distillation of the alcohol which normally reaches about 12% in the fermentation process. The distillation produces 95% alcohol. The residual 5% water is removed either by the addition of a third component to form a two-phase azeotrope or by passing the wet alcohol through a 3A molecular sieve which removes the water leaving 99.9 + % alcohol.

There is, however, enzymes which can still function up to almost 20% ethanol. The energy balance in ethanol production which takes into account the energy used to grow the corn, the energy to ferment and distillate the alcohol, and the energy in the remaining mash results in a ratio for the input energy/output energy of 1.3 which varies somewhat depending on the efficiency of the various steps. It would appear that it makes little sense to spend 1.3 kJ of fossil fuels to obtain 1 kJ of grain alcohol fuel and to consume grain which is a badly needed food supply for an expanding world population. Also Vaclav Smil noticed that because of America's extraordinary high gasoline consumption and the inherently low power density of ethanol production (only about 0.25 W/m^2 of cultivated land), corn-derived ethanol can never supply more than a relatively small part of the overall demand for fuel in the USA. If America's entire corn harvest—just over 280 million tons in 2005—was converted to ethanol at the best conversion ratio of 0.4 kg/kg of grain, the country would produce fuel equivalent to 13% of the total gasoline consumption.

Improved efficiency may be achieved by using pervaporation to separate out the alcohol instead of distillation. This consists of a membrane through which water and ethanol permeate with greatly different rates. When fermentation is continuously conducted in a membrane reactor, it is possible to continuously remove the alcohol as it is formed enriching it by a factor of six- to tenfold. This can reduce the costs of distillation which is as much as half the production cost of the ethanol.

The separation of ethanol from water can also be effected by freezing. The effect of ethanol concentration on the freezing point is given in Table 5.4. Thus, a liter of fermented brew with 12.5% ethanol by volume was completely frozen and then allowed to thaw. The first 500 mL of solution was 17% ethanol. When this 500 mL solution was frozen and allowed to thaw, again the first 250 mL was 23% ethanol. Various freeze-thaw cycles can thus concentrate ethanol. Another process which has been studied extensively for more than 70 years is the conversion of cellulose from wood and straw to glucose by enzyme or by acid hydrolysis.

The reaction is

$$\underset{\alpha-\text{cellulose}}{(C_6H_{10}O_5)_x} + xH_2O \ \rightarrow \ x \ \underset{\text{glucose}}{(C_6H_{12}O_6)} \tag{5.18}$$

Table 5.4 Some properties of ethanol-water (E/W) solutions

[E] (Wt. %)	Density (g/mL)	[E] (Molarity)	F.P. (°C)
2.5	0.9953	0.539	−1.02
5.0	0.9911	1.974	−2.09
10.0	0.9838	2.131	−4.47
15.0	0.9769	3.175	−7.36
20.0	0.9704	4.205	−10.92
25.0	0.9634	5.21	−15.4
30	0.9556	6.211	−20.5
35	0.9466	7.145	−25.1
40	0.9369	8.120	−29.3
44	0.9286	8.853	−32.7
50	0.9155	9.919	−37.7
54	0.9065	10.607	−40.6
60	0.8927	11.606	−44.9
64	0.8786	12.565	−48.64
68	0.8739	12.876	−49.52

Cellulose is the most abundant naturally occurring biomass material on earth, and its utilization as a fuel has not as yet been fully exploited. Paper, though presently recycled to some degree, can be included with straw and wood for conversion to ethanol fuel. According to Vaclav Smil's information, the US Department of Energy invested to build six cellulosic plants with combined capacity to 0.1% of transportation fuel used in the USA (2005). The construction of these plants (one based solely on corn stover, one on waste wood, and the rest on a mixture of agricultural wastes and waste wood) should be completed in Silicon Valley by 2011.

Ethanol has been used as a gasoline additive in Brazil since the 1920s. The conversion to straight alcohol during World War II has continued to the present. Ethanol is used in its pure form in the Otto cycle engine for cars and light trucks and in ethanol-gasoline blends (22% ethanol) where it replaces lead as antiknock additive and octane enhancer. Ethanol is also used in diesel engines. This is accomplished by the addition of 4.5% of a diesel ignition improver such as isoamyl nitrate, hexyl nitrate, and the ethylene glycol dinitrate to the fuel which consumes 65% more fuel volume than regular diesel fuel and 2% less than an Otto cycle engine running on pure ethanol. In some cases, 1% castor oil is added to the ethanol to hold lubricate engine parts.

The ethanol fuel program in Brazil is made possible by the large sugarcane crop as well as other sugar crops such as cassava and sorghum. According to Vaclav Smil's information, the total area planted to sugarcane in tropical and subtropical countries was about 19 million hectares in 2005, and if all that were devoted to ethanol production, the annual yield would be equivalent to 6% of the world's 2005 gasoline consumption. To cover the entire demand, sugarcane would have to be planted on some 320 million hectares—that is, on 20% of the world's arable land.

Though liquid fuels are used for automotive fuels throughout the world, its precise choice is determined by local conditions, which vary considerably.

Exercises

1. The standard heat of formation is $\Delta H_f^o(NO) = 90.25$ kJ/mol $\Delta G_f^o = 86.57$ kJ/mol.

   (a) Calculate the value of the equilibrium constant Kp for the reaction
   $N_2 + O_2 \rightleftarrows 2NO$
   at 800, 1,000, and 1,200 K and determine the % NO in air at these temperatures (see Exercise 2.18).

Table 5.5 Selected thermodynamic values

Chemical formula	ΔH_f° (kJ/mol)	ΔG_f° (kJ/mol)	S° (JK^{-1} mol^{-1})
	ΔH_f° (kJ/mol)	ΔG_f°	S° JK^{-1} mol^{-1}
CO	−110.5	−137.2	197.6
H$_2$	0	0	130.6
O$_2$	0	0	205.0
CH$_3$OH$_{(g)}$	−201.6	−162.4	239.7
CH$_3$OH$_{(l)}$	−239.1	−166.4	126.8
C$_2$H$_5$OH$_{(g)}$	−234.4	−167.9	282.6
C$_2$H$_5$OH$_{(l)}$	−277.1	−174.9	160.7
H$_2$O$_{(g)}$	−241.8	−228.6	188.7
H$_2$O$_{(l)}$	−285.8	−237.2	70.0
CH$_{4(g)}$	−74.7	−50.8	186.2
CH$_{4(l)}$	−184.4	−114.2	266.5

(b) Determine the thermodynamic values of the constants **a** and **b** in the equation of Exercise 2.18.

2. Using the data in Table 5.5, calculate the equilibrium constant for the reaction
$$CO + 2H_2 \rightarrow CH_3OH_{(g)}$$
at 298, 400, and 600°K.

3. Using the data in Table 5.5, calculate the equilibrium at 298, 400, and 600 K constant for reaction (5.3)
$$CH_4 + H_2O_{(g)} \rightarrow CO + 3H_2$$

4. The air fuel ratio (AFR) = weight of air available for combustion/weight of fuel available for combustion
Show that for complete combustion (stoichiometric), (a) the AFR = 15 for gasoline or diesel fuel; (b) calculate the AFR for pure methanol fuel.

5. What are the reasons for pursuing the development of alternate fuels for the automobile?

6. How can NOx formation be minimized when using air (with the N$_2$ present) in a combustion process?

7. Explain why propane is a "clear" fuel.

8. Explain why the heat of combustion by weight of propane is greater than that of gasoline (see Table 4.9).

9. Why would butane, if available in large quantities, be a better fuel than propane?

10. Why is it ill-advised to use methanol as a fuel in a vehicle designed to run on gasoline?

11. Can you suggest a nonfossil fuel source for methanol?

12. Calculate the temperature reached in the adiabatic compression (CR = 10) of an air/fuel mixture. Note: The heat capacity of air C$_u$ (air) = 21.5 J/K mol.

13. Would it be an advantage to convert methanol to dimethyl ether for an SI-ICE?

14. Calculate the energy required (if any) to convert natural gas (CH$_4$) to ethanol by the reaction
$$2CH_{4(g)} + O_{2(g)} \rightarrow C_2H_5OH_{(l)} + H_2O_{(l)}$$

15. Discuss the energetics and feasibility of reactions (5.14) and (5.15) to produce ethanol from methanol.

Further Reading

1. Ramadhas AS (ed) (2011) Alternate fuels for transportation, CRC Press, Taylor & Francis Group
2. Smil V (2010) Energy myths and realities. Bringing science to the energy policy debate. The AEI Press, Washington, DC

3. Kitasei S, Mastny L (2010) Powering the low-carbon economy: the once and future roles of renewable energy and natural gas. Worldwatch report 184. Worldwatch Institute, Washington, DC
4. (http://en.wikipedia.org/wiki/2008_Toronto_explosions).
5. (1983) Propane carburetion, EMR Canada
6. (1981) Switching to propane. MTC Energy Ontario
7. (1991) Focus on propane. Superior Propane Inc.
8. Ecklund EE, Mills GA (1989) Alternate fuels: progress and prospects. Part 1 and 2. Chemtech Sep p. 549 Oct. p. 626
9. Semanaitis D (1989) Alternate fuels. Road Track Nov. p 72
10. Chang TY et al (1991) Alternative transportation fuels and air quality. Environ Sci Technol 25(7):1190
11. Morton L et al (1990) Methanol: a fuel for today and tomorrow. Chem Ind 16:457
12. Gray CL, Alson JA (1989) The case for methanol. Sci Am p 108–114
13. Lincoln JW (1976) Methanol and other ways around the gas pump. Garden Way Publ, Charlotte Vermont
14. Marsden SS Jr (1983) Methanol as a viable energy source in today's world. Ann Rev Energy 8:333
15. Kohl WL (ed) (1990) Methanol as an alternative fuel choice: an assessment. John Hopkins University, Washington
16. Goodger EM (1981) Alternate fuels for transport, vol 1, Alternate fuels technology series. Cranfield Press, UK
17. Willkie HF, Kalachov PJ (1942) Food for thought—a treatise on the utilization of farm products for producing farm motor fuel as a means of solving the agricultural problem. Indiana Farm Bureau Inc., USA
18. Van Koevering'TE et al (1987) The energy relationships of corn production and alcohol fermentation. J Chem Educ 64:11
19. Biomass Energy Institute (1893) Canadian alternate fuels. Biomass Energy Institute, Winnipeg
20. World Bank (1980) Alcohol production from biomass in the developing countries. World Bank, Washington, DC
21. Anderson EV (1992) Ethanol from corn. Chem Engin News 2:7

Chapter 6
Gaseous Fuels

6.1 Introduction

The combustion process is usually an oxidation reaction involving oxygen, and the oxidant and is in many cases in the gaseous state. Thus, the burning candle and the alcohol burner are examples of solid and liquid fuels burning in the gaseous state. Coal however does not volatilize, and the combustion of many solids, including some plastics, occurs at the surface. By gaseous fuels, consideration is given to the storage state at ambient temperatures even though the fuel in usage can be in the liquid state.

The gaseous fuels which will be considered are methane (or natural gas), synthesis gas (which is a mixture of carbon monoxide and hydrogen), and pure hydrogen.

6.2 Natural Gas

Natural gas is a clean burning fuel which is easily transported by pipeline or as liquefied natural gas (LNG). Figure 6.1 shows the structure of a ship for transporting LNG. The production, consumption, and reserves of natural gas for 2006–2007 for various countries is given in Table 6.1. It can be noticed that some countries such as Norway are big producers with a large reserve in offshore gas but use little or no gas because in Norway, hydroelectricity is the preferred type of energy used. Likewise, Japan has very little natural gas and no significant reserves, but it is a major consumer bringing in the LNG from Australia, Saudi Arabia, Brunei, Qatar, Malaysia, and Indonesia. In the year ending April 1, 2000, Japan imported 52 Mtonnes of LNG. It is stored underground, as shown in Fig. 6.2. At the present rate of consumption, the world resources of natural gas will last about 60 years or until about 2060.

The composition of natural gas varies considerably from one country to another and from one well to another in the same locale. The representative composition of natural gas for various countries is given in Table 6.2. Some extreme values are riot shown and include 70% CO_2 in some gas in Indonesia and 40% C_2H_6 for some Siberian gas. Dry natural gas is gas freed of liquids of butanes and heavier hydrocarbons which are called natural gas liquids.

Natural gas is associated with petroleum deposits and usually accompanies the oil as it is drawn from the well. Often, such gas cannot be conveniently stored or utilized, and as a result, it is often wasted by venting or flaring. The proportion of this wasted gas is increasing every year as shown

O.V. Roussak and H.D. Gesser, *Applied Chemistry: A Textbook for Engineers and Technologists*,
DOI 10.1007/978-1-4614-4262-2_6, © Springer Science+Business Media New York 2013

Fig. 6.1 (**a**) Cross section of an LNG tanker and its detailed cell-type structure. (**b**) LNG transport ship

in Fig. 5.3. It is this gas which can often be converted directly into methanol. The gas in large wells is cleaned, stripped of liquids, and pumped into the pipeline for distribution.

The major contaminant in the gas is H_2S which is removed by the Claus process. Part of the H_2S is oxidized to SO_2 and then reacted by

$$2H_2S + SO_2 \rightarrow 2S + 2H_2O \tag{6.1}$$

The sulfur is stockpiled for sale, primarily to producers of sulfuric acid.

Methane mixed with carbon dioxide, usually in a 1:1 ratio, is formed by the anaerobic digestion of garbage and manure by microorganisms. Thus, landfill gas can be used as a source of heat when collected. Similarly, hog manure has been digested, forming biogas.

The anaerobic digestion is a two-stage biological process involving "acid-forming" bacteria which convert the fats, carbohydrates and proteins into simple acids such as acetic and propionic acid. At the same time, the "methane-forming" bacteria convert the acids by decarboxylation into CO_2 and CH_4. The optimum temperature range is 30–45°C and economically between 25°C and 35°C to reduce heating costs.

The biogas is composed of 60–70% CH_4 and 30–40% CO_2. Small amounts of H_2, H_2S, and NH_3 are also formed. The gas has an energy content of 22–26 MJ/m^3. The H_2S at about 10 g/m^3 can be

Table 6.1 World annual natural gas production, consumption, and reserves (IEA 2006)[a]

	Production (2006) ft$^3 \times 10^{-12}$	Consumption (2006) ft$^3 \times 10^{-9}$	Reserves (2007) ft$^3 \times 10^{-12}$
Canada	6.5	3,307	57.9
Mexico	1.7	2,200	14.5
USA	18.5	21,653	211.1
France	0.04	1,759	0.4
Germany	0.7	3,524	9.0
Italy	0.4	2.984	5.8
Netherlands	2.7	1,690	50.0
Norway	3.2	230	82.3
Russia	23.2	16,598	1,680.0
Ukraine	0.7	2,560	39.0
Turkmenistan	2.2	639	100.0
Uzbekistan	2.2	1,769	65.0
Indonesia	2.0	802	97.8
Japan	0.2	3,247	1.4
Algeria	3.1	904	161.7
Saudi Arabia	2.6	2,594	240.0
UK	2.8	3,202	17.0
Venezuela	0.9	918	152.4
Iran	3.8	3,839	974.0
Iraq	0.06	64	112.0
Argentina	1.7	1,475	16.1
Qatar	1.8	693	910.5
United Arab Emirates	1.7	1,522	214.4
Other	21.28	26,252	977.1
Total	**103.98**	**104,425**	**6,189.4**

[a]From the International Energy Annual (2006)

Fig. 6.2 Distribution scheme for natural gas from stored LNG

reduced to acceptable levels (<1.5 g/m^3) by passing the gas through iron oxide sponge which can be regenerated by exposing it to air.

Landfill gas and biogas can be readily converted to syngas either catalytically or by heated wires

$$CH_4 + CO_2^- + 2H_7 + 2CO \tag{6.2}$$

Table 6.2 Composition of various natural gas fields (% vol.)

	Methane	Ethane	Propane	Butane	Pentane	N$_2$He	H$_2$S	CO$_2$
Europe								
Lacq, France	69.0	3.0	0.9	0.5	0.5	1.5	15.3	9.3
S. Oldenburg, Germany	89.6	1.7	–[a]	–	–	8.2	–	0.5
Groningen, Netherlands	81.3	2.8	0.4	0.2	–	14.4	–	0.9
Ravenna, Italy	99.5	0.1	–	–	–	0.4	–	–
Hewett, UK	92.6	3.6	0.9	0.4	0.3	2.2	–	–
EKOFISH, Norway	90.9	5.9		1.1[b]		0.6	–	1.5
Urengoy, USSR	85.3	5.8	5.3	2.1	0.2	0.9	–	0.4
Africa								
Hassi R'Mel, Algeria	83.7	6.8	2.1	0.8	0.4	6.0	–	0.2
Umuechem, Nigeria	79.6	7.6	5.2	2.7	3.3	0.5	–	1.1
Amal, Libya	62.0	14.4	11.0	5.5	2.6	3.4	–	1.1
Middle East								
Ghawar, Saudi Arabia	59.3	17.0	7.9	2.6	1.1	0.4	1.6	10.1
Kirkouk, Iraq	55.7	21.9	6.5	3.9	1.7	–	7.3	3.0
Agha Jari, Iran	76.1	11.1	6.1	2.2	1.1	–	0.3	3.1
Murban OffS, Abu Dhabi	76.4	8.1	4.7	2.6	1.8	0.1	1.7	4.5
Asia								
Sui, Pakistan	88.5	0.9	0.3	0.4	–	2.5		
India								
Assam	83.3	10.9	1.5	0.3	0.9	0.4	7.4	
Bombay High (offshore)	87.0	6.4	4.5	0.7	0.4	0.01	–	2.6
Oceania								
North Bank, Australia	88.7	5.6	1.8	0.6	0.3	0.7		
Arun, Indonesia	71.9	5.6	2.6	1.4	3.6	0.4	–	2.3
Kapuni, New Zealand	44.2	11.6		11.6[b]		–	–	14.5
America								
California, USA	88.7	7.0	1.9	0.3	–	1.5		
Alberta, Canada	91.9	2.0	0.9	0.3	–	4.9	–	0.6
Tampico, Mexico	46.0	0.6		0.6[b]		2.4	–	–
Maracaibo, Venezuela	82.0	10.0	3.7	1.9	0.7	1.5	2.2	48.8
Camp Duran, Argentina	88.8	5.9	1.8	0.3	0.1	1.4	–	0.2

[a]Insignificant amounts
[b]Includes C$_3$ + C$_s$ fractions

The syngas can then be used to form methanol or gasoline by catalytic reaction after readjusting the H$_2$/CO ratio to 2.

Another potential source of CH$_4$ is primordial gas (nonfossil origin) which Thomas Gold has proposed exists 10–20 km below the earth's surface—just below the earth's crust. This gas is believed to exist throughout the earth and in sufficient quantities to last for several centuries. Experimental drilling has been conducted in Sweden and could determine if the Gold hypothesis is correct.

Still another source of methane is the gas hydrate. At high pressure, methane forms a clathrate (cage) complex with water which is stable at temperatures below 20°C and at pressures greater than 20 atm. This natural gas hydrate is present on the ocean floor, in the sediment below the seafloor, as well as in the cold permafrost of the Arctic.

The structure of the methane hydrate is illustrated in Fig. 6.3 where part of a unit cell is shown for 20 water molecules with one methane molecule. The unit cell contains two CH$_4$ molecules associated

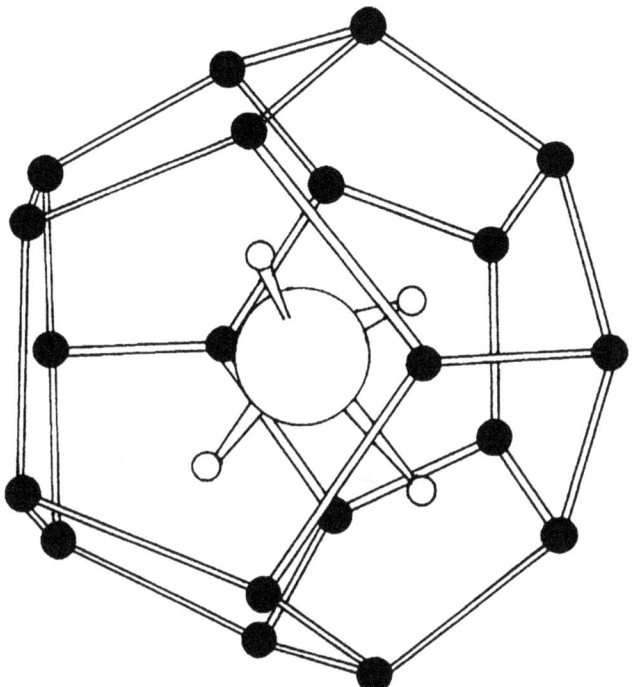

Fig. 6.3 Model of part of a methane-hydrate cell showing the oxygen atoms (•) of the water molecules with the CH$_4$ in the center

with 46 H$_2$O molecules to give about 3.7% CH$_4$ by weight. The estimated amounts of CH$_4$ bound as the hydrate in permafrost and deep ocean waters are 10 teratons or about 53% of the earth's total combustible carbon. Such a dilute system will require new technology and innovative engineering for the successful exploitation of this unique energy source.

Natural gas is usually saturated with water as it leaves the well, and as a result, the gas hydrates can form in the pipeline restricting the flow of gas. To destroy such plugs in the cold Siberian pipeline, large amounts of methanol are introduced at the approximate rate of 1 kg/1,000 m^3 of gas. This represents more methanol than can be produced by a standard methanol plant.

6.3   Natural Gas Uses

Methane—the major constituent of natural gas—is not only an excellent fuel but an important chemical. This is illustrated by the various reactions in Fig. 6.4 where each of the products are important reactants themselves. The thermodynamics of methane pyrolysis reactions is shown in Fig. 6.5 for equilibrium conditions. If, however, the hydrogen produced is removed from the reaction system, then the product yields can be substantially higher. This has been demonstrated using thin palladium membranes which allow only H$_2$ to pass through. Similarly, when the pyrolysis occurs on a hot wire, the hydrogen produced can diffuse out of the reaction zone faster than the other heavier product, and therefore, equilibrium conditions do not prevail. Thus, in Fig. 6.5, though reaction #1 is preferred thermodynamically, it is found that experimentally reaction #4 occurs on a hot wire forming aromatic oil.

Fig. 6.4 Chemical reactions and uses of methane

6.4   Natural Gas as a Fuel

Natural gas is an excellent fuel which burns cleanly with little or no residue. The main problem with its use is its low energy density—it is not readily stored in sufficient quantities to power a vehicle for the normal distance of 300–400 km. The simplest storage method is as a compressed gas in cylinders. Buses in Vancouver have such storage cylinders on the roof. It has been shown that by filling a cylinder with active carbon, it is possible to double the amount of CH_4 which can be stored in the cylinder. However, this still limits the use of compressed natural gas (CNG) to fleet vehicles which can routinely refill the cylinders with compressed gas. Buses run on natural gas in Vancouver (British Columbia, Canada) during the last 2010 Winter Olympic Games. This is shown in Fig. 6.6.

Natural gas can be stored in insulated containers as a liquid at $-161\,^{\circ}$C, and as a liquid fuel, the LNG has sufficient energy density. Its use has been tested in airplanes and locomotives. However, the use of natural gas as an automotive fuel is still limited unless higher densities can be achieved.

Recent designs of thin-walled polyethylene carbon fiber over wrapped lightweight high-pressure cylinders encapsulated with energy-absorbing foam in a protective casing have enabled a Geo Prizm sedan to achieve a driving range of 315 miles (500 km) when fueled with natural gas. This is the most promising new approach to the clean-fueled automobile.

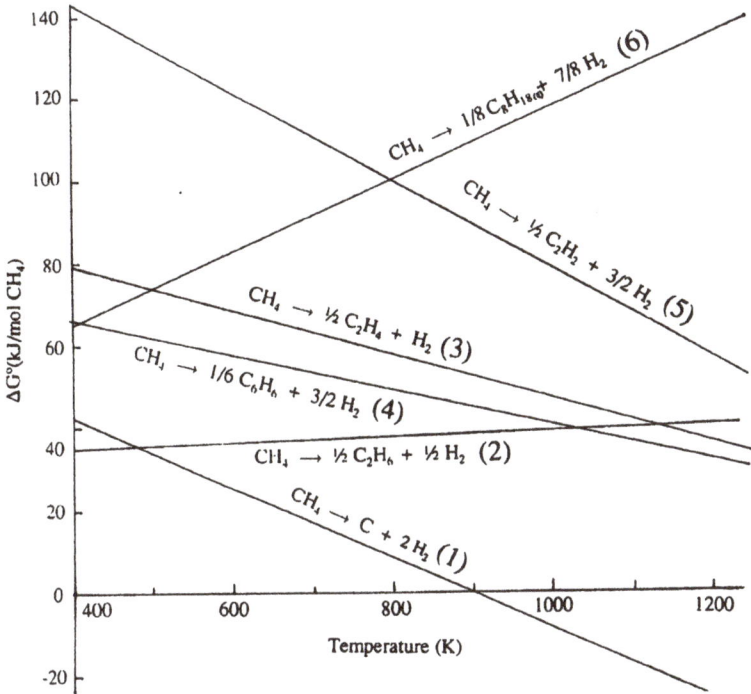

Fig. 6.5 The thermodynamics of six CH₄ pyrolysis reactions

Fig. 6.6 Buses run on natural gas in Vancouver (British Columbia, Canada) during the last 2010 Winter Olympic Games

6.5 Other Carbon-Based Fuel Gases

Natural gas is not the only carbon-based gas which can be used as a fuel. One of the simplest is derived from coal. The reaction of coal, coke, or charcoal with insufficient air for complete combustion forms producer gas which is a mixture of CO and N_2 in a 1:2 ratio:

$$2C + O_2 + 4N_2 \rightarrow 2CO + 4N_2 \tag{6.3}$$

Any moisture in the air results in the formation of hydrogen by the water gas reaction

$$C + H_2O_{(fg)} \rightarrow CO + H_2 \tag{6.4}$$

to form water gas. The water shift reaction can also occur to produce some CO_2

$$CO + H_2O_{(fg)} \rightarrow CO_2 + H_2 \tag{6.5}$$

When coal is heated to about 500°C, the CH_4 trapped in the pores is released. At higher temperature (1,000°C), the organic matrix in the coal is decomposed, forming mostly H_2 with some CO_x. This coal gas is also formed during the coking process.

The addition of oil to the water gas process at approx. 30 L of oil per 100 m^3 of gas increases the calorific value of the gas produced threefold by cracking the oil to $C_2 + C_3$ hydrocarbons.

The gasification of coal with oxygen and steam under pressure is called the *Lurgi process* and the gas is called *Lurgi gas*. As the pressure is increased from 5 to 20 atm, the CH_4 and CO_2 increases, while the H_2 and CO decrease. This gas, once popular as a town gas, is seldom used today.

The approximate composition and heating values of the various fuel gases are given in Table 6.3. The flame speeds relative to that of H_2 are also shown and account for the need to change burner nozzles when fuel gas composition is significantly changed.

The TLV value of CO is 50 ppm, and hence, a level of 20 ppm is considered to be safe. However, it has been shown that such low levels of CO can inhibit the learning process in children and reduce the response time of adults. Thus, apparently safe levels may not be healthy if long-term exposure is anticipated. A catalyst of gold supported on metal oxide surfaces has recently been shown to convert CO in air to CO_2 at room temperature.

Table 6.3 Approximate composition and calorific values of various fuel gases (%vol.)

Gas comp.	Natural gas	Producer gas	Water gas	Coal gas	Lurgi[a] gas	Landfill gas	Bio gas	Sewage gas	Flame[b] speed	ΔH^0 (MJ/m^3) combustion
CO	–	33	46	7	24	1	–	–	18	11.75
CO$_2$	1	1	2	2	26	58	63	60	–	–
H$_2$	–	4	48	53	45	–	–	1	100	11.88
CH$_4$	88	2	1	30	4	40	35	35	14	37.2
C$_2$ + C$_3$	10	–	–	3	–	1	–	–	16	60
N$_2$	1	60	3	5	1	–	2	2	–	0
Calorific value, MJ/m^3	39	5	11	20	16	22				

[a]2 atm pressure

[b]Relative to H_2 + 2 atm pressure

6.6   Explosion Limits

All combustible gases and vapors can be made to explode if mixed with the proper proportion of air or oxygen. There are two limits of gas concentration which are referred to as explosion limits. Below the lower limit (LEL), there is insufficient gas to carry the chain reactions which results in the rapid release of energy due to the exothermic heat of reaction. Above the upper limit (UEL), there is insufficient oxygen to permit the chain reaction to proceed. This is illustrated in Fig. 6.7 for methane and shows how the limits vary with pressure and temperature and how they differ for air and oxygen. The explosion limits and autoignition temperatures of various gases and vapors are listed in Table 6.4 which shows some wide ranges of limits for H_2, CO, and acetylene. Mixtures of gases such as water gas (CO + H_2) have two sets of limits. Gas which can explode over a wide range of concentrations presents a major hazard when these gases leak into air. However, light gases such as H_2 diffuse in air very rapidly, and hence, the danger of an explosion is reduced. Combustible solids in the form of dust can also explode. This is illustrated by coal dust, grain dust, and manufacturing dust which have all caused explosions. The very narrow explosion limits of methyl bromide are of interest since this substance can reduce the explosive characteristics of dust explosions.

6.7   Hydrogen

Hydrogen is receiving a lot of attention lately as a fuel of the future. We are all aware of the difficulty the world will be facing as our nonrenewable energy resources become depleted. Oil is expected to last, by various estimates, into the early twenty-second century and coal perhaps for another century

Fig. 6.7 Explosion limits for CH_4 in air (circles, ● and ○) and in O_2, filled points, ● and ▲ at 20°C and for CH_4 in O_2 at 300 °C (● and ■)

Table 6.4 Autoignition temperatures and explosive limits of some gases and vapors in air as vol. %

Autoignition			Autoignition		
Substance	Temp.°C	LEL/UEL	Substance	Temp.°C	LEL/UEL
Acetone	465	2.6–12.8	Hexane	225	1.1–7.5
Acetylene	305	2.5–8.0	Hydrogen	400	4.1–74.0
Ammonia	651	16.0–26.0	Isopropanol	456	2.0–12.0
Benzene	562	1.3–7.1	Jet fuel		1.3–80.0
Butane	405	1.8–8.4	Kerosene	210	1.3–80.0
Carbon disulfide	125	1.3–50.0	Methane	650	5.3–14.0
Carbon monoxide	609	12.5–74.0	Methyl bromide	537	13.5–14.5
Cyclo propane	500	2.4–10.4	Naphthalene	277	0.9–6.0
1,2-Dichloroethylene	460	9.7–12.8	Naphthalene	568	0.9–5.9
Dimethyl ether	350	3.4–27.0	n-Octane	220	0.8–3.2
Ethane	515	3.2–12.5	Propane	450	2.4–9.5
Ethanol	423	3.3–19.0	Toluene	480	1.2–7.1
Ethylene	490	3.0–36.0	Water gas		6.0–9.0, 55.0–70.0
Fuel oil		0.7–5.0	Xylene	530	1.0–6.0
Gasoline	280–456	1.4–7.6	Dust		~50 mg/L

or two depending on the alternate energy sources utilized. Natural gas will probably replace oil, and its reserves will not last much longer than oil. Hence, hydrogen, which is readily made from water, is being considered as a replacement for natural gas and even as a substitute for gasoline. However, there are some problems and difficulties which should be considered, since hydrogen is not a primary fuel but must be made using another energy source such as hydroelectricity, coal, nuclear energy, or solar energy to name but a few. Hydrogen is thus classified as an energy currency like electricity which can be pumped from one location to another and stored. It can also be used as a fuel for the automobile and so can replace gasoline when our fossil fuels are depleted.

What do we know about hydrogen?

1. It is easily made by the electrolysis of water or appropriate thermochemical cycles.
2. It is lighter than air and was used in lighter-than-air dirigibles such as the Zeppelin.
3. It is highly explosive and with the destruction of the Hindenburg in 1937, commercial travel in H_2-filled dirigibles ended.

However, it must be pointed out that the Hindenburg disaster resulted in the death of only 36 people, most from flaming diesel fuel from the engines and from jumping to the ground. The Hindenburg had made 54 flights, 36 of which were across the Atlantic.

The physical properties of hydrogen are given in Table 6.5, and the general methods of preparation are outlined in Table 6.6.

6.8   Methods of Preparation of H_2

6.8.1   Electrolysis

The most attractive method of producing hydrogen is by hydroelectricity at off-peak load. The minimum voltage required for water splitting at 25°C and 1 atm is 1.23 V (reversible voltage). Under these conditions, hydrogen would be produced only if heat is added. If no heat exchange with the surroundings takes place, the cell cools down. The thermoneutral potential at 25°C is 1.48 V.

Table 6.5 Selected physical properties of hydrogen (H_2)

Molecular weight (g/mol)	2.01594
Natural isotopic abundance*	
^1H	1.00794 g/mol 99.985 %
^2H (deuterium)	2.01355 g/mol 0.015 %
^3H (tritium)[a]	3.01550 g/mol 0
Melting point (K)	14.2
Boiling point (K) para-H_2	20.27
Boiling point (normal-H_2; 25 % p-H_2) (K)	20.39
Triple point (K, n-H_2)	13.96
Triple point liquid density (kg/m^3)	77.20
Triple point solid density (kg/m^3)	86.71
Triple point vapor density (kg/m^3)	0.131
Boiling point liquid density (kg/m^3)	71.0
Boiling point vapor density (kg/m^3)	1.33
Critical temperature (K, n-H_2)	33.19
Critical pressure (atm, n-H_2)	12.98
Critical volume (kg/m^3, n-H_2)	30.12
Latent heat of fusion at T.P.; p-H_2 (J/mol)	117.6
Latent heat vaporization at B.P. (J/mol, n-H_2)	897.3
Heat of combustion (gross) liquid H_2O (kJ/mol)	285.8
Heat of combustion (net) gaseous H_2O (kJ/mol)	241.8
Limits of flammability in air (vol.%)	4.0–75.0
Limits of detonability in air (vol.%)	18.0–59.0
Burning velocity in air (m/s)	Up to 2.6
Burning velocity in oxygen	Up to 8.9
Limits of flammability in oxygen (vol.%)	4.0–95.0
Limits of detonability on oxygen (vol.%)	15.0–90.0
Detonation velocity	
15 % H_2 in O_2 (m/s)	1,400
90 % H_2 in O_2 (m/s)	3,600
Nonflammable limits, air-H_2	<8 % H_2
Nonflammable limits, O_2-H_2	<5 % H_2
Maximum flame temperature at 31 % H_2 in air (K)	2,400
Autoignition temperature in air (K)	847
Autoignition temperature in oxygen (K)	833

*The radioactive decay constant $\lambda_{1/2} = 12.26$ year; n-H_2 refers to normal H_2 (25% p-H_2 and 75% o-H_2)

Table 6.6 Preparation of H_2

1. Electrolysis of water	$H_2O \rightarrow H_2 + \frac{1}{2}O_2$
2. Thermal methods	$2H1 \rightarrow H_2 + I_2$
3. Natural gas	$CH_4 \rightarrow C + 2H_2$; $CH_4 + H_2O \rightarrow CO + 3H_2$
	$CH_4 + \frac{1}{2}O_2 \rightarrow CO + 2H_2$
4. Thermal, nuclear, electrical	$CH_4 + H_2O \rightarrow CO + 3H_2$
5. Photoelectrolysis of water	$H_2O + hv + \text{catalyst} \rightarrow H_2 - + \frac{1}{2}O_2$

The electrolysis cells draw thousands of amperes (current densities of 1–4 kA/m^2) (100–400 A/ft^2) at high temperatures (up to 300°C) and pressures (above 200 atm). The electrolyte is usually potassium hydroxide (KOH), but acidic solutions (H_2SO_4) are also used occasionally.

Table 6.7 Three schemes for the thermal generation of hydrogen by closed-cycle processes with thermal efficiency (TE) and Carnot efficiency (CE)

Agnes		
TE = 41–58%	$3FeCl_2 + 4H_2O = Fe_3O_4 \; 4\text{-}6HCl + H_2$	450–750°C
CE = 58%	$Fe_3O_4 + 8HCl = FeCl_2 + 2FeCl_3 + 4H_2O$	100–110°C
	$2FeCl_3 = 2FeCl_2 + Cl_2$	300°C
	$Cl_2 + Mg(OH)_2 = MgCl_2 + \frac{1}{2}O_2 + H_2O$	50–90°C
	$MgCl_2 + 2H_2O = Mg(OH)_2 + 2HCl$	350°C
Behiah		
TE = 53–63%	$2Cu + 2HCl = 2CuCl + H_2$	100°C
CE = 63%	$4CuCl = 2CuCl_2 + 2Cu$	30–100°C
	$2CuCl_2 = 2CuCl + Cl_2$	500–600°C
	$Cl_2 + Mg(OH)_2 = MgCl_2 + H_2O - \frac{1}{2}O_2$	80°C
	$MgCl_2 + 2H_2O = Mg(OH)_2 + 2HCl$	350°C
Catherine		
TE = 64–83%	$3I_2 + 6LiOH = 5LiI + LiIO_3 + 3H_2O$	100–190°C
CE = 83%	$LiIO_3 + KI = KIO_3 + LiI$	0°C
	$KIO_3 = KI + 1\frac{1}{2}O_2$	650°C
	$6LiI + 6H_2O = 6HI + 6LiOH$	450–600°C
	$6HI + 3Ni = 3NiI_2 + 3H_2$	150°C
	$3NiI_2 = 3Ni + 3I_2$	700°C

At 25°C, $\Delta G^0 = -237.13$ kJ for water, but at 900°C, $\Delta G^0 = -182.88$ kJ for steam. Hence, the minimum voltage E_{th} (900°C) = 0.906 V at the higher temperature and the electrolysis of steam at high temperatures and pressures are also being studied as a means of producing hydrogen efficiently. At 200°C, the voltage of 1.3 V is the minimum voltage at which electrolysis would occur. Above this voltage, heat is produced due to the IR drop in the cell and due to the overvoltage on the electrodes. At 1 ¢/kWh, the cost of 1 GJ of H_2 is 3.8 ¢ at 1.3 V and 4.9 ¢ at 1.8 V. Hence, great care is required in designing cells, electrodes, and electrode separators in order to reduce costs.

The production of oxygen which accompanies the hydrogen is a surplus product since oxygen is readily separated from nitrogen in liquid air or by the use of molecular sieves or membranes. Hence, by using a carbon anode, the following reactions would occur:

$$H_2O_{(1)} + C \rightarrow H_2 + CO \qquad \Delta G^0 = 100.0 \text{ kJ/mol} \qquad (6.6)$$

$$H_2O_{(a)} + C \rightarrow H_2 + CO_2 \qquad \Delta G^0 = 40 \text{ kJ/mol} \qquad (6.7)$$

This can be compared with ΔG_j^0 $(H_2O)_1 = -237.2$ kJ/mol and represents a substantial energy saving. However, when hydrogen is prepared by this method, it no longer is an environmentally friendly fuel since CO and CO_2 will be formed.

The main problem involved in hydrogen production by electrolysis is the materials of construction since reliability is essential if explosions are to be avoided.

6.8.2   Thermal Methods

Several closed-cycle thermal processes have been developed whereby hydrogen can be produced from water using, e.g., coal as the source of heat. Three common schemes are presented in Table 6.7.

The corrosive nature of some of the products implies rather elaborate precautions in construction and design. The total energy required for the reaction

$$H_2O \rightarrow H_2O_{(g)} \rightarrow H_2 + \tfrac{1}{2}O_2 \qquad \Delta H^0 = 285.8 \text{ kJ} \tag{6.8}$$

is 285.8 kJ if one starts with liquid H_2O and 241.8 kJ if steam is available. In either case, this energy can be supplied in a series of steps as shown in Table 6.7. The efficiencies given represent the thermal efficiencies of the sum of the steps as well as the Carnot (reversible) efficiency.

6.8.3   Natural Gas

The major use of hydrogen today is in the production of ammonia via the Haber reaction:

$$\frac{1}{2}N_2 + \frac{3}{2}H_2 \xrightarrow[\text{[500°C,300 atm]}]{\text{Catalyst}} NH_3 \qquad \Delta H^0 = -46.2 \text{ kJ} \tag{6.9}$$

The nitrogen is prepared by the fractional distillation of liquid air (b.p. $N_2 = -196°C$, b.p. $O_2 = -183°C$), whereas the hydrogen is usually prepared by the thermal cracking of natural gas:

$$CH_4 \rightarrow C + 2H_2 \qquad \Delta H^0 = 74.8 \text{ kJ} \tag{6.10}$$

Hydrogen, as a component of synthesis gas, is also made from carbon and methane by the following reactions:

$$C + H_2O \rightarrow CO + H_2 \qquad \Delta H^0 = 131.3 \text{ kJ} \tag{6.11}$$

$$CH_4 + H_2O \rightarrow CO + 3H_2 \qquad \Delta H^0 = 206 \text{ kJ} \tag{6.12}$$

$$CH_4 + \tfrac{1}{2}O_2 \rightarrow CO + 2H_2 \qquad \Delta H^0 = -36 \text{ kJ} \tag{6.13}$$

$$CH_4 + 2H_2O \rightarrow CO_2 + 4H_2 \qquad \Delta H^0 = 165 \text{ kJ} \tag{6.14}$$

The CO_2 can be separated from H_2 by its solubility in water at 25 atm.

6.8.4   Thermal-Nuclear-Electrical

Water is readily dissociated into its elements H_2 and O_2 by an electrical discharge or by thermal energy. The separation of H_2 from O_2 is then required. An interesting proposal presented several years ago was to detonate a nuclear explosive device in a landfill (garbage dump). A calculated yield of the products H_2, CO, CO_2, and CH_4 showed that as the organic content of the garbage reaches 80%, the H_2 and CO each approach 50%. The obvious problem in such a scheme is that the gases cannot be used until the radioactivity of the gases has decayed to acceptable levels.

6.8.5   Photoelectrolysis

About 25 years ago, it was first shown that ultraviolet light can dissociate water on a suitable semiconductor catalyst. Since then, it has been possible to photocatalytically dissociate water into H_2 and O_2 at separate electrodes (hence, not requiring a separation of the gases) or to generate electrical energy for direct use. This rapidly expanding field will eventually achieve what nature does in plants, i.e., using visible light in multiple steps to convert CO_2 and water into cellulose. Efficiencies are usually low, and long-term stability has not as yet been achieved, but with continued efforts, it will be possible to produce hydrogen from sunlight and water.

6.9   Transportation and Storage of H_2

The cost of pipeline transportation of hydrogen is higher (20–30%) than that for natural gas for equivalent energy flow. Another major difference is due to the small molecular size of H_2 (diameter = 0.289 nm) compared to CH_4 (diameter = 0.38 nm). Hence, H_2 pipelines are more susceptible to leaks.

Another problem anticipated with hydrogen is the hydrogen embrittlement of metals, usually at elevated pressures and at ambient temperatures. Hydrogen has a tendency to diffuse through metals along grain boundaries and, in some cases, forming metal hydrides. This makes the metal brittle, causing fractures and failure. This was first noted about 100 years ago, and special alloys have been made specifically for H_2 service. Thus, the use of natural gas pipelines to transport hydrogen may not be the safest thing to do.

The storage of hydrogerras, a liquid (LH$_2$), is well known in the nuclear field where the liquid hydrogen bubble chamber (5–1,500 L) is used to detect nuclear particles. Liquid hydrogen is also used as a fuel.

The hydrogen molecule exists in two forms: ortho-H_2 and para-H_2. In ortho-H_2, the nuclear (proton) spins of the two hydrogen atoms are parallel, whereas in para-H_2, the nuclear spins are antiparallel. It is possible to catalytically convert hydrogen to the pure lower energy p-H_2 at low temperatures ($T < 20$ K), but at higher temperatures ($T > 200$ K), only an equilibrium mixture of 25% p-H_2 and 75% o-H_2 can be obtained. The conversion of o-H_2 to p-H_2 is an exothermic process which at 10 K, $\Delta H^0 = -1.062$ kJ/mol and at 300 K, $\Delta H^0 = 55.5$ J/mol.

The liquefaction of hydrogen produces 25% p-H_2 + 75% o-H_2, and the slow conversion of o-H_2 to p-H_2 adds an additional heat source to the storage system. Hence, it is desirable to convert the o-H_2 to p-H_2 either completely in the liquid state (by adding charcoal) or preferably partially in the pre-cooled gas phase (at 77°K, liquid N_2, the equilibrium mixture of H_2 vapor, is 60% p-H_2) followed by complete conversion in the liquid state. This reduces the losses on storage from 25%/day for 25% p-H_2 to 0.02%/day for 98% p-H_2.

Liquid hydrogen, like liquid helium, is usually stored in double Dewar flasks, an outer Dewar containing liquid nitrogen (b.p. = 77 K) into which is placed the second Dewar containing the liquid hydrogen (b.p. 20.3 K). However, in transport by truck or rail, single horizontal cylindrical Dewars are used with multilayer insulation, resulting in typical boil-off losses of 0.25%/day.

Table 6.8 lists some thermal conductivities of evacuated powders used in large Dewars for LH$_2$. Since the dielectric powders are transparent to room temperature radiation, the effectiveness of the powders can be increased significantly by the addition of metal powders (e.g., Al, Cu, Ag) which reflect the radiation. Since the metal powders conduct, their addition is an optimum at certain levels.

It is obvious that liquid hydrogen is practical only for large continuous users such as ships, trains, and even aircraft but impracticable for intermittent users such as automobiles. Hydrogen is not very

Table 6.8 Thermal conductivities (λ) of some evacuated insulating powders

Powder	Particle size	Density (g/mL)	Vacuum (Torr)	λ (μW/cm, K)
Perlite, expanded	+30 mesh	0.06	$<10^{-4}$	21
Perlite, expanded	−80 mesh	0.14	$<10^{-4}$	10
Silica aerogel	250 A	0.10	$<10^{-4}$	21
Diatomaceous earth	1–100 μm	0.29	$<10^{-4}$	10
Cab-O-Sil	200–300 Å	0.04	$<10^{-6}$	30
Cab-O-Sil + 50% Metal powder <44 μm	200–300 Å	–	$<10^{-6}$	2.0

Fig. 6.8 Equilibrium pressure-temperature curves (isotherms) for the alloy LaNi$_5$H. where n changes from 0 to 6 as the pressure of H$_2$ increases from 0 to over 2 atm. Note the hysteresis effect

soluble in liquids but dissolves in some metals forming hydrides which readily dissociate reversibly back to the metal and hydrogen at elevated temperatures. The formation and dissociation cycle of a typical system is shown in Fig. 6.8 for a LaNi$_5$ alloy which reacts with three molecules of H$_2$ to form LaNi$_5$H$_6$

$$\text{LaNi}_5 + 3\text{H}_2 \rightleftharpoons \text{LaNi}_5\text{H}_6 \qquad (6.15)$$

though usually more hydrogen can be "dissolved" if the pressure is increased. Several metal systems can be used in this way to store hydrogen. A comparison is given in Table 6.9 where cryoadsorption of H$_2$ on active carbon at liquid nitrogen temperatures (-196°C) is included along with encapsulated H$_2$ in zeolite. (Zeolites are natural and synthetic minerals of silicon and aluminum which have

Table 6.9 A comparison of some H_2 storage systems

Storage medium	Storage capacity		Energy density	
	Wt.%	Vol. g/mol	Wt. kJ/g	Vol. kJ/mL
MgH_2	7	0.10	9.93	14.3
Mg_2NiH_4	3.16	0.08	4.48	11.5
VH_2	2.07	0.095	2.93	13.9
$FeTiH_{1.95}$	1.75	0.096	2.48	13.6
$LaNi_5H_7$	1.37	0.089	1.94	12.6
H_2 (liquid) $-253°C$	100	0.07	141.8	9.93
H_2 (gas) 100 atm	100	0.008	141.8	1.13
H_2 (carbon)[a] $-196°C$	6.8	0.024		
H_2 (zeolite)[b]	1.5	0.009		
H_2 (NiO; 2.5 SiO_2)$-196°C$	3.4	0.024		

[a]At 40 atm, mass and volume of liquid nitrogen is not included
[b]Encapsulated by high pressures greater than 1,000 atm but stored at ambient temperatures (25°C)

Table 6.10 Approximate comparative costs for alternative hydrogen storage systems

Facility	Cost ($/kWh)
Cryoadsorber (carbon)	0.025
Metal hydride (FeTl alloy)	0.028
Compressed gas (11 atm)	0.028
Liquid H_2	0.035

interconnected channels and cages.) By modifying the size of the opening of a type 3A molecular sieve (approx. 3 Å cage size) by replacing Na^+ ions by the larger Cs^+ ion, it is possible to force H_2 into the cage at 250°C and 10,000 psi (500–4,000 atm). When quenched to room temperature, the gas remains in the cage even when the pressure is released. The loss is insignificant for short periods (approx. 1 week) and could be reduced by further modification of the zeolite. In the case of methane encapsulation, it has been shown that up to about 7% by weight CH_4 can be stored in an unmodified type 3A molecular sieve for 5 months without noticeable loss.

It should be pointed out that the density of hydrogen in the metal hydride is similar to that of liquid H_2, implying that hydrogen in the metal is in an atomic form. This has been verified by other methods. The relative cost of the various storage systems is estimated in Table 6.10, but this too will greatly depend on the quantities involved.

Hydrogen has been used in surface vehicles such as automobiles and buses, and plans are in progress to convert a Lockheed L-1011 to burn LH_2 and fly between Pittsburgh, Montreal, Birmingham, Frankfurt, and Riyadh.

6.10   Safety

It is constantly reported that hydrogen is a fuel which is so safe—it requires no chimney since its only product is water (H_2O). This is not always correct since it has been shown that when used as a fuel in an internal combustion spark, engine hydrogen also forms hydrogen peroxide (H_2O_2) which was

Table 6.11 Ignition energy
of H_2-air mixtures

%H_2 in air	10	15	20	30	40	50	60
Ignition energy, mJ	0.2	0.06	0.025	0.02	0.03	0.07	0.3

initially shown to be at about 225 ppm in the exhaust. The TLV value for H_2O_2 is 1 ppm, and though it is easy to destroy H_2O_2, its removal to below 1 ppm is not a simple task. A more recent determination of the H_2O_2 formed in the exhaust of a hydrogen-fueled engine showed that very little H_2O_2 is present and that the previously reported high levels were due to the method of analysis which was affected by NO_x in the exhaust.

Hydrogen is also one of the easiest explosives to ignite when suitably mixed with air or oxygen. For example, the ignition energy of hydrogen (H_2), methane (CH_4), and propane (C_3H_8) are 0.02, 0.45, and 0.25 mJ, respectively. This value depends on the composition of the mixture, and for hydrogen, the values are given in Table 6.11.

Though hydrogen is extremely dangerous because of its wide range of explosive limits and low ignition energy, it has a low mass and therefore diffuses very rapidly away from a source. Thus, a spilled tank car full of LH_2 caused no problem. With reasonable precautions we will be able to handle hydrogen with as much ease as we now manage gasoline.

6.11 Helium

The composition of natural gas (Table 6.1) indicates the presence of up to 0.5–1% helium. The concentration of helium in air is much less than 0.001%, and it is impractical to extract helium from air. Argon, another noble gas, is present at up to 1% in air, and it is obtained as a by-product in the liquefaction of air and the production of O_2 and N_2. Thus, since He is only available economically from natural gas, it is obvious that when natural gas is exhausted, so will be the inexpensive supply of He. Hence, it is imperative that wherever possible, He should be extracted from natural gas before it is put into pipelines, burned, or used for chemical processes. He is usually stored in abandoned mines (storage in cylinders as a compressed gas is too costly). It must be stressed that if natural gas with even 0.5% He is burned, the He is lost and will only be recoverable from air at a very great expense.

Exercises

1. How can the heat value of hydrocarbon gases be determined?
2. What volume of (a) air and (b) oxygen is required to burn 1 L of octane (density = 0.7025 g/mL at 25°C)?
3. Explain how a ratio of $H_2/CO = 1$ can be increased to 2 without the addition of hydrogen from an external source.
4. If the heats of combustion of carbon monoxide, hydrogen, and methane are 11.75, 11.88, and 37.28 MJ/m^3, respectively, calculate the heating value of the water gas and biogas given in Table 6.3.
5. What is the theoretical weight of steam necessary to convert 1 tonne of coke containing 2% ash into water gas at 900°C?
6. Comment on the statement "Natural gas is more explosive than manufactured gas."
7. Write equations for the removal of sulfur from natural gas.
8. What burner adjustments would have to be made in changing from a bottled gas to natural gas?

9. Hydrogen is claimed to be the ideal fuel since it is made from water and burns to form water. Comment on the validity of this statement.
10. When helium is exhausted, hydrogen would be ideal for lighter-than-air ships. Do you agree?
11. Comment on the prevalence of explosions caused by dust.
12. Some prepared gases such as syngas have two sets of explosion limits. Explain.
13. How can the calorific value of a producer gas be increased?
14. Vehicles in New Zealand are required to run on at least two types of fuels which include CNG, propane, methanol, gasoline, gasohol, and diesel. Comment on the two fuels you would select for your vehicle if living in New Zealand.
15. Some metal oxides can be reduced with hydrogen to form an activated metal $MO_x + H_2 = M + H_2O$. The activated metal can be reacted with water to produce hydrogen and the metal oxide, i.e., the reverse of the above reaction. This means that water is effectively the fuel that produces the combustible hydrogen. The metal oxide can be regenerated to metal similarly to the charging of a battery. Iron sponge has recently been proposed and tested as a typical metal for such a system. Calculate the amount of iron and water needed for a trip of 35 km using H_2 in an SI-ICE vehicle with a thermal efficiency equivalent to a gasoline engine which uses 10 L/100 km.
16. Consider a small vehicle fueled by hydrogen in an SI-ICE. The efficiency of the engine is assumed to be thermally equivalent to gasoline, which uses 10 L/100 km. The hydrogen is prepared by the electrolysis of water during the evening (7 p.m.–7 a.m.) and stored for use to drive 35 km. What current is required to produce this hydrogen in one cell (1.5 V) electrolysis unit?

Further Reading

1. Canadian Standards Association (2007) Liquefied natural gas (LNG): production, storage, and handling, 8th edn. Canadian Standards Association, Mississauga
2. Murphy MJ (2005) Emergency response procedures for natural gas transit vehicles. National Academy Press, Washington, DC
3. Canadian Standards Association (2010) Natural gas and propane installation code, 8th edn. Canadian Standards Association, Mississauga
4. Natural Gas Potential in Canada 2001: a report. University of Calgary. Canadian Gas Potential Committee. Calgary: The Committee. (2001) G. D. Berry, Hydrogen as a Transportation Fuel: Costs and Benefits. Bus. Tech. Bks. (1997)
5. Berry GD, Pasternak AD, Rambach GD, Smith JR, Schock RN (1996) Hydrogen as a Future Transport Fuel Energy 21(4):289–303
6. Willett RW (1996) The 1996 natural gas handbook. Wiley, New York
7. Pohl H (1995) Hydrogen and other alternate sources of energy for air and ground transportation. Wiley, New York
8. Bell SR, Sekar R (eds) (1994) Natural gas and alternate fuels for engines, vol 21. American Society of Mechanical Engineers, Fairfield, NJ
9. Brewer GD (1991) Hydrogen aircraft technology. CRC Press, Boca Raton, FL
10. Billings RE (1991) The hydrogen world view. International Academy of Science Independence, Missouri
11. Marinescu-Pasoi L et al (1991) Hydrogen metal hydride storage with integrated catalytic recombiner for mobile application. Int J Hydr Energy 16(6):407–412
12. Swain MR et al (1990) Hydrogen peroxide emissions from a hydrogen fueled engine. Int J Hydr Energy 15(4):263–266
13. Melvin A (1989) Gas taps its natural flare. New Sci 59
14. EMR Ottawa (1988) Natural gas: an alternative transportation fuel. EMR Ottawa, Ontario
15. Klass DL (1984) Energy from biomass and wastes: 1983 update. Institute of Gas Technology, Chicago, IL
16. Klass DL (1984) Methane from anaerobic fermentation. Science 223:1021
17. Bockris JO'M (1980) Energy options. Real economics and the solar-hydrogen system. Taylor and Francis Ltd., London
18. Bell C et al (1980) Methane: fuel of the future. Prism Press, Dorset
19. Gold T, Soter S (1980) The deep-earth gas hypothesis. Sci Am 242(6):154
20. Reilly JJ, Sandrock GD (1980) Hydrogen storage in metal hydrides. Sci Am 242:118–129

21. Kenneth EC, Williamson KD Jr (1977/1979) Hydrogen: its technology and implications, vol 5. CRC Press, Cleveland, OH
22. Woolley RL, Hendrickson DL (1977) Water induction in hydrogen powered I.C. engines. Int J Hydr Energy 1:401–412
23. Woolley RL (1976) Performance of a hydrogen powered transit vehicle. In: Proceedings of 11th intersociety energy conversion conference, Lake Tahoe
24. Billings RE et al (1974) Ignition parameters for hydrogen engines. In: Proceedings of 9th intersociety energy conversion engineering conference, San Francisco, CA
25. Finegold JG et al (1973) The U.C.L.A. hydrogen car: design, construction, and performance. Paper SAE 730507. Society of Automotive Engineers, New York
26. Billings RE, Lynch FE (1973) Performance and nitric oxide control parameters of the hydrogen engine. Publication No. 73002, Energy Research, Provo, Utah
27. King RO (1958) The hydrogen engine: combustion knock and related flame velocity. Trans EIC 2:143
28. King RO, Rand M (1955) The oxidation, decomposition, ignition, and detonation of fuel vapors and gases, XXVII. The hydrogen engine. Can J Technol 33:445
29. Biomass to gases and liquid fuels. http://www.gpc.peachnet.edu/
30. National Fuel Gas Co. http://www.natfuel.com/
31. Pan American Enterprises Inc. (NG, H2). http://www.panent.com/
32. US geological survey resources. http://www.usgs.gov/
33. HyWeb-H2 Fuel Cell. http://www.hydrogen.org/index-e.html
34. Gaseous explosions. http://www.directedtechnologies.com/energy/safety.html
35. Helium source and reserves. http://members.truepath.com/gr/helium.html

Chapter 7
Nuclear Energy

7.1 Introduction

Nuclear energy represents a viable choice as an alternate source of energy to fossil fuels and as a means of supplying the energy needed to produce some of the alternate fuels and fuel currencies which are environmentally acceptable. Though the hazards and dangers of nuclear energy are known to be enormous, a world with a population growth of 2–3% per annum (the world population was 2.56×10^9 in 1950, 4.48×10^9 in 1980, 6.0×10^9 in 2000, and 6.8×10^9 in 2010. It is estimated to be 7.5×10^9 in 2020) will require more energy at a reasonable cost in order to raise the standard of living of the developing nations, to meet the needs of the growing population, and at the same time to preserve our environment. Though solar energy has the potential of meeting the energy needs of the world, the diffuseness of its power makes the present and foreseeable costs too great to be competitive with nuclear energy on a global basis.

Several countries have already embarked on a program to develop their nuclear energy. For example, France and Belgium produce 77% and 60% of their electricity by nuclear energy. A map showing the location of nuclear power stations in Europe is given in Fig. 7.1. In 1993, there were 430 nuclear power reactors producing 337 GWe of electrical power. Canada has 22 power reactors and the USA 109 power reactors, producing about 17% and 21%, respectively, of these nations' electricity. Due to recent accidents in Ukraine and Japan, a major concern has been raised about the safety of the reactors. More will be discussed later.

In spite of a moratorium in some countries on the construction of new nuclear power plants, predictions indicate that the present world nuclear energy production of 22.7 EJ will reach about 30 EJ by 2020. This growth in the use of nuclear power is primarily due to its attractiveness as a "clean" nonfossil energy source.

7.2 Basis Theory of Nuclear Energy

Nuclear energy is based on the Einstein equation

$$E = mc^2. \tag{7.1}$$

During some nuclear reactions, a small loss in mass, Δm, means that a corresponding liberation of energy, ΔE, occurs. During a chemical reaction, some heat may be liberated or absorbed, but there is no measurable change in mass. The law of conservation of mass is valid. The simplified model of the atom is that of a nucleus surrounded by electrons. The nucleus is composed of protons, Z,

O.V. Roussak and H.D. Gesser, *Applied Chemistry: A Textbook for Engineers and Technologists*, DOI 10.1007/978-1-4614-4262-2_7, © Springer Science+Business Media New York 2013

Fig. 7.1 Maps showing the location of sites of nuclear power stations with the number of sites/total number of operating power stations as of 1996. (**a**) World. (**b**) Europe

Table 7.1 Nuclear components

Component	Symbol	Mass kg	Mass amu	Charge C	Charge Atomic
Electron	e, β^-	9.10939×10^{-31}	5.48580×10^{-4}	-1.60218×10^{-19}	-1
Proton	p	1.67262×10^{-27}	1.00728	$+1.60218 \times 10^{-19}$	$+1$
Neutron	n	1.67493×10^{-27}	1.00866	0	0
Positron	β^+	9.10939×10^{-31}	5.48580×10^{-4}	$+1.60218 \times 10^{-19}$	$+1$
Neutrino	ν	0	0	0	0

1 amu $= 1.6605402 \times 10^{-27}$ kg $= 931.4874$ MeV

Table 7.2 An isotopic composition of natural uranium and zirconium

(a) Uranium

Mass number	Atomic mass	Abundance %
234	234.11379	0.006
235	235.11704	0.720
238	238.12493	99.274

(b) Zirconium (Zr)

Mass no.	Natural Abundance %	Thermal cross section Barns[a]
91	11.27	1.58 ± 0.13
92	17.17	0.26 ± 0.08
94	17.33	0.08 ± 0.04
96	2.78	0.3 ± 0.1
(91.224) Average		0.185

[a]1 Barn $= 10^{24}$ cm^2

and neutrons N = (A–Z), where A is the mass number of the atom and represents the sum of protons and neutrons (nucleons) in the nucleus. The mass and charges of these fundamental particles are given in Table 7.1. Isotopes of an element have the same number of protons (and electrons) but different numbers of neutrons. Thus, natural uranium consists of three isotopes, whereas there are five isotopes of zirconium which are shown in Table 7.2.

The notation used to represent the nucleus of an isotope is in terms of the symbol of these elements, E, which has the appropriate number of protons Z and the mass number A where A = Z + N and N is the number of neutrons in the nucleus, for example, $^A_Z E$, $^{238}_{92} U$, $^{87}_{38} Sr$, $^{39}_{18} Ar$, $^{94}_{40} Zr$.

The mass of the components of the nucleus is invariably (except for H) greater than the actual mass of the nucleus. The difference (a mass defect) represents the binding energy of the nucleus, ΔE, and represents the energy needed to break a nucleus into its individual components (Fig. 7.2).

There are about 300 stable isotopes and over 1,000 unstable (radioactive) isotopes which have been characterized. Nuclei with the same mass number A but having different nuclear charges Z are called isobars; there are 59 with stable isobar pairs starting A = 36, $^{36}_{16} S$ $^{36}_{18} Ar$ and ending with

A = 204, $^{204}_{80} Hg$, $^{204}_{82} Pb$, as well as five isobaric triads such as A = 50, $^{50}_{22} Ti$, $^{50}_{23} V$, $^{50}_{24} Cr$.

The nucleus is very small, having a radius, R, which is approximated by the relation

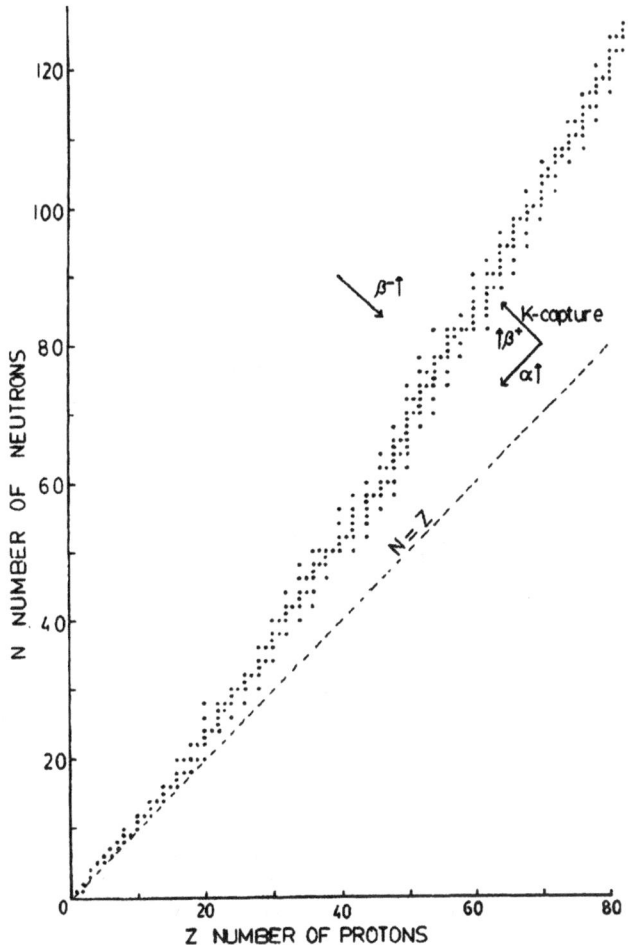

Fig. 7.2 Plot showing the number of neutrons, N, and number of protons, Z, in the nucleus of naturally occurring isotopes of the elements from Z = 1 to Z = 80. The N = Z line is shown to illustrate the deviation of the stable isotopes. Elements above the stability line can become more stable by β^- emission or neutron emission. Elements below the stability line can become more stable by β^+ emission, K-capture, or α particle decay

$$R = R_o A^{1/3} \tag{7.2}$$

where $R_o = (1.5 \pm 0.2) \times 10^{-13}$ cm. The density of the nucleus is very high, approximately 10^{14} g/cm^3 or about 10^8 t/cm^3

The binding energy, ΔE_B, is given by

$$\Delta F_B(A, Z) = \{(Z\, M_p + (A - Z)M_n) - M_N\}c^2 \tag{7.3}$$

where ΔE_B (A,Z) is the binding energy of a nucleus with A, Z values for the nucleons expressed in MeV units of energy.

M_N is the isotopic mass of the particular nucleus.
M_p is the mass of the proton.
M_n is the mass of the neutron.

$$\Delta E\ (A, Z) = 931.4874(1.00728Z + 1.0086(A - Z) - M_N) \tag{7.4}$$

If the mass of the atom is utilized, then p + e = 1.00783 is used instead of the simple proton mass. The specific nuclear binding energy is $\Delta E_B/A$ and represents the average binding energy per nucleon.

The binding energy of the helium nucleus can be calculated from this (7.4)

$$\begin{aligned}
\Delta E &= 931.4874[(2 \times 1.00728 + 2 \times 1.00866) - 4.00151] \\
&= 931.4874(4.03188 - 4.00151) \\
&= 931.4874(.03037) \\
&= 28.29\ \text{MeV}\ \text{or}\ \ 7.07\ \text{MeV/neucleon}
\end{aligned} \tag{7.5}$$

The atomic mass of He is 4.00260 amu. The nuclear mass is

$$4.00260 - 2 \times M_e = 4.00260 - 2 \times .00054864$$
$$M_N = 4.00151\ \text{amu}$$

7.3 Nuclear Model and Nuclear Reactions

Several models of the nucleus exist and attempt to account for the stability of various nuclei. The shell model predicts stable nuclei with N or Z equal to 2, 8, 20, 28, 50, 82,126, and 152.

For nuclei with low values of Z, n $\approx$ p. However, as Z increases n $>$ p, and for the largest nuclei, n = 1.5 p. Nuclei which are not in the stability region of the N/Z curve try to achieve stability by a change in the nuclear composition, that is, they are radioactive and undergo a transformation which depends on whether the isotope is above or below the stability line.

Isotopes above the stability line can become stable by emitting a β particle thereby converting a neutron into a proton (the neutrino, n, is produced in most of these reactions, but not discussed here)

$$n \rightarrow p + \beta - \tag{7.6}$$

$$_{90}^{234}\text{Th} \rightarrow _{91}^{234}\text{Pa} + _{-1}^{0}e \tag{7.7}$$

$$_{6}^{14}\text{C} \rightarrow _{6}^{14}\text{N} + _{-1}^{0}e \tag{7.8}$$

Another reaction which will move a nucleus closer to the stability line would be the emission of a neutron. This occurs only for nuclei in excited states and therefore is primarily seen as a result of a reaction

$$_{4}^{9}\text{Be} + _{2}^{4}\text{He} \rightarrow \ _{6}^{14}\text{C} + _{0}^{1}n \tag{7.9}$$

where the α particles can be supplied by a radium source.

$$_{88}^{226}\text{Rn} \rightarrow \ _{86}^{222}\text{Rn} + _{2}^{4}\text{He} \tag{7.10}$$

Isotopes which are below the stability line can achieve stability (a) by positron emission of β^+,

$$^{11}_{6}C \rightarrow {}^{11}_{5}B + {}^{0}_{-1}e \tag{7.11}$$

$$^{39}_{19}C \rightarrow {}^{39}_{18}B + {}^{0}_{-1}e \tag{7.12}$$

(b) by electron capture (K-capture) in which the nucleus absorbs an electron from the external structure of the nucleus, the K shell of the atom (n = 1, s = 0, m_l = 0),

$$^{81}_{37}Rb + {}^{0}_{-1}e \rightarrow {}^{81}_{36}Kr \tag{7.13}$$

(c) by emitting an α particle, $^{4}_{2}He$,

$$^{238}_{92}U + {}^{234}_{90}Th + {}^{4}_{2}He \tag{7.14}$$

and (d) by proton emission which seldom occurs.

7.4 Radioactive Decay Rates

The radioactive nucleus decays spontaneously at a rate which is proportional to the number of active (unstable) nuclei

$$-\frac{dN}{dT} \propto N \quad \text{or} \quad -\frac{dN}{dT} = \lambda N \tag{7.15}$$

$$dN/N = \lambda dt; \quad \ln N/N_0 = -\lambda t \tag{7.16}$$

where N_0 is the activity at $t = 0$.

This is identical to a first-order reaction with λ representing the decay constant. The half-life $t_{1/2}$ is given by $\ln 2/\lambda$ or $0.693/\lambda$ and is an important characteristic of each radioactive nucleus. Some values of half-lives of selected nuclei are given in Table 7.3.

The $^{14}_{6}C$ isotope is formed by the reaction of neutrons formed by cosmic radiation

$$^{1}_{0}n + {}^{14}_{7}N \rightarrow {}^{14}_{6}C + {}^{1}_{1}p \tag{7.17}$$

A shorthand notation for this reaction is $^{14}_{7}N$ (n, p) $^{14}_{6}C$.

The $^{14}_{6}C$ reacts with oxygen and ends up as CO_2 which is incorporated into living matter.

The rate at which the $^{14}_{6}C$ decays is balanced by the rate at which it is formed, and thus, an equilibrium is established and $^{14}_{6}CO_2$ is constantly replaced as it decays. When the living matter "dies," it ceases to absorb CO_2, and thus, the $^{14}_{6}C$ begins to decay without being replaced.

Table 7.3 Characteristics of some natural radioactive isotopes

Isotope	% Abundance	$\lambda_{1/2}$ (years)	Type of decay	Energy, MeV
$^{14}_{6}C$	Trace	5,730	β^-	0.156
$^{49}_{19}K$	0.0117	1.25×10^9	β^-	1.36
$^{50}_{23}V$	0.25	24×10^{17}	e-Cap	
$^{87}_{37}Rb$	27.83	4.5×10^{10}	β^-	0.273
$^{115}_{49}In$	95.77	6×10^{14}	β^-	0.5
$^{124}_{50}Sn$	6.1	1.5×10^{17}	β^-	
$^{147}_{62}Sm$	15.0	1×10^{11}	α	2.2
$^{187}_{75}Re$	62.6	4.5×10^{10}	β^-	0.0025
$^{238}_{92}U$	99.274	4.5×10^9	α	4.0
$^{235}_{92}U$	.720	7×10^8	α	4.68

Prior to 1945 when atomic bombs began to contaminate the atmosphere, the rate of $^{14}_{6}C$ production and its decay had reached equilibrium. In 1960, Willard Libby received the Nobel Prize for his work showing that it was possible to date when natural materials ceased to grow by incorporating CO_2 into its matter. The following example will illustrate the application of C14 dating.

In 1950, a sample of linen in which some of the Dead Sea Scrolls were wrapped was burned to CO_2 and analyzed for ^{14}C. It was found that the activity was 12.05 Bq/g. Living matter had a ^{14}C activity of 15.30 Bq g^{-1}. What is the age of the linen?

From (7.16),

$$\ln \frac{N_0}{N} = kt = \ln \frac{15.30}{12.05} = kt \tag{7.18}$$

since $t_{1/2} = 5,730$,

$$k = \frac{0.693}{t_{1/2}} = \frac{0.693}{5,730} \text{ years}^{-1}$$

$\ln \dfrac{15.30}{12.05} = \dfrac{0.693}{5730} t$ $t = 1974 \pm 200$ years (with estimated error).

Hence, the linen was made in 1950–1974 = 24 B.C. $\pm$ 200 years.

Radiation is measured by various methods which depend on the type of radiation—α, β, and γ—and the activity of the source. Early measurements were made using photographic films or electroscopes. Modern methods include ionization chambers and fluorescing plastics (scintillation counters) and solutions in which the emitting photons are counted individually. The Geiger counter is one of the simplest means of detecting and measuring β and α particles of sufficient energy to penetrate the counter's thin window. A diagram of a Geiger tube is shown in Fig. 7.3.

Fig. 7.3 A Geiger counter tube. Radiation which passes through the thin window ionizes the gas in the tube. The electrons are accelerated by the high voltage creating (500–1,200 V), creating a pulse which is detected electronically

7.5 Radioactivity Units

The radioactive nucleus decays by a spontaneous process. The activity is measured by several methods: (1) By the activity of the material. The basic unit was the curie (Ci) which was the equivalent of 3.7×10^{10} disintegration/sec for 1 g of radium. The present unit is the becquerel (Bq) which is one disintegration per sec. (2) The energy of the radiation in terms of the absorbed energy. The Roentgen (r) is 83.8 ergs of absorbed energy defined as the quantity of x or γ (as well as α and β) radiation which on passage through 1 mL of dry air at 0°C and 1 atm (STP) (0.001293 g/mL) will produce ions carrying 1 esu of charge (either positive or negative). This is limited to radiation with energy less than 3 MeV. A dose of 1 r produces 2.08×10^9 ion pairs corresponding to the absorption of 0.114 ergs. (2.58×10^{-4} coulombs of negative charges in 1 kg of dry air at STP).

The dose of radiation absorbed by a unit of mass per unit of time is called the absorbed dose rate in units of joules/kg. The rad (radiation absorbed dose) is 10^{-2} J/kg. A rad of α rays is more harmful than a rad of β rays. Hence, the RBE (relative biological effectiveness) of radiation can normalize the damage to tissue by different types of radiation. This is illustrated by the equation

$$\text{Rem} = (\text{Rad})(\text{RBE}) \tag{7.19}$$

where Rem is the Roentgen equivalent to man. The values of some of the parameters are given in Table 7.4. Radiation units are given in Table 7.5. Radiation damage at about 400–500 Rem is fatal to humans (see Table 7.6). Some evidence exists which indicates that small doses may be beneficial (called hormesis) and implies that there is a repair mechanism in living man and animals exposed to small doses of radiation.

The human body contains some naturally occurring radioactive isotopes besides ^{14}C. These are all listed in Table 7.7. The energy of the radiation is very low, and the biological damage due to these isotopes is of minor importance. The normal exposure levels to man is given in Table 7.8.

Table 7.4 The approximate values of RBE for various types of radiation

Radiation	RBE
β, γ	1
α, p	10
n (Thermal)	5
n (Fast, E < 40 MeV)	10
Heavy ions (cosmic ray)	20

Table 7.5 Units of radiation

1 Curie (Ci) = 3.7×10^{10} disintegrations/s[a]
1 Becquerel (Bq) = 1 disintegration/s
1 Gray (Gy) = 100 Rads = 1 J/kg
1 Sievert (Sv) = 100 Rem

[a]The activity of 1 g of pure radium

Table 7.6 Radiation damage for the whole body exposure

0–25	Rem	No clinical effects
25–50	Rem	Decrease in white cells in blood
100–200	Rem	Nausea, less white blood cells
500	Rem	LD_{50}

Table 7.7 The naturally occurring radioactive isotopes in a 70-kg human body

Source	Amount of isotope (g)	Number of atoms	Total activity (Bq)	Particle energy MeV	γ
$^{3}_{1}$H	8.4×10^{-15}	1.7×10^{9}	3	0.018	
$^{14}_{6}$C	1.9×10^{-8}	8.1×10^{14}	3.1×10^{3}	0.158	
$^{40}_{19}$K	8.3×10^{-2}	1.2×10^{21}	4×10^{7}	1.36	1.46

7.6 Nuclear Reactors

The first nuclear chain reaction occurred naturally about two billion years ago in Gabon, Africa, where a uranium deposit moderated by water spontaneously became critical. In 1942 as a result of the war efforts, a sustained chain reaction was achieved by E. Fermi in Chicago working on the Manhattan Project which eventually led to the atomic bomb.

Of the naturally occurring isotopes, only ^{235}U is fissionable with a neutron absorption cross section (σ) of 582 barns with thermal neutrons at 0.025 eV. This cross section decreases as the energy of the neutrons increase ($\sigma \propto E^{-\frac{1}{2}}$), and in the MeV range, σ is about 2 barns. ^{238}U reacts with neutrons ($\sigma \sim 0-1/4$ for E < 0.9 MeV) to form plutonium (Pu) by the reaction sequence:

$$^{238}_{92}U\ (n,\ \gamma)\quad ^{239}_{92}U\ \xrightarrow{\beta^-}\quad ^{239}_{91}Np\ \xrightarrow{\beta^-}\quad ^{239}_{90}Pu \qquad (7.20)$$
$$\qquad\qquad\quad [23.5\ min]\qquad [2.33\ days]$$

Table 7.8 Yearly radiation
exposure

Natural radiation	
Cosmic rays sea level at 2,500 m	40 mRem
(Banff AB)	90 mRem
Terrestrial exposure due to radium and other isotopes in ground and buildings	40 mRem
Internal radiation mainly due to potassium 40	18 mRem
Cosmic radiation during 10,000-km flight at 10-km altitude	4 mRem
Man-made radiation	
1 Chest X-ray	40 mRem
1 Dental X-ray	20 mRem
Fallout	3 mRem
Misc.—TV, monitors, etc.	2 mRem

Plutonium is radioactive with $t_{1/2} = 2.4 \times 10^4$ years and is fissionable like ^{235}U. It is also possible to convert ^{232}Th to fissionable ^{233}U which has a half-life of 1.63×10^5 years by the reaction sequence.

$$^{232}_{94}\text{Th (n, }\gamma\text{)} \xrightarrow{\beta^-} \;^{233}_{93}\text{Pa} \xrightarrow{\beta^-} \;^{233}_{92}\text{U} \qquad (7.21)$$
$$[23.6\,\text{min}] \qquad [27\,\text{day}]$$

The process of fission in ^{235}U proceeds by the absorption of a thermal neutron and formation of the compound nucleus ^{236}U in an excited state with about 6.5 MeV. During the fission process, the compound nucleus is distorted and splits into two fission fragments which by virtue of coulombic repulsion achieve a kinetic energy equivalent to about 80–90% of the fission energy (200 MeV).

The fragment nuclei decay by neutron and γ emission leading to an average of 2.42 neutrons released for each neutron captured. The stabilization of the fragments and the radiation results in the thermal energy produced by the chain reaction.

The neutron flux can also be controlled by control rods made of cadmium or boron which are strong absorbers of neutrons and which are raised or lowered into the reactor to maintain the chain reaction at the appropriate level.

The enrichment of $^{235}_{92}U$ from natural uranium was first performed by the differential diffusion of gaseous UF_6 through porous barriers. This very energy-intensive method was replaced by the simpler gas centrifuge. With the development of the tunable laser, it became possible to photochemically excite one isomeric uranium atom by virtue of its hyperfine splitting of the isotopic lines and thus enrich one isotope. Though the cost of the enrichment process has thus been greatly reduced, ^{235}U is still a very expensive fuel.

The yield of the fusion products as a function of the atomic mass is shown in Fig. 7.4 and indicates that the major elements formed are in the range A = 85–105 and 130–150. The major elements formed and their fission yields are listed in Table 7.9.

Reactors can be classified in terms of the fuel, (^{235}U, ^{233}U, ^{239}Pu) and its containment alloy, zircaloy, the moderator which is needed to slow down neutrons, and the heat exchanger used to generate steam. The most common type of classification is in terms of the coolant which can be either a liquid or gas.

The four categories are (1) the light water reactor (LWR) of which are 81% of all reactors, (2) gas-cooled reactors (GR) make up 11%, (3) 7% use heavy water (HWR), and (4) the remaining reactors use molten metals (such as sodium). (See Fig. 7.5c.)

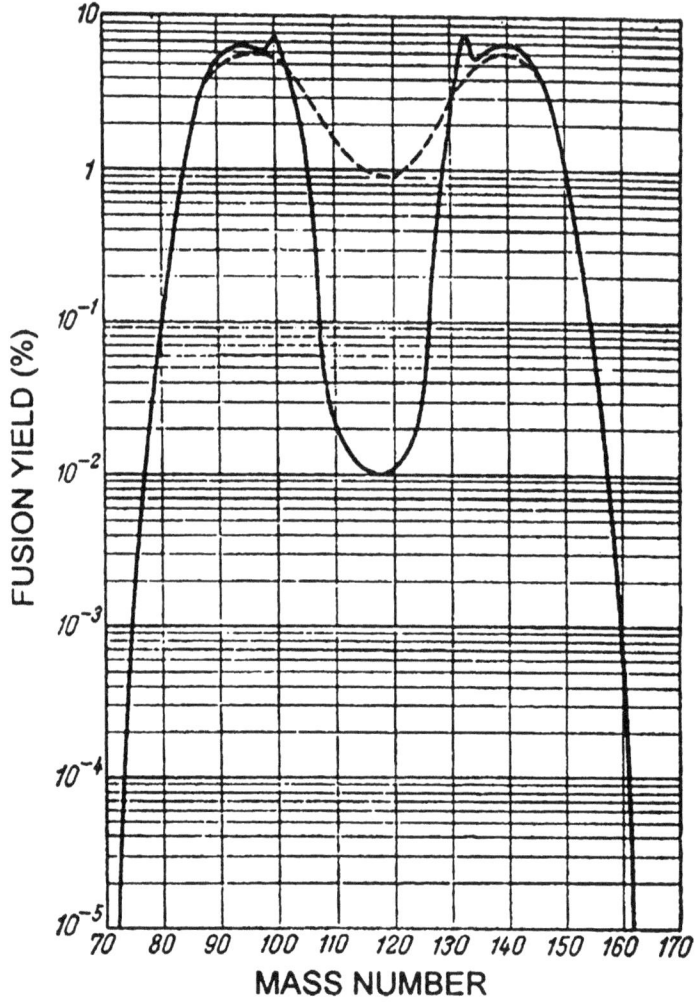

Fig. 7.4 The distribution of fission products from the reaction of $^{235}_{92}U$ with slow neutrons

Successful operation of current light water reactors and implementation of advanced nuclear energy systems is strongly dependent on the performance of fuels and materials. A typical light water reactor (LWR) contains numerous types of materials that all must perform successfully. A majority of the LWRs in the USA are extending their operating licenses from a 40-year period to a 60-year period, with initial discussions about 80-year lifetimes now underway. Many proposed advanced systems (also known as Generation IV systems) anticipate operation at temperature and radiation exposures that are beyond current nuclear industry experience, as well as most previous experience with developmental systems. Table 7.10 summarizes the expected environments during normal operation for the six Generation IV systems. For comparison, the operating conditions for a pressurized water reactor (PWR is a type of light water reactor) are also listed. The Generation IV systems are expected to operate at high temperatures and in some cases with coolants that present more challenging corrosion problems than current LWRs. Generation IV systems are expected to operate for at least 60 years.

Table 7.9 Most important products of thermal-neutron fission of uranium-235

Isotope	Half-life	Fission yield %	Isotope	Half-life	Fission yield %
^{137}Cs	33 years	6	^{95}Zr	65 days	6
^{90}Sr	19.9 years	5	^{95}Nb	38.7 days	6
^{140}Ba	13.4 days	5.7	^{99}Tc	2.1×10^5 years	6.2
^{91}Y	61 days	5.9	^{129}Te	35.5 days	0.2
^{140}La	1.65 days	5.7	^{131}I	8.1 days	3
^{141}Ce	33 days	5	^{129}I	1.7×10^7 years	–
^{144}Ce	282 days	3.6	^{103}Ru	39.8 days	3.7
^{143}Pr	13.5 days	5.3	^{106}Ru	290 days	0.5
^{147}Nd	11.9 days	2.6	^{105}Rh	1.54 days	0.5
^{147}Pm	2.26 years	2.6	^{133}Xe	5.3 days	6
^{155}Eu	1.7 years	0.03			

The LWR is further classified into the pressurized water reactor (PWR) which operates at about 150 atm and 318°C with a thermal efficiency of about 34%. The other type of reactor is the boiling water reactor (BWR) which operates at 70 atm pressure and 278°C with a thermal efficiency of 33%. These reactors require fuel with enriched ^{235}U to about 3% to have a sufficient neutron flux for the chain reaction. The fuel, as UO_2, is in the form of pellets enclosed in a zirconium alloy, Zircaloy-2.

The gas-cooled reactors were developed in Great Britain using CO_2 as the gas coolant and graphite as the moderator with natural uranium metal as the fuel. The thermal efficiency is about 25%. With uranium enriched to 2.2% as the oxide (UO_2) fuel, the thermal efficiency increased to 41% and is called the advanced gas reactor (AGR). Helium is also used in a high-temperature version.

The heavy water reactor was developed in Canada and is known as the CANDU reactor. The D_2O is used as both coolant and moderator. The relative moderating efficiency of various materials is given in Table 7.11. Because of the superior moderating property of D_2O, it is possible to use natural uranium as the fuel in the form of UO_2 pellets in zircaloy tubes. This makes the CANDU one of the best designed reactors in the world. The coolant cycle and the moderator are separate flow circuits shown in Fig. 7.6. The fuel elements in the pressure tubes and the D_2O flow is shown in Fig. 7.7 where the coolant is at about 293°C and 100 atm pressure. The moderator is at lower temperature. The efficiency is rated at 29%.

The average composition of deuterium (^2_1D) in water (H_2O) is about 150 ppm which varies from place to place, as shown in Fig. 7.8 for Canada.

Heavy water is extracted from water usually in two steps: the first being a dual temperature deuterium transfer process and the second through vacuum distillation.

The dual temperature process is based on the atomic exchange of hydrogen and deuterium between hydrogen sulfide gas (H_2S) and fresh water with a deuterium concentration of approximately 148 parts per million.

When hot water is in contact with hot H_2S gas, the deuterium atom migrates from the water into the gas; when both are cold, it migrates in the reverse direction.

$$HOD + H_2S \rightleftarrows H_2O + HSD \qquad K = 0.99(25°C) \qquad (7.22)$$

By repeating this process in successive stages, the deuterium oxide is increased to a concentration of 15%. The ten extraction towers of the plant contain sieve trays. Water flows downward across each tray in succession. H_2S gas is forced upward, bubbling through the sieve holes for contact with the water. This is shown in Fig. 7.9.

Fig. 7.5 As with the boiler in a coal, oil, or gas burning oil plant, a nuclear power reactor produces steam to drive a turbine which turns an electric generator. Instead of burning fossil fuel, a reactor fissions nuclear fuel to produce heat to make steam. (**a**) The *PWR* shown here is a type of reactor fuelled by slightly enriched uranium in the form of uranium oxide pellets held in zirconium alloy tubes in the core. Water is pumped through the core to transfer heat to the steam generator. The coolant water is kept under pressure in the core to prevent boiling and transfers heat to the water in the steam generator to make the steam. (**b**) The *BWR* shown here is a type of reactor fuelled by slightly enriched uranium in the form of uranium oxide pellets held in zirconium alloy tubes in the core. Water is pumped through the core, boils and produces steam that is piped to the turbine. (**c**) The *HTGR* shown here is a type of reactor fuelled by uranium carbide particles distributed in graphite core. Helium gas is used as a coolant to transfer the heat from the core to the generator. (**d**) In an *LMFBR*, molten sodium in the primary loop is pumped through the reactor core containing the fuel. This sodium collects the heat and transfers it to a secondary sodium loop in the heat exchanger from which it is carried to the steam generator. In addition to producing electricity, this reactor also produces more fissionable material than it consumes, which is why it is called a "breeder reactor." In the reactor, uranium-238 is transmuted to fissionable plutonium-239 which is extracted periodically and fabricated into new fuel

Table 7.10 Approximate operating environments for Gen IV systems

Reactor type	Coolant inlet temperature (°C)	Coolant outlet temperature (°C)	Maximum dose (dpa[a])	Pressure (Mpa)	Coolant
Supercritical water-cooled reactor (SCWR)	270	500	15–67	25	Water
Very high temperature gas-cooled reactor (VHTR)	600	1,000	1–10	7	Helium
Sodium-cooled fast reactor (SFR)	370	550	200	0.1	Sodium
Lead-cooled fast reactor (LFR)	600	800	200	0.1	Lead
Gas-cooled fast reactor (GFR)	450	850	200	7	Helium/SC carbon dioxide
Molten salt reactor (MSR)	700	1,000	200	0.1	Molten salt
Pressurized water reactor (PWR)	290	320	100	16	Water

[a]dpa is displacement per atom and refers to a unit that radiation material scientists used to normalize radiation damage across different reactor types. For one dpa, on average, each atom has been knocked out of its lattice site once

Table 7.11 The relative moderating character of various materials

Substance	Moderating ratio	Average slowing down length from "fast" to thermal (cm)
Light water	72	5.3
Organic liquids	60–90	
Beryllium	159	
Graphite	160	19.1
Heavy water (99.8 %)	2,300	
Pure D_2O	12,000	11.2

Fig. 7.6 The simplified flow diagram for the CANDU reactor

Fig. 7.7 The coolant around the fuel elements in the calandria tubes in D_2O

The 15% enriched heavy water is delivered to a finishing unit where, through vacuum distillation, the concentration is upgraded to 99.8% (reactor grade) deuterium oxide. The distillation unit consists of two towers containing special packing.

Deuterium is also enriched during the electrolysis of water since $_1^1H$ is more readily liberated as H_2 than $_1^2D$ as HD. Hence, the first samples of heavy water were obtained from Norway where water electrolysis is the method of producing H_2 and O_2 since electricity is inexpensive there—even today where it is the major method of heating homes.

It is possible to increase the fuel efficiency by selecting the zirconium isotope $_{40}^{94}Zr$ which has the lowest neutron absorption cross section (0.08 barns; see Table 7.2) for the fabrication of the zircaloy. Several attempts have been made since the high cost of an enrichment process is a major capital investment which could be of continuous benefit since the $_{40}^{94}Zr$ can be recycled from spent fuel and reused with little reprocessing.

7.7 The Hazards of Nuclear Energy

It is at present somewhat uncertain if very low levels of radiation are harmful. We cannot avoid all radiation since there is a natural radiation background (with approximate yearly exposure) due to, for example, the cosmic rays (40 millirem at sea level, 250 Rem at 500 m elevation) radium and radon in ground and building material (40 Rem) and potassium 40 (18 Rem). In addition, we can add some man-made radiation sources such as one chest X-ray (40 Rem), one dental X-ray (20 Rem), fallout from nuclear explosions (5 Rem), as well as miscellaneous sources such as TV, CRT, etc., all of which total to 163 Rem/year (for sea level). The average annual radiation dose to a nuclear reactor worker in Ontario is 0.68 Rem with an annual limit of 5 Rem set by radiation protection regulations.

Thus, a nuclear power reactor worker receives about as much additional radiation as that of an office worker who moves from New York to Denver. Hence, nuclear energy presents an insignificant risk to its workers. It can be argued that the potential risk of a nuclear accident (such as the Three Mile Island accident, March 1979, or the Chernobyl accident, April 1986) is not worth the benefits. This must be viewed in a perspective which is tempered by history and time. The Chernobyl nuclear accident on Saturday, April 26, 1986, has been called the worst accident in the world and will probably slow down and, in some countries, end the development of nuclear energy. The Russian reactor was a graphite-moderated water-cooled enriched uranium reactor which was idling at 7% power, generating 200 MW of heat while maintenance work was being done. Human error appears to have caused a sudden increase in power output, 7–50%, in 10 s due to part of the reactor going critical. The resulting runaway high temperature converted water to hydrogen which exploded, setting the

Fig. 7.8 Map of Canada showing the deuterium concentration in ppm (µg/g) in various parts of the country. Also shown are the sites of heavy water plants and nuclear power plants

PRODUCTION OF HEAVY WATER (D₂O)

Fig. 7.9 Basic principles of heavy water production by the HDO + H$_2$S → HDS + H$_2$O equilibrium reaction

reactor on fire and releasing enormous amounts of radioactivity into the atmosphere. The radioactive cloud drifted across Europe and was first detected in Sweden, Monday morning, April 28, 1986. Based on the prevalence of radioactive cobalt, iodine, and cesium, it was concluded that a nuclear bomb test was not the source of the activity but that a nuclear accident had occurred. This was acknowledged at 9:00 p.m. by Moscow TV news.

The radioactivity contaminated the vegetables and meats of central Europe. Millions of dollars were used to compensate farmers for the loss they suffered. Though about 31 deaths were associated with acute radiation exposure and fire, it is estimated that thousands of delayed cancers will result from the fallout.

Another nuclear disaster rated equal to Chernobyl happened at the Fukushima nuclear plant in Japan on the 11th of March 2011. Nuclear reactors were damaged by strong earthquakes and tsunami. Crisis level at the Fukushima nuclear plant was raised to level 7 after 1 month of the disaster. Level 7 is classified by IAEA as the worst level on an international scale.

The accident at Three Mile Island released 15 curies and caused no direct fatalities. It has been estimated that not more than 1 delayed cancer victim will result.

However, we accept the construction of dams, some of which have failed and others, if they do fail, would cause hundreds of thousands of deaths. A list of some hydroelectric dam disasters is given in Table 7.12.

We accept the occasional mine disaster and continue to mine coal. Since a tonne of uranium ore yields 300 times more energy than a tonne of coal, it is obvious that mining uranium leads to less mining deaths per energy unit (1/300) than mining coal. Again, it could be argued that the radon and radiation exposure in uranium mining is an additional hazard not encountered in coal mining. However, black lung disease kills thousands of miners every year. It has been estimated that for every 10^{15} We (watts of electricity) generated there are 1,000 deaths by black lung among coal miners and 20 deaths by lung cancer among uranium miners.

Table 7.12 Loss of lives due to hydroelectric and dam disasters

Date	Place	Lives lost
1923	Santa Paula, California, USA	450
1923	Gleno, Italy	600
1926	St. Francis, USA	430
1959	Malpasset, France	412
1961	Kiev, USSR	145
1963	Vaiont, Italy	2,000
1967	Kayna, India	180
1972	Buffalo Creek, WV, USA	118

Table 7.13 Average annual risk of fatality by various causes (USA)

Accident type	Total number	Individual chance per year
Motor vehicle	55,791	1 in 4,000
Falls	17,827	1 in 10,000
Fuels and hot substances	7,451	1 in 25,000
Drowning	6,181	1 in 30,000
Firearms	2,309	1 in 100,000
Air travel	1,778	1 in 100,000
Falling objects	1,271	1 in 160,000
Electrocution	1,148	1 in 160,000
Lightning	160	1 in 2,000,000
Tornadoes	91	1 in 2,500,000
Hurricanes	93	1 in 2,500,000
All accidents	111,992	1 in 1,600
Nuclear reactor Accidents (100 plants)	5,000[a]	1 in 200,000

[a]The British accident at Windscale in October 1957—a fire in the atomic pile—resulted in the release of radioactive gas $^{131}_{-50}$I (20,000 Cu), $^{210}_{82}$Po (37 Cu), as well as $^{103}_{44}$Ru, $^{106}_{44}$Ru, $^{95}_{40}$Zr, $^{95}_{41}$Nb, $^{137}_{55}$Cs (600 Cu), $^{89}_{38}$Sr (80 Cu), $^{90}_{38}$Sr (9 Cu), Ce-144, and Te-132

It has only recently been estimated that the number of deaths from leukemia and cancer which could be directly due to this accident is from 1,000 to 2,000 with the predominant effect to children. Hence, we must distinguish between direct and indirect or more remote fatalities. If we add the direct deaths and the leukemia and cancer deaths for the Three Mile Island and Chernobyl nuclear accidents, we would estimate a total of about 5,000 deaths. This gives a risk factor of 1 in 200,000 which is still better than air travel $^{103}_{44}$Ru, $^{106}_{44}$Ru, $^{131}_{-50}$I, $^{210}_{82}$Po, $^{95}_{40}$Zr, $^{95}_{41}$Nb, $^{137}_{55}$Cs, $^{89}_{38}$Sr, $^{90}_{38}$Sr

An indication of relative average risk of fatality by various causes is given in Table 7.13. The automobile remains the major cause of accidental deaths in our society, yet we accept this with minor token complaints. It is possible to predict that there will be 500 deaths in the USA due to automobile accidents (50 in Canada) during the Labor Day weekend holiday in September.

A comparison of man-days lost per megawatt-year output by various energy sources has been given by Inhaber and shown in Fig. 7.10. Coal and oil show the highest losses, whereas natural gas shows the lowest loss. It may be surprising to note that solar energy sources are relatively hazardous. Edward Teller—the father of the H-bomb—once pointed out that solar energy is not free of dangers because the ladder, which would be required to clean solar cells on the roof of our homes, causes more accidents in the home than any other device.

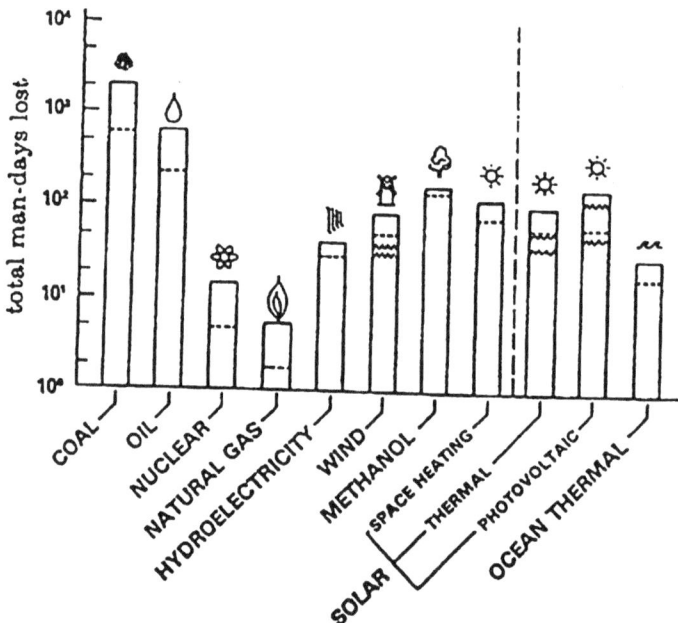

Fig. 7.10 Total man-days lost per megawatt-year net output. Public and occupational deaths are combined. The *top* of the bars and the *dotted lines* indicate the upper and lower part of the range, respectively. Bars to the *right* of the *vertical dotted lines* indicate technologies less applicable to Canada. Note the logarithmic scale. Deaths are assumed to contribute 6,000 man-days lost per fatality and are combined with accidents and illness on that basis. *Solid* and *dashed jagged lines* on the three of the bars indicate maximum and minimum values, respectively, when no backup (or low-risk backup) energy is assumed

7.8 Nuclear Waste

The opponents of nuclear energy besides being concerned with the hazards of accidents in power are also anxious about the handling and storage of nuclear waste—the end product of nuclear fuel. This waste consists of a multitude of radioactive nuclei having a wide range in activity as well as lifetime. In general, about 2,000 m^3 of uranium ore must be mined to produce 1 t of nuclear fuel grade material which yields 91 L of UO_2 reactor fuel. The spent fuel is stored in water pools to permit the radioactive decay of short-lived fission products. The discharged fuel is reprocessed resulting in about 1,200 L of high-level waste (HEW) which is stored and evaporated to about 570 L and calcined (heated) to 80 L of oxides. When fixed in glass, the waste occupies 70 L or 1/3 of a canister 30 cm in diameter × 3 m long which is destined for disposal and storage underground. After 450 years, the residual activity is equivalent to that of natural uranium and presents the same radiation hazards. A list of the major waste components in spent fuel is given in Table 7.14. As illustrated in Table 7.14, the major activity is due to plutonium and strontium.

Decommissioning and immediate dismantlement of nuclear power reactors at the end of their useful life (about 30–40 years) is both a challenge and expensive process because of the need to dispose large volumes of low-level wastes (LLW) in many countries today. For example, the volumes and activities estimated for immediate dismantlement of reference nuclear power reactors PWR (1,175 MWe) and BWR (1,155 MWe) are below:

(a) Waste Stream (PWR, 1,175 MWe): activated metal—97,100 ft^3 of total activity 4,841,300 Ci; activated concrete—25,000 ft^3 of total activity 2,000 Ci; contaminated metal—192,000 ft^3 of

Table 7.14 Activities of selected fission products and actinides in irradiated fuel[a]

Radionuclide	Mode of decay	Half-life (years)	Activity (curies/kg U)		
			At discharge	After 1 year	After 10 years
Fission products					
Tritium (H-3)	β	12.3	0.17	0.16	0.10
Krypton-85	β, γ	10.7	2.22	2.19	1.23
Strontium-89	β, γ	0.14	443	3.95	9.8×10^{-20}
Strontium-90	β	29	17.5	16.0	12.9
Yttrium-91	β, γ	0.16	578	77	1.1×10^{-16}
Zirconium-95	β, γ	0.18	825	17.3	1.3×10^{-14}
Niobium-95	β, γ	0.10	802	36.6	2.9×10^{-14}
Technetium-99	β, γ	2.1×10^5	3.4×10^{-3}	3.4×10^{-3}	3.4×10^{-3}
Ruthenium-106	β	1.0	182	101	0.21
Iodine-129	β, γ	1.6×10^7	7×10^{-6}	7.9×10^{-6}	7.9×10^{-6}
Iodine-131	β, γ	0.02	525	1.2×10^{-11}	0
Cesium-134	β, γ	2.17	16.9	11.3	0.55
Cesium-135	β	2.3×10^6	4.5×10^{-5}	3.8×10^{-5}	3.8×10^{-5}
Cesium-137	β, γ	30.2	25.3	24.8	20.2
Cerium-144	β, γ	0.78	424	181	0.06
Promethium-147		2.6	58.9	50.7	4.7
Actinides					
Neptunium-237	α, γ	2.1×10^6	2.1×10^{-5}	2.1×10^{-5}	2.2×10^{-5}
Plutonium-238	α, γ	87.7	7.2×10^{-2}	8.3×10^{-2}	8.0×10^{-2}
Plutonium-239	α, γ	2.4×10^4	0.15	0.15	0.15
Plutonium-240	α, γ	0.24	0.24	0.24	0.24
Plutonium-241	β, γ	14.7	22.9	21.8	14.2
Americium-241	α, γ	432	11.5×10^{-3}	4.7×10^{-2}	0.3
Americium-243	α, γ	7,380	5.3×10^{-3}	5.3×10^{-4}	5.3×10^{-4}
Curium-242	α, γ	0.45	2.58	0.44	8.9×10^{-6}
Curium-244	α, γ	18.1	1.6×10^{-2}	1.5×10^{-2}	1.1×10^{-2}

[a]Based on Pickering fuel irradiated to 7.5 MWd/kg U (Courtesy of AECL)

total activity 900 Ci; contaminated concrete—374,700 ft^3 of total activity 100 Ci; dry solid waste (trash)—50,600 ft^3; spent resins—1,100 ft^3 of total activity 42,000 Ci; filter cartridges—300 ft^3 of total activity 5,000 Ci and evaporator bottoms—4,700 ft^3

(b) Waste Stream (BWR, 1,155 MWe): activated metal—4,900 ft^3 of total activity 6,552,300 Ci; activated concrete—3,200 ft^3 of total activity 200 Ci; contaminated metal—549,200 ft^3 of total activity 8,600 Ci; contaminated concrete—59,200 ft^3 of total activity 100 Ci; dry solid waste (trash)—119,500 ft^3; spent resins—1,500 ft^3 of total activity 200 Ci and evaporator bottoms—18,300 ft^3 of total activity 43,800 Ci

The storage and handling of nuclear waste by the various countries all include immobilization of either the reprocessed fuel waste or the unreprocessed spent fuel by glass or mineral formation followed by encapsulation. Storage varies due to facilities and geological formation within a country. Thus, salt mines have been used in Germany, the USA, and the Netherlands. Salt mines by their existence testify to a guaranteed isolation from the environment by ground water. The abandoned salt mine Asse near Brunswick in Germany has since 1967 stored more than 100,000 special drums of nuclear waste. However, this has been stopped in 1979 due to political pressure. An alternate salt mine at Gorleben is now under consideration and testing.

Some alternate proposals for waste disposal include the following: (a) the space option, (b) Antarctic ice shelf, (c) ocean dumping, (d) ocean burial, and (e) nuclear transformation.

(a) The soft-landing approach for the storage of nuclear waste on the moon has been evaluated but until such flights become routine, it must be considered to be too risky and expensive (the launch costs are presently about US $20,000,000 per tonne for low earth orbit).
(b) Disposal under the Antarctic ice is also too risky due to the unknown stability of the underlying shelf which requires predicable stability for thousands of years.
(c) Ocean dumping has been carried out since 1943 with disastrous results. Evidence of leaking drums has been obtained and long-term storage is not feasible for high-level waste.
(d) Burial in the ocean floor is attractive if deep holes can be drilled and sealed with the sediment. This is being studied actively and the environmental impact evaluated.
(e) The conversion of radioactive nuclei by nuclear reactions and transformation into stable nuclei is possible. The cost, however, is prohibitive, and no concerted effort is being made to make this a viable alternative.

Canada is very fortunate in having the Canadian Shield of Precambrian rock formation in Central Canada around Hudson's Bay. Some hard rock formations, called pluton, within the shield are ideal sites for the location of an underground nuclear waste storage facility.

Studies now in progress in an underground laboratory will establish the permeability and water movement in such formations.

Another possible location of an underground nuclear storage facility is on the Novaya Zemlya island of Russian Federation surrounding with huge resources of natural adsorbents (the Kovdor vermiculite; the Karelian amphibole asbestos, shungite, bentonite, and clays) for development of engineering barriers and localization of nuclear wastes in a geological repository. With the end of the Cold War, Russia stopped nuclear weapon testing program on the Novaya Zemlya and started reduction of nuclear weapon stockpiles in accordance to a signed treaty committing agreement to the USA. The USA Federation and Russian Federation are going to reduce number of strategic nuclear warheads below 1,700–2,000 by 2012.

It is interesting to note that the natural nuclear reactor discovered in Oklo, Gabon, Africa, operated for about 7×10^5 years and released about 15 GW years of energy about 1.8×10^9 years ago. The study showed that the fission products could lie immobile in the ground at the original reactor site during these many years. It is evident that responsible disposal of nuclear waste is feasible and will eventually be accomplished.

7.9 Nuclear Fusion

In contrast with nuclear fission where a large nucleus is split into two more stable nuclei, fusion relies on the formation of larger stable nuclei from small nuclei. The main difference is that fusion requires an initial high temperature of millions of degrees to overcome the energy repulsion barrier of the nuclei. In the fusion H-bomb, the high temperature (10^8 K) is achieved by a fission bomb.

Various systems have been designed and tested to achieve these high temperatures and to initiate the fusion reaction. Such reaction has been maintained for very short time intervals, which is slowly being extended to longer times.

Some fusion reaction and the corresponding energy liberated are given below:

1. DT: Deuterium-tritrum reaction

$$\ {}^2_1\text{D} + {}^3_1\text{T} \rightarrow {}^4_2\text{He} \ (3.5 \text{ MeV}) + {}^1_0\text{n}(14.1 \text{ MeV}) \tag{7.23}$$

This reaction can occur at the lowest temperature and produces the highest fusion—power density. The highly energetic neutrons present technical problems regarding material of construction.

2. DD: Deuterium-deuterium reactions

$$_{1}^{2}D + _{1}^{2}D \rightarrow _{2}^{3}He \ (3.5 \ MeV) + _{0}^{1}n \ (2.45 \ MeV) \tag{7.24}$$

$$_{1}^{2}D + _{1}^{2}D \rightarrow _{1}^{3}T \ (1.01 \ MeV) + _{1}^{1}H \ (3.02 \ MeV) \tag{7.25}$$

The tritium and $_{2}^{3}He$ can react further to give

$$3 \ _{1}^{2}D \rightarrow 2 \ _{2}^{4}He \ (3.5 \ MeV) + _{1}^{1}H + _{0}^{1}n + 21.6 \ MeV \tag{7.26}$$

Recent experiments on the electrolysis of LiOD in D_2O on a palladium cathode has been claimed to result in "cold" fusion. The detection of helium, neutrons, and even tritium has been reported. However, there is considerable doubt about the validity of the claims and "cold" fusion, like polywater will soon be buried and its obituary published.

3. Other reactions

The following reactions do not involve neutrons:

$$_{1}^{1}H + _{3}^{6}Li \rightarrow _{2}^{3}He + _{2}^{4}He \ (4.0 \ MeV) \tag{7.27}$$

$$_{2}^{3}He + _{3}^{6}Li \rightarrow _{1}^{1}H + 2 \ _{2}^{4}He(16.8 \ MeV) \tag{7.28}$$

$$_{2}^{3}He + _{2}^{3}He \rightarrow 2 \ _{1}^{1}H + _{2}^{4}He \ (12.9 \ MeV) \tag{7.29}$$

$$_{1}^{1}H + _{5}^{11}B \rightarrow _{6}^{12}C \rightarrow _{4}^{8}Be + _{2}^{4}He$$
$$\downarrow \tag{7.30}$$
$$2 \ _{2}^{4}He$$

The overall reaction is

$$_{1}^{1}H + _{5}^{11}B \rightarrow 3 \ _{2}^{4}He \tag{7.31}$$

Reactions involving $_{2}^{3}He$ are not really suitable because $_{2}^{3}He$ must be artificially produced. The amount of deuterium in the oceans which is available for fusion can supply enough energy to satisfy earth's energy requirements for as long as the sun shines.

7.10 Summary

Radioactivity can be harmful or beneficial. As tracer isotopes, it can be used to map the fate of a chemical or the location of a malfunctioning organ. It is used in radiation therapy for cancer patients and in the irradiation of food to prevent or reduce spoilage. It is also dangerous in significant dosage causing cancer and death. Nevertheless, it is an efficient source of energy which does not contribute to the greenhouse effect, and thus, it helps preserve our climate. The need to deal with the problem of waste management remains but appears to be soluble. The real question is "can we survive without nuclear energy?"

Exercises

1. Uranium is present in seawater at about 3.3 mg/tonne. Calculate the amount of uranium in the oceans, ($V = 1.5 \times 10^9$ km^3).
2. Deuterium is present in water to about 150 ppm. Calculate the amount of deuterium present in earth's water.
3. Calculate the binding energy in the nitrogen nucleus.
4. Calculate the energy change in reactions (7.9) and (7.17).
5. A sample of wood from a cross was tested for ^{14}C and found to have 13.2 Bq/g of carbon. What was the age of the cross?
6. A sample of a radioactive mineral had an activity of 40,000 Bq. Three months later, the activity was 32,000 Bq. What is the half-life of the active component in the mineral?
7. A radioactive tracer with a half-life of 12 days was administered to a patent weighing 75 kg. Its initial activity was 1,600 Bq/mL of solution. When testing the patient's blood a week later, what would be the expected activity of the blood (assuming the volume of blood in the patient is 5 L)?
8. Explain how the breeder reactor can be used to extend the energy convertibility of nuclear fuel.
9. One of the major concerns about nuclear energy is the disposal of radioactive waste from the reactors. Discuss the problem and the various solutions which have been proposed.
10. From Table 7.2, calculate the average molar mass of uranium.
11. The expected value of an event (EV) is equal to the risk (R) times the benefit (B), EV = RB. Calculate the expected value for each of the following:

    (a) Throw a 6 in a cubic dice with 1–6 on the sides and win $400.
    (b) Pick heads on the toss of a coin and win $100.
    (c) Pick the ace of spades out of a full deck of cards (52) and win $3,500.
    (d) Pick any ace out of a full deck of cards (52) and win $950.
    (e) Pick the month of birth of a total stranger and win $800.

12. Calculate the energy liberated in the fusion reaction (7.31).
13. Write the shorthand notation for the reaction

$$^1_1\text{H} + {}^6_3\text{Li} \rightarrow {}^3_2\text{He} + {}^4_2\text{He}$$

14. Write the nuclear reaction for $^{63}_{29}\text{Cu}(\alpha, \text{ n}) \rightarrow {}^{64}_{31}\text{Ga}$.
15. Calculate the energy liberated when 1 g of U-235 is split by a slow neutron to form three neutrons, Ba-141 and Kr-92.

Further Reading

1. Allen T, Busby J, Meyer M, Petti D (2010) Materials challenges for nuclear systems. J Mater Today 13(12):14–23
2. Alexander WR, McKinley LE (2007) Deep geological disposal of radioactive waste. Elsevier, Amsterdam, 273pp
3. Chapman N, McCombie C (2003) Principles and standards for the disposal of long-lived radioactive wastes. Pergamon, Amsterdam, 277pp
4. Phylippovsky VI, Roussak OV (1997) Use of natural inorganic adsorbents to increase safety of long-term storage of radioactive wastes. Journal "Problems of Material Science" No. 2. Research Institute of Construction Materials "Prometej". St. Petersburg, 1997
5. Berlin RE, Stanton CC (1990) Radioactive waste management. A Wiley-Interscience Publication Company, New York, pp 46–47
6. International Nuclear Energy guide (1987) 15th edn. Editions Technip. Paris
7. Cohen BL (1983) Before it's too late. Plenum Press, New York
8. Wolfson R (1991) Nuclear choices. MIT Press, Cambridge, MA
9. Hunt SE (1980) Fission, fusion and the energy crisis, 2nd edn. Pergamon Press, New York
10. Heppenheimer TA (1984) The man-made sun. Omui Press, Boston
11. Berger JJ (1976) Nuclear power—the unviable question. Ramparts Press, Pala Alto
12. Haissinsky M (1964) Nuclear chemistry and its applications. Addison-Wesley Publishing Company, New York
13. Kessler G (1983) Nuclear fission reactors. Springer, New York
14. Orensen B (1974) Renewable energy. Academic, New York
15. Berlin RE, Stanton CC (1990) Radioactive waste management. A Wiley-Interscience Publication Company, New York
16. http://ca.news.yahoo.com/fire-japan-nuclear-reactor-heightens-radiation-20110315-171920-237.html

Chapter 8
Lubrication and Lubricants

8.1 An Introduction to Tribology

Tribology, from the Greek word *tribos* meaning rubbing, is the science of friction, lubrication, and wear. The basic laws of tribology are concerned with the general behavior of surfaces interacting during sliding and to a first approximation are given by

$$F_{kin} \propto W \ or \ F_{kia} = \mu W \tag{8.1}$$

$$Z \propto W \ or \ Z = \kappa W \tag{8.2}$$

where F_{kin} is the sliding force of friction, W is the load applied normal to the sliding direction, Z is the wear rate (volume of surface removed per unit sliding distance), μ is the coefficient of friction and κ is the wear factor. F_{kin} and Z are independent of the apparent area of contact.

The factors of proportionality in laws (1) and (2) above are not necessarily related to the same operating parameters and it is possible to obtain heavy wear rates associated with low frictional forces.

Friction is not always undesirable. For example, the stopping of an automobile relies on friction between the tires and the road and on friction within the braking mechanism.

The function of a lubricant is to (1) prevent the moving interacting surfaces from coming into direct contact, (2) to provide an easily sheared interfacial film, (3) to remove the heat evolved in the process, and (4) to reduce wear of the surfaces. Solid lubricants can only satisfy (1), (2), and (4), but only liquids and gas can also satisfy (3). Lubricants can thus conveniently be classified into gaseous, liquids, and solids.

8.2 Gaseous Lubricants

Gas-lubricated bearings have special applications, and its use is often neglected. The gases used include air, helium, nitrogen, and hydrogen—though air is the most commonly employed. It must be pointed out that in gas-lubricated bearings, the coefficient of friction approaches zero. The viscosity of air under atmospheric conditions is 0.018 cP (centipoise, a poise $= 1$ g cm^{-1} s^{-1}), and its temperature coefficient is small, allowing both low temperature and high temperature operations. An example of a hydrostatically lubricated gas thrust bearing is shown in Fig. 8.1.

O.V. Roussak and H.D. Gesser, *Applied Chemistry: A Textbook for Engineers and Technologists,* DOI 10.1007/978-1-4614-4262-2_8, © Springer Science+Business Media New York 2013

Air out at Final Pressure P$_f$

Fig. 8.1 Schematic cross section of a hydrostatic (gas-lubricated) cone step

For hydrodynamic conditions, the minimum film thickness for a gas thrust bearing is given by the relation

$$h_o = \sqrt{0.66\eta \frac{Vl}{P_{av}}} \tag{8.3}$$

where η = the viscosity, V = mean runner velocity, l = the length of shoe in rotation direction, and P_{av}
– average thrust pressure. The viscosity of a gas η is given by

$$\eta = \frac{N'm\,\bar{c}\lambda}{3} \tag{8.4}$$

where N' = the concentration of molecules in the gas, molecules per unit volume, m = mass of the gas molecule, c = average molecular speed, and λ = mean free path.

The mean free path of a gas λ is the average distance a molecule moves between collisions and is given by

$$\lambda = \frac{1}{\sqrt{2}\pi\sigma^2 N'} \tag{8.5}$$

where σ is the molecular diameter and the average molecular speed, c, is given by

$$\bar{c} = \sqrt{\frac{8RT}{\pi Nm}} \tag{8.6}$$

where R = the gas constant, T = the absolute temperature, and N = Avogadro's number (6.022×10^{23}).

Thus, by combining Eqs. 8.4, 8.3, and 8.4, we get

$$\eta = \frac{2}{3\pi^{3/2}\sigma^2} \sqrt{mkT} \tag{8.7}$$

where k is Boltzmann's constant.

$$k = \frac{R}{N} \tag{8.8}$$

Thus, $\eta \propto \sqrt{T}$ to a first approximation

$$\text{fluidity } \varphi = \frac{1}{\eta} \tag{8.9}$$

The ratio λ/h_o is known as Knudsen number (K_n) and determines the conditions under which the gas can be considered a continuum, that is, when $K_n < 0.01$, the gas can be considered as a fluid; when, however, $0.01 < K_n < 1.0$, the gas is a continuum if slip at the boundary is corrected. For air at 0°C and 1 atm, $\lambda \_ 6 \times 10^{-6}$ cm; hence, for film thicknesses of 6×10^{-4} cm or more, $K_n < 0.01$ and proper lubrication without slip can be expected.

Gas lubrication is best suited to continuous operation at extreme temperatures and in radiation fields which would normally degrade oil. It is non-contaminating and "free." Starting and stopping gas-lubricated bearings can be harmful unless special design and precautions are taken. Other disadvantages include limited load capacity, susceptible to instability, and require precise machining.

8.3   Liquid Lubricants

Like gas lubrication, liquids lubricate by keeping the two moving surface apart. The bearing is one of the most common lubricated system. There are three types of bearing: the journal bearing, the thrust bearing, and the slider bearing.

8.3.1   Journal Bearings

This consists of a rotating axel or journal which contacts the bearing for support. Rotations and the small angle between the two surfaces cause a buildup of pressure in the oil layer which supports the load during rotation. This is called hydrodynamic lubrication and is shown in Fig. 8.2. The important characteristic of the bearing is the dimensionless number C where

$$C = \frac{ZN}{P} \tag{8.10}$$

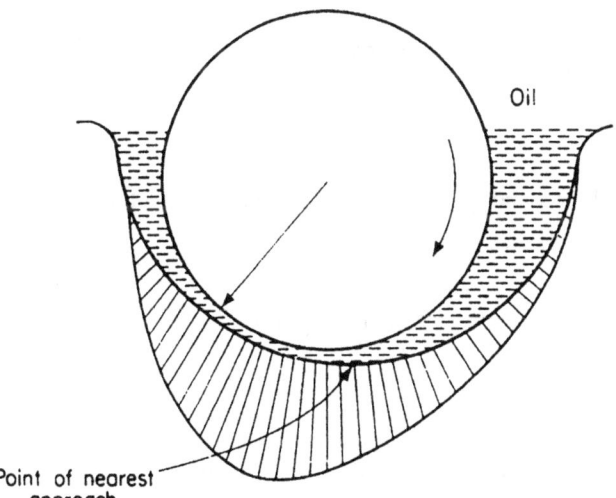

Fig. 8.2 The journal bearing showing the pressure distribution and oil forces supporting the journal

Point of nearest approach

Fig. 8.3 (**a**) Fine ground
surface showing high spots
called asperities. (**b**) Polished
surface showing bumpy
contour

$Z =$ the viscosity of the lubricant, $N =$ the speed of the shaft, and $P =$ the pressure or load divided by the projected area. Values of $C > 35$ imply that hydrodynamic lubrication is effective. When $C \leq 35$, the oil wedge responsible for hydrodynamic lubrication is too thin, leading to an increase in friction, that is, a reduction in lubrication.

The true surface area of a polished surface can be many times the nominal area. Thus, for example, the true area of a polished plate glass surface is about ten times the nominal area. This implies the presence of many hills and valleys which can interact as shown in Fig. 8.3. If the speed of rotation N and/or the viscosity of the lubricant is reduced or if the pressure is increased, then although the coefficient of friction may increase, the distance between the journal and the bearing will eventually decrease to less than the height of the largest surface irregularity, and wear will result. There are thus three possible causes of lubrication failure: high pressures, low speeds, and of course, high temperature which reduce the viscosity of the lubricant oil. The effect of temperature on the viscosity of a liquid is discussed in Appendix B.

The variation of viscosity (η) with temperature (T), in its simplest form, is given by Reynolds relation

$$\eta = k\ e^{-at} \tag{8.11}$$

where k and a are constants depending on the liquid. This can also be expressed as

$$\eta = k\ e^{Evis/RT} \tag{8.12}$$

A more exact relationship is given

$$\log_{10}\log_{10}(v + 0.8) = n\ \log_{10}T + C \tag{8.13}$$

where $v = \frac{\eta}{\rho}$ (kinematic viscosity with unites of stokes) in which n and C are constants for a given oil or lubricant and ρ is the density of the liquid.

Fig. 8.4 A schematic illustration of a monomolecular iron stearate boundary film. Cohesive forces between the iron stearate molecules result in a closely packed and difficult to penetrate film (Courtesy of Texaco Inc.)

To reduce wear during low speeds, long-chain fatty acids or soaps are added to the oil. These form chemical bonds to the metal surfaces and thus prevent direct metal-metal contact or welding of the two surfaces. This type of lubrication is called boundary lubrication and is illustrated in Fig. 8.4.

Two surfaces which are in relative motion with an oil film between them are hydrodynamically lubricated. The viscosity of the oil determines the friction of motion. When the thickness of oil film is less than the surface irregularities, the asperities, then surface wear will eventually bring the two surfaces sufficiently close so as to rely on boundary layer lubrication. This is illustrated in Fig. 8.5 for a moving piston in a car engine.

8.3.2 Thrust Bearings

Thrust bearings are meant to keep a rotating shaft from motion parallel to the axis of rotation. Various configurations are possible, but the most common is one in which the load is supported by the bearing in a vertical shaft system.

8.3.3 Slider Bearings

Slider bearings are also called *ways or guide* bearings and are meant for rectilinear and curvilinear motion.

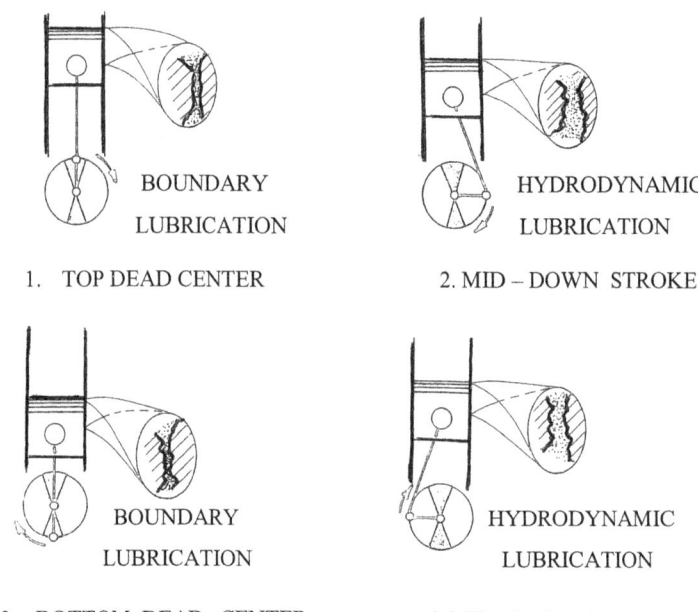

Fig. 8.5 Schematic diagram describing piston-to-cylinder wall and piston skirt-to-cylinder wall lubrication regime changes during an engine revolution (Courtesy of Texaco Inc.)

8.3.4 Ball Bearings

Elastohydrodynamic Lubrication (EHL): In journal bearings or other surface area interactions under lubricating conditions, deformation of the materials or the influence of pressure on viscosity of the lubricant can be ignored. However, in roller and ball bearings, the point of contact is deformed elastically under the high load pressure. This also increases the pressure on the film lubricant which increases in viscosity according to the equation

$$\eta_p = \eta_o e^{\alpha p} \tag{8.14}$$

where η_o is the viscosity of a liquid at 1 atm pressure
η_p is the viscosity of the liquid at pressure P
α is the pressure coefficient of viscosity

The film thickness is about 10–50 μm and increases as the speed is increased. An increase in load has the effect of deforming the metal surfaces more, causing an increase in contact area rather than a decrease in film thickness.

An important characteristic of EHL is the specific film thickness (λ) which is the ratio of the film thickness (h) to the composite roughness (σ) of the two surfaces, that is,

$$\lambda = \frac{h}{\sigma} \tag{8.15}$$

Fig. 8.6 Schematic representation of elastic deformation of touching curved surfaces with the accompanying pressure increase on the lubricating oil

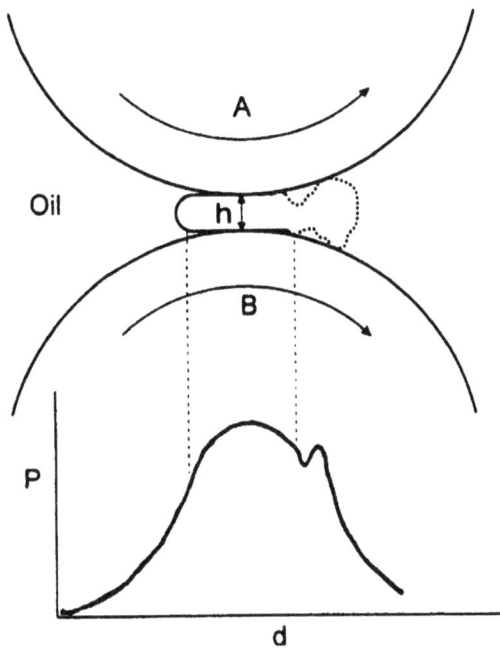

The composite roughness is determined from

$$\sigma = \sqrt{\sigma_1^2 + \sigma_2^2} \tag{8.16}$$

where σ_1 and σ_2 are the root mean square roughness

$$h = B\sqrt{\eta_0 VR} \tag{8.17}$$

where η_0 = viscosity of oil at atm pressure
$V = \frac{1}{2}(V_1 + V_2)$, mean speed
R = mutual radius of curvature of the two contacting surfaces

and where B is a constant depending on the units used, for example, $B = 8 \times 10^{-6}$ cm for c:g.s units. The elastic deformation and oil pressure profile is shown in Fig. 8.6.

8.4 Extreme Pressure Lubrication

At higher pressures and higher sliding speeds or at temperatures above about 150°C, the surface boundary layer breaks down or "melts," and boundary lubrication ceases to operate. Such conditions can occur in hypoid and worm gears, and special oil-soluble substances called E.P. (extreme pressure) additives are used. These materials react with the metal surface at a specific temperature and pressure and form a layer of low shear strength solid which are normally sulfides, chlorides, or phosphorus compounds in the form of sulfurized fats, chlorinated esters, and tricresyl phosphates of the base metal-iron. The reaction is a form of controlled corrosive attack which is preferred to uncontrolled abrasive wear. The solids formed by E.P. lubricants have high fusion temperatures, making them more effective than boundary lubricants. Though corrosive wear is slow, it cannot be eliminated by E.P. lubricants, and the alternate use of solid lubricants such as graphite or molybdenum disulfide is employed. These lubricants will be discussed later.

8.5 Wear

Not too many years ago, a new car owner was required to "run the car in" by driving at lower speeds for a few months to wear off the high spots. The oil and lubricants were then changed, and the car was allowed to run normally. The initial operation removes the asperities of the bearings and results in wear.

The analysis of lubricating oils for trace metals can give an early indication of excessive wear. Over 40 years ago, the New York Central Railway routinely analyzed locomotive lubricating oils for trace metals and frequently predicted, and eventually prevented, engine failure. The development of modern automatic-computer-controlled instrumental methods can routinely analyze for as many as 20 elements at a rate of about 200 samples per hour. The methods include atomic absorption spectrophotometry (AAS), inductively coupled argon plasma spectrometry (ICAPS), and atomic fluorescence spectrophotometry (AFS). The method is commonly referred to as SOAP (Spectrographic Oil Analysis Program). Table 8.1 shows the elements commonly found in oil from a diesel engine, the typical values, and the source. The data in Table 8.2 show the results of consecutive analysis for several elements in a diesel engine over a few months of operation. The high levels of lead in Engine B clearly indicate a potential problem.

This approach (SOAP) is now used for fleet vehicles, aircraft engines, as well as locomotives and can prevent breakdowns by scheduling maintenance in relation to wear at predictable parts of an engine.

8.6 Oil Additives

Straight chain aliphatic (alkanes) hydrocarbons do not have optimum lubricating properties. However, the addition of certain substances can even improve a paraffin oil.

The general purposes of lubricant additives are the following:

(a) *To Extend the Useful Range of Conditions of the Lubricant by Increasing the Viscosity Index (VI):*
    One of the major factors affecting the conditions of a lubricant is its temperature which usually

Table 8.1 Sources of trace elements in oil

Element	Wear or contamination source	Typical values[a] μg/g
Si	Dirt due to faulty air filter, also found in antifoamers	5
Fe	Wear from engine block, gears, rings, camshaft, oil pump, and/or crankshaft	20
Cu	Wear of bushings, valve guides, connecting rods, or piston pins	6
Ni	Wear of plating on gears and some bearings	3
Sn	Wear of some bearings and coated surfaces	2
Pb	Wear of bearings in diesel engines	5
Cr	Wear of rings and cooling system if chromates are used as corrosion inhibitors	4
Al	Wear of pistons and some bearings	4
Mo	Wear of some alloy bearings and oil coolers	3
Na	Possible leakage of antifreeze	30
B	Coolant leakage if boron inhibitors are used	2
Ba	Normally used in rust and corrosion inhibition and in oil detergents	2
Zn	Detergents	<2
P	Normally used as an antiwear additive in oil	<2
Ca	Normally used in detergents and antirusting additives	3,000

[a]Under normal running conditions

Table 8.2 Typical concentrations ($\mu g/g$) of trace elements in engine oil as a function of time of use

	Week	Al	B	Cr	Cu	Fe	Pb	Si	Sn
Engine A	1	3	2	2	4	10	4	4	2
	6	3	3	2	4	21	6	5	3
	12	2	2	2	5	22	12	5	2
	13	4	3	2	11	31	22	6	4
			Ni						
	0	<5	1.	<1	1	31	4		<2
Engine B	25	12	1.7	<1	75	11	120		0
Engine C	25	<5	1.0	<1	1	11	0		<2
					16		18		0
									<2
									0

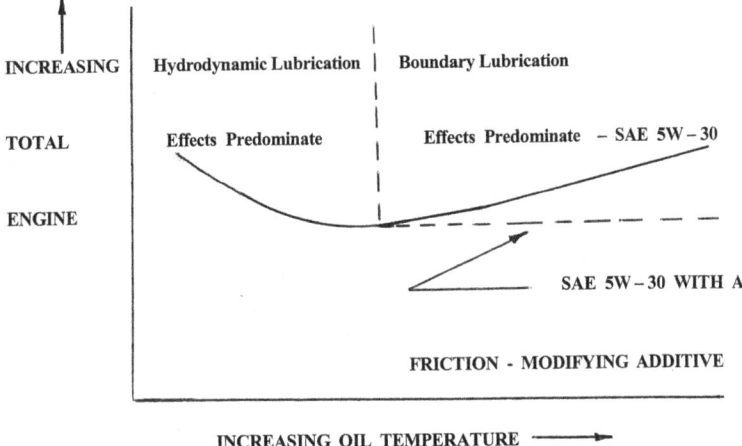

Fig. 8.7 The effect of a friction-modifying additive on the total engine friction response to oil temperature for an SAE 5W-30 engine oil (Courtesy of Texaco Inc.)

starts low (cold) and increases with running time. Thus, it is most desirable for the temperature effect on viscosity to be a minimum. This is determined by the viscosity index (VI) which is related to E_{vis} and determined by a comparison with standard oils. A high VI (80) implies a low value in E_{vis} and a small dependence of viscosity on temperature, whereas the converse is true for a lubricant with a low VI (30).

Common VI improvers which are added to lubricants are high molecular weight (~20,000 g/ mol) polymers. These tend to increase the viscosity of the oil to a greater extent at high temperatures than at low temperatures. Examples of polymers added include polymethacrylates, polyacrylates, polyisobutylenes, and alkylated styrenes. VI improvers are used in engine oils and automatic transmission fluids. The effect of a VI additive to an oil is shown in Fig. 8.7.

(b) *To Prevent or Retard the Deterioration of the Lubricant*: The most common additive designed to preserve the function of the lubricant is the antioxidant which reduces or inhibits the rate of oxidation of the oil. In oxidation, the oil can form peroxides which eventually decompose to acids causing corrosion of the metallic components. Zinc dithiophosphate acts as both an antioxidant and a corrosion inhibitor. Other additives include sulfur compounds, amines, and metal salts such as calcium salts of alkylphenol formaldehyde.

(c) *To Protect the Surfaces in Contact with the Lubricant*: Surface protectors include boundary layer additives and rust inhibitors such as sulfonates, phosphates, and fatty acid salts. Extreme pressure additives also preserve metallic surfaces at lower temperatures.

(d) *To Depress the Pour Point and Break Foam*: The pour point (PP) is the temperature below which the oil will not flow due to the crystallization of the heavier wax components. Paraffinic type oils have higher pour points than the naphthenic oils. Hence, the reduction in the PP temperature is achieved by either removing the wax components or adding PP depressants. Such depressants include the additions of high molecular weight polymers of polymethacrylate, polyalkyl-naphthalenes, and polyalkylphenol esters. An oil can have its PP depressed by $30°$ upon the addition of 1% depressant. The cloud point, the temperature at which the oil becomes cloudy on cooling, is due to the separation of wax or moisture. Thus, low cloud points are a desirable property of a lubricant.

Foam and froth in the oil is caused by the entrapment of air during the pumping and movement of oil through various parts of the engine. Foam reduces the effectiveness of the lubricant and thus can result in increased friction and wear. Antifoaming agents are usually silicones which at 5 ppm or less can lower the surface tension of the oil and destroy the tendency to form bubbles.

(e) *To Improve the Wetting Properties of the Lubricant*: A good lubricant must be able to wet the metal surface that requires lubrication. The discussion of wetting is given in Appendix C. Additives which lower the contact angle of oil on steel include fluorinated polymers. Other additives include dispersants and detergents which tend to keep solid particles (that pass through the filter) in suspension, preventing them from forming larger agglomerates.

The modern lubricating oil has many additives, and further research and developments continue to improve oils; extending their useful life and effectiveness will be discussed later.

8.7 Synthetic Lubricants

Lubricants which are not based on petroleum are classed as synthetic and include silicones, polyglycols, polyphenyl ethers, fluoro compounds, chlorofluorocarbon polymers, and phosphate esters. The major advantage of these fluids is that they are tailor-made for a specific function and usually require few additives.

Silicone oils are especially useful for high temperature applications and can be polymerized to give any desired viscosity. However, their major disadvantage is that when they decompose, they invariably form solid silicates and silica (SiO_2) which can damage lubricated joints and bearings. Their low surface tension makes them less desirable for thin film lubrication and boundary conditions. However, the high viscosity index, good chemical stability, high shear resistance, low volatility, and antifoam characteristics make silicone oils a good lubricant in special situations such as torque converters, ball bearings, etc., in spite of the high cost.

Polyglycols are water insoluble, but by incorporating ethylene oxide into the polymer, a water-soluble ethylene oxide-glycol polymer is obtained which is water soluble. They have very high viscosity index (160), low volatility, and good compatibility with rubber. The water-soluble type, with as much as 40% water content, is considered to be fire-resistant and therefore used in mines and ships. Their degradation products are soluble in the fluid or volatile, thus leading to very low sludge formation.

Fluorolube consists of fluorinated hydrocarbons and polyethers. They can function effectively from $-90°C$ to over $250°C$, have a high density, are thermally stable, have low volatility, and are nonexplosive. They are usually used in highly corrosive environments such as in the preparation of hypochlorite (ClO^-). The costs of such specialized lubricants are very high (about $2,000/kg), but since they do not degrade, they almost last forever.

8.8   Solid Lubricants

Solids in the form of soap have long been used as a lubricant. More recently, the three most prominent solids which have lubricating properties are graphite, molybdenum disulfide (MoS_2), and Teflon. A more complete list is given in Table 8.3. In graphite and MoS_2, the solid is a layer lattice (lamellar) in which the distance between the parallel planes is 3.40 Å for graphite and 3.49 Å for MoS_2. These large values are to be compared with 1.42 Å for the in-plane C-C distance and 2.41 Å for the in-plane Mo-S distance. Thus, the large interplanar distance is considered to be bonded by relatively weak van der Waal forces which permit the planes to slide past each other.

Lamellar solid lubricants function by being interspersed between the asperities of the two moving surfaces. This is illustrated in Fig. 8.8 where the solid adhering to the surfaces clear at the planes when the metal-lubricant bonding forces are greater than the interplanar forces. This is the case with MoS_2 even under vacuum conditions. However, in the presence of water vapor and to a lesser extent in O_2 or N_2, the lubricating effectiveness of MoS_2 is reduced due to a reduction in the metal-MoS_2 attractive forces. With graphite, the converse applies, that is, graphite is a poor lubricant in vacuum but much better in nitrogen and somewhat better in oxygen but best in the presence of water vapor. This is explained by the adsorption of O_2 and water to the dangling valences at the edge of the graphite planes, thereby reducing the shear strength required to move the interlaminar planes. With MoS_2, no such residual valency is present, and the vapors adsorbed onto the metal surfaces reduce its adhesion to MoS_2. Other lamellar solids such as boron nitride(BN) or talc do not have similar lubricating

Table 8.3   Selected solid lubricants

	Electrical resistivity	
	ohm—cm	Air stability°C
Teflon	10^{18}	320
$MoS_2(n)$	8.5×10^2	350
$MoSe_2$	1.8×10^{-2}	400
WS_2	1.4×10^1	440
WSe_2	1.1×10^2	350
NbS_2	3.1×10^{-3}	420
$NbSe_2$	5.4×10^{-4}	350
TaS_2	3.3×10^{-3}	600
$TaSe_2$	2.2×10^{-3}	580
Graphite	1.4×10^{-3}	450
ReS_2		230

SURFACE BONDING FORCES

IRON OXIDE SURFACE

Fig. 8.8   Schematic illustration of how microcrystalline plates of solid materials such as molybdenum disulfide can form strong surface coatings. The plates slide easily over each other thereby providing lubrication (Courtesy of Texaco Inc.)

Fig. 8.9 Schematic diagram of the mode of action of a lamellar solid lubricant when used in dispersion in a liquid such as oil. (**a**) Orientation of lubricant particles between sliding surfaces, (**b**) prevention of welding of asperities by the lubricant particle (From Hickman J, Middleton K (1970). Some surface chemical aspects of lubrication, advancement of science, June, p 442, with permission from the British Association for the Advancement of Science)

properties because the adhesion to surfaces is less than the shear energy of the interlaminar planes. The mode of action of lamellar solid lubricants is shown in Fig. 8.9.

These solids, when dispersed in lubricating oils, increase the efficiency of lubricants and decrease wear during engine start-up. Particle size must be less than 2 μm in order to pass through the filter and to remain in suspension.

A unique solid lubricant has been described which can function at over 800°C in air or other oxidizing environments and maintain its low coefficient of friction of 0.1–0.2. At very high temperatures, normal solid lubricants such as MoS_2 or WSe2 revert to abrasive metallic oxides, while graphite is oxidized to CO and CO_2, thereby losing its lubricating surface. The lubricant is a blend of WSe_2 or WS_2 and gallium, with 10–30% Ga. Under dynamic bearing pressures of 1,500 psi (100 Atm), a wear rate of less than 5 mg/h is obtained.

Teflon is a polymer of tetrafluoroethylene, often referred to as PTFE. It is well known as a non-sticking surface on the frying pan and other cooking utensils. It has a low coefficient of friction and is chemically inert and thermally stable to over 300°C.

Teflon is a thermosetting plastic, but it cannot be used as such because it is too viscous above its glass transition temperature of 325°C. It is therefore formed by the high pressure-high temperature compression of its powder. Teflon is often impregnated with powdered glass or other solids to increase its yield point since it tends to flow and deform under pressure. Other disadvantages of Teflon is that it is a poor conductor of heat and has a high coefficient of expansion. During wear, the polymer breaks down resulting in fluorocarbon free radicals and C_1 and C_2 fluorocarbons. The sliding friction (60 cm/s, loaded at 0.7 k Pa) of steel on Teflon (filled with 25% carbon) in a high vacuum ($P = 5 \times 10^{-8}$ torr, coefficient of friction = 0.17) was shown to give rise to fluorocarbon gaseous fragments which were identified in a mass spectrometer to be CF, CF_2, CF_3, C_2F_2, C_2F_3, and C_2F_4. Thus, the wear of Teflon can give rise to dangerous toxic gases, and its use in spacecraft is not recommended.

The effectiveness of Teflon as a means of reducing friction of a ski was compared to a conventionally waxed ski on crystalline snow at 0°C over a 213-m slope. With a 76-kg weight on the skis, the normal ski took 63 s compared to 42 s for a Teflon-coated ski. When the weight was reduced to 64 kg, the relative times for the decent of the skis were 83 s and 54 s, respectively. Teflon has many uses, but its application is limited by the difficultly to bond it to other surfaces and to itself (see Chap. 12). Another fluorocarbon polymer, perfluoropoly-propylene (FEP), is similar to Teflon in its lubricating properties though much more transparent in thin films. However, its lower melting point (290°C) allows it to be fabricated by conventional methods when copolymerized with PTFE.

These polymers are more expensive than other plastics such as polyethylene, but in special applications, they fulfill exceptional needs.

8.9 Greases

Grease consists of an oil which is thickened by the addition of a finely ground highly dispersed solid and which are used over a temperature range of $-70°$ to $300°C$. The ancient Egyptians ($\sim$ 1400 BC) prepared lubricating grease by mixing olive oil with lime. In modern greases, the solids include metallic soaps such as aluminum, barium, and lithium. Greases serve to (1) act as a barrier in bearings to water and abrasive materials, (2) protect surfaces against corrosion and wear, (3) lubricate bearing by virtue of the oil present in the grease, and (4) greases usually have a smaller temperature coefficient of viscosity than the base oil.

As much as 14% lithium soap is added to the base oil in a lithium grease. The oil is held in the gel structure by molecular and polar forces, by capillarity, and by mechanical entrapment. Greases are thixotropic gels which flow under shearing forces such as stirring or motion. The oil in the grease thus lubricates the bearing during motion.

Lithium soap greases with added MoS_2 are generally superior to the base lithium greases in wear reduction, lubrication, and rust prevention. In systems where the lithium soap acts as an EP lubricant, the MoS_2 has little effect on the grease. The amounts of MoS_2 added vary from 0.25% to up to 25%.

Silicone greases are readily made by adding finely ground or fumed silica (SiO_2) to silicone oil. The amount added determines the thickening effect. Greases are used in bearings and often preferred to lubricating oils in cases were explosions and fires are possible from the vapors. For example, coal mining is reported to use a 1 kg of grease for every 15 tonnes of coal mined.

Exercises

1. Gas under pressure can act as a lubricant. What factors affect the lubrication qualities of a gas?
2. What are the advantages and disadvantages of gas lubrication compared to normal liquid lubricants?
3. The analysis of trace metals in oils can be an indication of excessive wear. Explain.
4. Boron nitride forms a layer lattice—would you expect this substance to have lubricating properties?
5. Explain why Teflon-coated surfaces are considered to be self-lubricated in some cases.
6. Determine the mean free path of CO_2 at 2 atmospheres pressure and $300°C$. The diameter of the CO_2 molecule is 520 pm.
7. The value of E_{vis} of a liquid is from 20 % to 50 % of the enthalpy of vaporization of the liquid. Why is this not unexpected?
8. The boiling point of an oil is $325°C$. Using Trouton's Rule ($\Delta H_{vap}/T_{bp}=88$ kJ/mol) and the ratio $\Delta H_{vap}/E_{vis}=5$, determine E_{vis} for the lubricant.
9. Suggest a lubricant which is compatible with (a) liquid oxygen, (b) concentrated HNO_3 (70 % by wt.), (c) concentrated HF (36 % by wt.), (d) 50 % NaOH solution, and (e) molten sodium at $250°C$.
10. Based on the results in Table 8.2 what faults would you expect in the 3 engines, A, B, and C?
11. What lubricants would you use for operating in a spacecraft intended for a journey of 6 months?

12. Distinguish between hydrodynamic lubrication, boundary lubrication, and extreme pressure lubrication.
13. How can a viscous lubricant be converted into a grease?

Further Reading

Wikipedia: Lubrication & Friction

B.P. Trading Ltd. (1969) Lubrication theory and its applications. B.P. Trading Ltd., London
Bowden F, Tabor D (1950/1964) Friction and lubrication of solids I and II. Clarendon Press, Oxford
Cameron A (1976) Basic lubrication theory. Wiley, New York
Gunther RC (1971) Lubrication. Chilton Book Company, New York
Halling J (1976) Introduction to tribology. Wykeham Publications Ltd., London
Kogelsky IV, Allison VV (eds) (2005) Tribology, lubrication, friction & wear. Wiley, New York
Moore DF (1975a) Principles and application of tribology. Pergamon Press, Oxford
Moore DF (1975b) Principles and applications of tribology. Pergamon Press, New York
Quinn TFJ (1971) The application of modern physical techniques to tribology. Van Nostrand Reinhold Company, New York
Wills JG (1980) Lubrication fundamentals. Marcel Dekker Inc., New York

Chapter 9
Electrochemistry, Batteries, and Fuel Cells

9.1 Introduction

Electrochemistry is concerned with the effect of electrical voltages and currents on chemical reactions (*ionics*) and chemical changes which produce the voltages and currents (*electrodics*). This is illustrated in Table 9.1 where ionics is governed by Faraday's laws, whereas electrodics is determined by the Nernst equation.

9.2 Ionics

Faraday's laws of electrolysis are:

1. The mass of material formed at the electrodes is proportional to the quantity of electricity passed through the solution.
2. For a fixed quantity of electricity passed through a solution, the masses of different materials formed (or dissolved) at the electrode are proportional to their equivalent weight, which is the atomic mass/electron charge, Z.
3. The charge on the electron is 1.60219×10^{-19} coulombs. Hence, one mole of electrons (6.02214×10^{23}) represents $1.60219 \times 10^{-19} \times 6.02214 \times 10^{23} = 9.6486 \times 10^4$ coulombs/mol which is called a Faraday ($\mathscr{F}$) and usually rounded off to 96,500 coulombs. A coulomb, Q, is equal to 1 amp × 1 s, and 26.80 amp-h $= 1.\mathscr{F}$

$$Q = It \tag{9.1}$$

where I is the current in amperes and t is the time in seconds.

Example 9.1

How much silver would be electrodeposited by a current of 6.0 amp for 3 h from a solution of $AgNO_3$?

Answer: The atomic mass of silver is 107.87 g/mol. The charge on silver is +1 (Z = 1), and this means that $1.\mathscr{F}$ would deposit 107.87 g. We must now determine the number of coulombs, Q, which were passed through the solution.

O.V. Roussak and H.D. Gesser, *Applied Chemistry: A Textbook for Engineers and Technologists*, DOI 10.1007/978-1-4614-4262-2_9, © Springer Science+Business Media New York 2013

Table 9.1 The classification
of electrochemical systems

Ionics	Electrodics
Faraday's law	Nernst equation
Current flow causes reaction to occur	Reactions cause voltage to develop and current to flow
Electrolysis	Batteries
Electrodeposition	Fuel cells
Electrochemical machining	Corrosion protection
Battery charging	

$$Q = It = 6.0 \text{ amp} \times 3 \text{ h} \times 60 \text{ min/h} \times 60 \text{ s/min} = 64,800 \text{ amp.s} \qquad (9.2)$$

96,500 C would deposit 107.87 g of Ag and 1 C would deposit 107.87/96,500 g of Ag. Hence, 64,800 C would deposit $64,800 \times 107.87/96,500$ g of Ag, or 72.43 g of silver would be deposited.

Example 9.2

During the 4.0 h electrodeposition of copper from a copper sulfate ($CuSO_4$) solution, 140.0 g of Cu was deposited. What was the current flowing through the cell?

Answer: The atomic mass of copper is 63.55 g/mol. Since the charge on the copper in the solution, Cu^{2+}, is +2, then the equivalent weight is 63.55/2 g. Hence, $1\mathscr{F}$ would deposit 63.55/2 g Cu. Since 140.0 g of Cu was deposited, the quantity of current which flowed through the cell is 140.0/63.55/2 or $4.406\,\mathscr{F}$.

$$Q = It$$
$$I = \frac{Q}{t} = \frac{4.406 \times 96,500}{4 \times 3,600} \text{amp} = 29.53 \text{ amp.} \qquad (9.3)$$

Example 9.3

An alloy of tin and lead is deposited from a solution of $Sn(NO_3)_2$ and $Pb(NO_3)_2$. What is the percentage of tin in the alloy if 35.00 g of alloy is deposited when 3.00 amp of current is passed through the solution for 4.00 h?

Answer: The quantity of electricity Q passed through the solution is given by:

$$Q = It = 3.00 \times 4.00 \times 3,600 = 43,200 \text{ C}$$

96,486 C liberates 1 equivalent weight, 1 C liberates 1/648 equivalent weights, and 43,200 C liberates 43,200/96,486 equivalent weights, that is, a total of 0.448 equiv. If M is the mass of Sn in the deposit, then 35.00 M is the mass of Pb.

The equivalent weight of Sn is

$$\frac{118.7}{2} = 59.35 \text{ g/eq.}$$

The equivalent weight of Pb is

$$\frac{207.2}{2} = 103.6 \text{ g/eq.}$$

Hence, the number of equivalents of Sn in the deposit is

$$\frac{M}{59.35},$$

and the number of equivalents of Pb in the deposit is

$$\frac{35.00 - M}{103.6}.$$

Thus,

$$\frac{M}{59.35} + \frac{35.00 - M}{103.6} = 0.448.$$

Solving for M

$$103.6M + 2{,}077 - 59.35M = 2{,}754$$
$$44.2M = 677$$
$$M = 15.3 \text{ gSn}$$
$$35.00 - M = 19.7 \text{ gPb}$$

Percentage of Sn in the alloy is

$$\frac{15.3}{35.00} \times 100 = 43.7\%.$$

Example 9.4

High purity gases are often prepared by electrolysis. It is desired to produce O_2 at 25°C and 800 torr at a flow rate of 75.0 cm³/min. What current would be required for the electrolysis cell which consists of aqueous NaOH solution and nickel electrodes?

Answer: Using $PV = nRT$, we see that a flow rate of 75.0 cm³/min corresponds to a flow rate of

$$n = PV/RT = \{800 \text{ torr}/760 \text{ torr}/\text{atm}\}$$
$$\times 75.0 \text{ mL/min} \times (82.06 \text{ mL.atm/K.mol})^{-1} \times [298 \text{ K}]^{-1}$$
$$n = 3.228 \times 10^{-3} \text{ mol/min}$$

For O_2 gas, 1 mole corresponds to 4 equivalents. Thus, this flow rate corresponds to

$$4 \times 3.228 \times 10^{-3} \text{ equiv./min}$$

or

$$\frac{4 \times 3.228 \times 10^{-3}}{60} \text{ equiv./s,}$$

that is, $\mathscr{F}/s$. Thus,

$$\text{current required} = \frac{4 \times 3.228 \times 10^{-3} \times 96{,}485}{60} \text{C/s}$$

$$I = 20.77 \text{ C/s}$$

$$I = 20.77 \text{ amp}$$

One mole of H_2 prepared by the electrolysis of water would require $2\mathscr{F}$, whereas 1 mole of O_2 would require $4\mathscr{F}$. The electrolysis of Al_2O_3 in cryolite (Na_3AlF_6) requires $3\mathscr{F}$ for each mole of Al (27 g). This is very energy intensive, and a tonne of aluminum uses over 15,000 kW. Some effort (though unsuccessful) has been made to convert Al^{3+} to Al^{+1} in order to save on electrical energy by electrolyzing Al^+ instead of Al^{3+}.

9.3   Electrolysis and Electrodeposition of Metals

When an increasing DC voltage is applied to a solution of metal ions, the metal, in some cases, will begin to deposit the deposition material at the cathode at a minimum voltage. This is illustrated as D in Fig. 9.1. The deposition potential depends on the metal, the surface, the current density, the concentration of the metal, and other ions in solution. The electrodeposition or electroplating reaction

$$M^{+n} + ne^- \rightarrow M \tag{9.4}$$

is most common for metals such as Au, Ag, Pb, Cd, Zn, Cr, Ni, Cu, and Sn. Two or more metals can also be deposited to form alloys. If the voltage is too high, then hydrogen will be evolved:

$$2H^+ + 2e^- \rightarrow H_2 \tag{9.5}$$

The theoretical minimum voltage required for the electrolysis of water is 1.23 V at 25°C. However, the process has an activation energy which is referred to as the overvoltage or polarization which depends on the current, temperature, and materials used for the electrodes. Some typical overvoltages are given in Table 9.2.

The industrial preparation of hydrogen by the electrolysis of water on nickel electrodes requires a voltage of more than 1.50 V (1.23 V + 0.210 V + 0.060 V) since it is necessary to add the RI drop (due to the internal resistance of the electrolyte). However, at very high current densities, the polarization is much higher, and higher temperatures are used to reduce the excess power

---

[1] The ground state electronic configuration of Al is $1s^2 2s^2 2p^6 3s^2 3p^1$, and it would be expected for Al to have a (+1) oxidation state. This is in fact the case, and at high temperatures, the reaction.

$$AlCl_{3(g)} + 2Al_{(3)} = 3AlCl_{(g)}$$

occurs. However, attempts to electrolyze AlCl have so far proved to be unsuccessful.

Fig. 9.1 Typical current–voltage plot for an electrolyte solution. *A* start of deposition, *A–B* linear segment of plot, *D* extrapolated deposition potential

Table 9.2 Overvoltage (mV) in water electrolysis, 25°C

Metal surface	Cathode polarization dilute H_2SO_4	Anode polarization dilute KOH
Platinized platinum	5	250
Smooth platinum	90	450
Palladium	~ 0	430
Gold	20	530
Silver	150	410
Nickel	210	60
Lead	640	310
Cadmium	480	530
Mercury	780	–

requirements. The overvoltage encountered in the electrodeposition of a metal can be associated with the various steps by which the metal in solution, for example, Cu^{2+}, becomes the atom in a copper lattice.

The various steps in the overall mechanism are:

1. Cu^{2+} (hydrated in solution) diffuses to the cathode.
2. Cu^{2+} (hydrated) at electrode is transferred to the cathode surface.
3. Cu^{2+} (partially hydrated and adsorbed onto the surface), the ad ion, diffuses across the electrode surface to a crystal building site.
4. Cu^{2+} (adsorbed at a crystal building site) becomes a part of the lattice.
5. $[Cu^{2+} + 2e^-]$ occurs, and Cu is part of the metal.

The sequence of relative importance is $2 > 3 > 4 > 1 > 5$ with step 2 being the slowest step or rate-controlling process.

9.4   Electrochemical Machining

Electrolysis, with an anode that dissolves under controlled conditions, is the basis of electrochemical machining (ECM). High-strength metals such as Nimonic alloys (nickel with Al, Ti, and Mo) used in the aircraft industry resist deformation even at high temperatures and are exceedingly difficult to machine by the normal cutting process because of the limitations and expense of tool materials. In ECM, the metal alloy does not determine the rate or characteristics of the dissolution process, and hard tough metals can be dissolved as readily as soft metals. Only the current density (amp/cm^2) determines the rate of machining where approx. 5×10^{-3} mm/min can be removed at a current density of 0.3 amp/cm^2. The additional advantage of ECM is the absence of mechanical and thermal stress usually associated with conventional machining.

A schematic representation of a typical ECM apparatus is shown in Fig. 9.2 and consists of a workpiece to be machined (the anode), a properly shaped cathode tool which is movable and maintains a constant gap with the workpiece. The electrolyte flows between the two electrodes, removing the products of electrolysis as well as heat. The power supply furnishes the high currents necessary to dissolve the anode.

9.4.1   The Cathode

The cathode is a tool shaped to conform to the desired cut in the workpiece. The tool is usually made from copper, steel, or alloy and insulated on the sides to give directed current lines. The tool is moved during electrolysis to maintain a constant small gap (about 0.25 mm) between the electrodes to reduce the voltage required.

Fig. 9.2 Schematic diagram of an ECM unit

Table 9.3 Composition of various iron-base alloys[a]

Iron-base alloy	Element (%)							
	C	Cr	Mo	Mn	Si	P	S	Ni
SCM 3	0.35	1.1	0.2	0.7	0.25	0.03	0.03	–
SKD 11	1.5	12.0	1.0	–	–	–	–	–
SNC2	0.3	0.8	–	0.5	0.2	0.03	0.03	2.7
SUS 304	0.08	17.5	–	20	1.0	0.045	0.03	9.5
Inconel 718	0.06	18.26	2.93	0.11	0.13	0.0001	0.003	51.85

[a]Iron is the remaining element making up the alloy's 100%

Fig. 9.3 Surface roughness of steels (see Table 9.3) obtained in various electrolytes. O SCM 3, ● SKD 11, Δ SNC 2, ▲ SUS 304, □ Inconel 718

9.4.2   The Electrolyte

The purpose of the electrolyte is to provide a conducting medium, and at the same time, it must not corrode the cathode tool. The cheapest material commonly used is sodium chloride (NaCl) at about 30% by weight. In some cases, additives such as alcohols, amines, thiols, and aldehydes are used to inhibit stray currents, which results in overcuts. Other electrolytes such as $Na_2Cr_2O_7$, $NaNO_3$, and $NaClO_3$ at 50–250 g/L have also been used, but the choice is limited primarily by cost.

The electrolyte is usually recirculated with the metal products removed or reduced before being reused. This minimizes cost and pollution and prevents the formation of a precipitate in the electrolysis gap.

The effect of electrolyte on surface roughness, H, was studied by Y. Sugie (1978) for five different iron alloys (characterized in Table 9.3) and is shown in Fig. 9.3. The roughness depends on electrolyte and alloy.

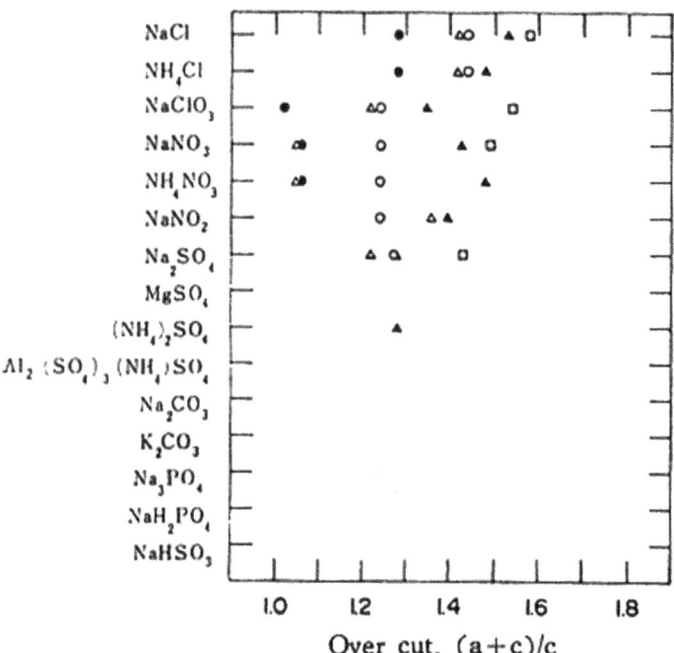

Fig. 9.4 Overcut obtained for various steels (see Table 9.3) in various electrolytes. *O* SCM 3, ● SKD 11, Δ SNC 2, ▲ SUS 304, □ Inconel 718

The accuracy of the machining, Figs. 9.4 and 9.5, was determined for the five alloys by measuring the overcut and machined angle (shown in Fig. 9.6) for the various electrolytes. Based on the results, it was concluded that $NaClO_3$ was most suitable for the low alloy steel and that Na_2SO_4 was best for the high nickel alloy steels.

The power supply and tool drive complete the apparatus. Current densities of 100–200 A/cm^2 at voltages from 10 to 50 V give cutting rates of about 1 mm/min with surface finishes of 5×10^{-4} cm. By pulsing the DC current, it is possible to reduce the required flow rate of the electrolyte. However, one difficulty, sparking, common in some continuous systems, occurs in intermittent electrolysis and is a decided disadvantage. Other disadvantages in ECM include high initial cost of the equipment, the hazards of hydrogen evolution at the cathode, sparking and "wild cutting" by stray current, and the need of a machinist who has some knowledge of chemistry.

Intergranular corrosion observed for some alloys can be reduced by using mixed electrolytes such as 15% NaCl with 20% $NaClO_3$.

It is possible to affect electrochemical grinding and polishing by using a cathodic wheel with an anodic workpiece separated by a flowing electrolyte. Tungsten carbide tools are usually prepared by such a process. An electrochemical saw has also been produced.

Other refractor materials such as TiC, ZrC, TiB$_2$, ZrB$_2$, TiC/Ni can also be subject to ECM using NaCl and KNO$_3$ as electrolytes. However, the number of equiv./mol (Z) is usually not a whole number because of multiple dissolution reactions, for example, $Z = 6.6$ for TiC because

$$TiC + 3H_2O \rightarrow TiO_2 + CO + 6H^+ + 6e^- \tag{9.6}$$

as well as

$$TiC + 4H_2O \rightarrow TiO_2 + CO_2 + 8H^+ + 8e^- \tag{9.7}$$

Machined angle obtained for various
steels in various electrolytes

○: SCM 3, ●: SKD 11, △: SNC2
▲: SUS 304, □: Inconel 718

Fig. 9.5 Machined angle obtained for various steels (see Table 9.3) in various electrolytes. O SCM 3, ● SKD 11, △ SNC 2, ▲ SUS 304, □ Inconel 718

Fig. 9.6 Schematic diagram of machined surface (see Figs. 9.3, 9.4, and 9.5). W working electrode (15 mmφ), C counter electrode (5 mmφ), θ machined angle

One interesting result of ECM is that the surface composition of an alloy may be significantly different from the bulk composition. Thus, a TiC/Ni alloy of Ni, Ti, C, O of 11, 44, 39, and 6 at.%, respectively, had a surface composition of 47, 22, 10, and 21 at.% after ECM. This difference disappears within 1 μm from the surface and reflects the preferential dissolution of one or more elements in the metal.

The major disadvantages of ECM over conventional machining are the high cost of the ECM machine and its high maintenance cost, the technical expertise required of the operator, and the lower accuracy achieved.

9.5 Electrodics

A metal in equilibrium with its ions in solution will establish a potential difference which depends on the concentration of the metal ion in solution and the temperature. This potential is best determined by comparison with a standard which is arbitrarily set to zero—the hydrogen electrode:

$$2H^+(aq) + e^- \rightleftharpoons H_2(g) \tag{9.8}$$

The electrical energy, $\mathcal{E}$, is directly related to the free energy, ΔG, by the relation

$$\Delta G = -n\mathcal{F}\mathcal{E} \tag{9.9}$$

where n is the number of electrons transferred in the process and $\mathcal{F}$ is the Faraday. For an equilibrium reaction,

$$\Delta G = \Delta G° + RT \ln Q \tag{9.10}$$

where $\Delta G°$ is the standard free energy and

$$\Delta G° = -RT \ln K_{eq} \tag{9.11}$$

R is the gas constant equal to 8.314 J/K.mol; T is the absolute temperature; Q is the ratio of the concentration of products to concentration of reactants, and K_{eq} is the value of the equilibrium constant at the specified temperature. For a chemical reaction, (9.9) becomes

$$\mathcal{E} = \mathcal{E}° - RT \ln Q \tag{9.12}$$

which is called the Nernst equation. For example, in the Daniell cell,

$$Zn(s) + Cu^{2+}(qa) \rightarrow Zn^{2+}(aq) + Cu(s) \tag{9.13}$$

$$\mathcal{E} = \mathcal{E}° - \frac{RT}{n\mathcal{F}} \ln \frac{[Zn^{2+}]}{[Cu^{2+}]} \tag{9.14}$$

Reaction (9.13) can be considered to be composed of two reactions called *half-cell* reactions:

$$Zn(s) \rightarrow Zn^{2+}(aq) + 2e^- \tag{9.15}$$

and

$$Cu^{2+}(aq) + 2e^- \rightarrow Cu(s) \tag{9.16}$$

Table 9.4 Standard reduction potentials for some common redox systems at 25°C (↑ signifies gas state, and ↓ represents solid state; in all other cases, state is liquid or solution)

Reduction half-reaction	$\mathcal{E}^o_{cell}(V)$	Reduction half-reaction	$\mathcal{E}^o_{cell}(V)$
$F_2 \uparrow +2e^- = 2F^-$	+2.87	$Cu^+ + e^- = Cu \downarrow$	+0.521
$O_3 \uparrow +2H^+ + 2e^- = O_2 \uparrow +H_2O$	+2.07	$Fe(CN)_6^{3-} + e^- = Fe(CN)_6^{4-}$	+0.356
$Co^{3+} + e^- = Co^{2-}$	+1.82	$Cu^{2+} + 2e^- = Cu \downarrow$	+0.337
$H_2O_2 + 2H^+ + 2e^- = 2H_2O$	+1.77	$Hg_2Cl_2 \downarrow +2e^- = 2Hg + 2Cl^-$	+0.2680
$Ce^{4+} + e^- = Ce^{3+}$ (in $HClO_4$ solution)	+1.70	$AgCl \downarrow +e^- - Ag \downarrow +Cl^-$	+0.2224
$MnO_4^- + 4H^+ + 3e^- = MnO_2 \downarrow +2H_2O$	+1.69	$Cu^{2+} + e^- = Cu^+$	+0.153
$2HClO + 2H^+ + 2e^- - Cl_2 \uparrow +2H_2O$	+1.63	$Sn^{4+} + 2e^- = Sn^{2+}$ (in HCl solution)	+0.14
$Ce^{4+} + e = Ce^{3+}$ (in HNO_3 solution)	+1.60	$S \downarrow +2H^+ + 2e^- = H_2S$	+0.14
$2HBrO + 2H^+ + 2e^- = Br_2 + 2H_2O$	+1.6	$S_4O_6^{2-} + 2e^- = 2S_2O_3^{2-}$	+0.09
$MnO_4^- + 8H^+ + 5e^- = Mn^{2+} + 4H_2O$	+1.51	$2H^+ + 2e^- = H_2 \uparrow$	0.0000
$2BrO_3^- + 12H^+ + 10e^- Br_2 + 6H_2O$	+1.5	$Pb^{2+} + 2e^- = Pb \downarrow$	−0.126
$Mn^{3+} + e^- = Mn^{2+}$	+1.49	$Sn^{2+} + 2e^- = Sn \downarrow$	−0.140
$Ce^{4+} + e^- = Ce^{3+}$ (in H_2SO_4 solution)	+1.44	$Ni^{2+} + 2e^- = Ni \downarrow$	−0.23
$Cl_2 \uparrow +2e^- = 2Cl^-$	+1.359	$Co^{2+} + 2e^- = Co \downarrow$	−0.28
$Cr_2O_7^{2-} + 14H^+ + 6e^- = 2Cr^{3+} + 7H_2O$	+1.33	$Cr^{3+} + e^- = Cr^{2+}$ (in HCl solution)	−0.38
$Ce^{4+} + e^- - Ce^{3+}$ (in HCl solution)	+1.28	$Cd^{3+} + 2e^- = Cd \downarrow$	−.402
$MnO_2 \downarrow +4H^+ + 2e^- = Mn^{2+} + 2H_2O$	+1.23	$Fe^{2+} + 2e^- = Fe \downarrow$	−0.440
$O_2 \uparrow +4H^+ + 2e^- = 2H_2O$	+1.229	$2CO_2 \uparrow +2H^+ + 2e^- - H_2C_2O_4$	−0.49
$ClO_4^- + 2H^+ + 2e^- = ClO_3^- + H_2O$	+1.19	$Cr^{3+} + 3e^- = Cr \downarrow$	−0.74
$2IO_3^- + 12H^+ + 10e^- = I_2 \downarrow +6H_2O$	+1.19	$Zn^{2+} + 2e^- = Zn \downarrow$	−0.7628
$Br_2 + 2e^- = 2Br^-$	+1.087	$SO_4^{2-} + H_2O + 2e^- = SO_3^{2-} + 2OH^-$	−0.93
$N_2O_4 \uparrow +2H^+ + 2e^- = 2HNO_2$	+1.07	$Mn^{2+} + 2e^- = Mn \downarrow$	−1.190
$NO_3^- + 3H^+ + 2e^- = HNO_2 + H_2O$	+0.94	$Al^{3+} + 3e^- = Al \downarrow$	−1.66
$2Hg^{2+} + 2e^- = Hg_2^{2+}$	+0.907	$H_2 \uparrow +2e^- = 2H^-$	−2.25
$2NO_3^- + 4H^+ + 2e^- = N_2O_4 \uparrow +2H_2O$	+0.80	$Mg^{2+} + 2e^- = Mg \downarrow$	−2.37
$Ag^+ + e^- = Ag \downarrow$	+0.7994	$Na^+ + e^- = Na \downarrow$	−2.7
$Hg_2^{2+} + 2e^- = 2Hg$	+0.792	$Ca^{2+} + 2e^- = Ca \downarrow$	−2.87
$Fe^{3+} + e^- = Fe^{2+}$	+0.771	$Sr^{1-} + 2e^- - Sr \downarrow$	−2.89
$O_2 \uparrow +2H^+ + 2e^- = H_2O_2$	+0.69	$Ba^{2-} + 2e^- = Ba \downarrow$	−2.90
$H_3AsO_4 + 2H^+ + 2e^- = HAsO_2 + 2H_2O$	+0.56	$K^+ + e^- = K \downarrow$	−2.925
$I_3^- + 2e^- = 3I^-$	+0.545	$Rb^+ + e^- = Rb \downarrow$	−2.93
$I_2^- + 2e^- = 2I^-$	+0.536	$Li^+ + e^- = Li \downarrow$	−3.03

The potential of these half-cell reactions can be determined by comparison with the hydrogen electrode[2] all under standard conditions, which is unit concentration for ions in solution and 1 atm pressure for gases at 25°C. The value of these standard electrode reduction potentials is given in Table 9.4. The standard cell potential of reaction (9.15) is

$$\mathcal{E}^o Zn/Zn^{2+} = -\mathcal{E}^o Zn^{2+}/Zn = 0.763 \text{ V}, \quad \mathcal{E}^o Cu^{2+}/C_U = 0.337 \text{ V}$$

and

$$\mathcal{E}^o{}_{cell} = 0.763 + 0.337 = 1.100 \text{ V}$$

---

[2] The standard potential of the hydrogen electrode (SHE) is defined as zero at all temperatures for $[H^+]aq = 1$ M and P $(H_2) = 1$ atm.

Fig. 9.7 The zinc–copper Daniell cell where the zinc dissolves at the anode (−) and copper is plated out at the cathode (+)

This is represented in shorthand notation as

$$Zn\big|Zn^{2+}(1\ M)\big\|Cu^{2+}(1\ M)\big|Cu \qquad\qquad (9.17)$$

where the single | line indicates a phase change and the double ‖ line represents a salt bridge or connection between the two half-cells. This is illustrated in Fig. 9.7.

The simplest cell is one in which the electrode material is the same in each half-cell, but the metal ion concentration is different in the two half-cells. This is called a *concentration cell*, an example of such a cell is

$$Cu\big|Cu^{2+}(0.01\ M)\big\|Cu^{2+}(0.10\ M)\big|Cu$$

The cell potential is

$$\mathcal{E}_{cell} = \mathcal{E}^{\circ}{}_{cell} - \frac{RT}{nF}\ \ln\frac{0.01}{0.10}, \quad \mathcal{E}_{cell} = 0 - \frac{0.2568}{2}\ \ln 0.10, \quad \mathcal{E}_{cell} = 0.0296\ V$$

The cell will continue to develop a potential until the concentration of Cu^{2+} in the two half-cells becomes equal.

The concentration inequality adjacent to metals is the basis of the corrosion of metals which is discussed in the next chapter. All redox reactions can be divided into two or more half-cells which can be combined into a full-cell. The voltage generated and the current which can be drawn determine its usefulness as a battery.

9.6 Batteries and Cells

The means by which chemical energy is stored and converted into electrical energy is called a *battery* or *cell*. We saw how the Daniell cell composed of a zinc electrode immersed in a 1 M zinc sulfate solution and a copper electrode dipping into a 1 M copper sulfate

$$Zn|ZnSO_4(1\ M)||CuSO_4(1\ M)|Cu \tag{9.18}$$

will develop 1.10 V.

This cell, however, has limited use because as soon as current is drawn from the cell, the voltage drops because of polarization which is primarily caused by the buildup of hydrogen at the copper electrode. This polarization can be minimized by the addition of a depolarizer which supplies oxygen readily and so removes the hydrogen from the electrode to form water. Such a cell, nonetheless, has a limited lifetime and restricted use.

Batteries are classified as primary or secondary. Batteries which are not rechargeable are referred to as primary batteries. Secondary batteries are rechargeable either by an electrical current or by a replacement of the electrode material (anode).

9.6.1 Primary Batteries

A dry cell, which has become a primary power source for transistorized electronic equipment, was developed over a hundred years ago in 1865, is called the *Leclanché* cell.

It consists of a zinc negative electrode, which acts as the container and which is slowly oxidized; a porous carbon rod as the positive electrode, which takes no part in the overall reaction but can act as a gas vent for the cell; and an electrolyte, which is ammonium chloride to which is added manganese dioxide. The MnO_2 acts as a depolarizer and is reduced at the carbon electrode by the following reaction:

$$MnO_2 + 4H^+ + 2e^- \rightarrow Mn^{2+} + 2H_2O \tag{9.19}$$

This is complicated by further reactions:

$$Mn^{2+} + MnO_2 + 2OH^- \rightarrow 2MnOOH \tag{9.20}$$

$$Mn^{2+} + MnO_2 + 4OH^- + Zn^{2+} \rightarrow ZnOMn_2O_3 + 2H_2O \tag{9.21}$$

The zinc ion is then complexed by the ammonia as follows:

$$Zn^{2+} + 4NH_3 \rightleftharpoons \left[Zn(NH_3)_4\right]^{2+} \tag{9.22}$$

The cell voltage is between 1.5 and 1.6 V. A D-type battery, commonly used in flashlights, has a capacity of about 4 amp-h. Although this type of cell contains about 20 g of zinc, only about 5 g is used. The overall reaction, although complex, can be represented as follows:

$$Zn + 4NH_4Cl + MnO_2 \rightarrow MnCl_2 + [Zn(NH_3)]Cl_2 + 2H_2O \tag{9.23}$$

Because the reduction reaction of MnO_2 is not a well-defined reaction, the Nernst equation cannot be applied successfully, and the cell voltage changes unpredictably with time and discharge condition.

Fig. 9.8 Plots showing the loss in capacity with storage time for the zinc–carbon, manganese–zinc–alkaline, and mercury batteries

The shelf life of this battery is rather poor, as shown in Fig. 9.8, because of the slow loss of water and because of side reactions due to impurities in the MnO_2 ore commonly used. When specially purified MnO_2 is used, the performance of the battery is greatly improved. The zinc–carbon dry cell is considered the workhorse of the battery industry: It provides power at a very low cost.

Another popular dry cell that is commonly associated with transistorized electronic components is the manganese–zinc alkaline cell. It utilizes refined MnO_2, as does the improved zinc–carbon cell, but in large excess. The electrolyte is 40% KOH presaturated with zinc (ZnO) to prevent the zinc electrode from dissolving while in storage. A steel can, instead of the zinc electrode, usually serve as the container; hence, the cell is highly leak resistant. However, the cost of the cell is about twice that of the zinc–carbon cell.

The most efficient dry cell is the mercury cell, which has an excellent stabilized voltage. Developed in 1942 by Ruben, it consists of zinc, which dissolves and becomes the negative electrode and mercuric oxide, which is reduced to mercury at the positive electrode. The overall cell, which has no salt bridge, can be represented as follows:

$$Zn|ZN(OH)_2|40\% \text{ KOH}|HgO|Hg \tag{9.24}$$

One half-cell reaction is

$$Zn \rightleftharpoons Zn^{2+} + 2e^-, \qquad \mathcal{E}°_{zn^{2+}/Zn} = -0.763 \text{ V} \tag{9.25}$$

This is followed by the reaction

$$Zn^{2+} + 2OH^- \rightleftharpoons Zn(OH)_2 \tag{9.26}$$

for which $K_{sp} = 4.5 \times 10^{-17}$.

The other half-cell reaction is

$$Hg^{2+} + 2e \rightleftharpoons Hg, \qquad \mathcal{E}°_{Hg2+/Hg} = 0.085 \text{ V} \tag{9.27}$$

This is preceded by the reaction

$$HgO + H_2O \rightarrow Hg(OH)_2 \rightleftharpoons Hg^{2+} + 2OH^- \tag{9.28}$$

Fig. 9.9 Typical discharge curves of voltage plotted against time for four different types of cells of comparable size (two AA penlight cells discharged at 250 mA)

for which $K_{sn} = 1.7 \times 10^{-26}$. The cell emf is given as follows:

$$\mathcal{E}_{cell} = \mathcal{E}_{Zn/Zn^{2+}} + \mathcal{E}_{Hg^{2+}/Hg} = -\left(\mathcal{E}_{Zn^{2+}/Zn}\right) + \mathcal{E}_{Hg^{2+}/Hg}$$

$$\mathcal{E}_{cell} = -\left(\mathcal{E}^{o}_{Zn} - \frac{RT}{n\mathcal{F}} \ln \frac{1}{[Zn^{2+}]}\right) + \left(\mathcal{E}^{o}_{hg} - \frac{RT}{n\mathcal{F}} \ln \frac{1}{[Hg^{2+}]}\right) \tag{9.29}$$

However,

$$[Zn^{2+}] = \frac{Ksp(Zn(OH)_2)}{[OH-]_2} \text{ and } [Hg^{2+}] = \frac{Ksp(Hg(OH)_2)}{[OH-]_2} \tag{9.30}$$

$$\mathcal{E}_{cell} = -\left(\mathcal{E}^{o}_{Zn} - \frac{0.02568}{2} \ln \frac{[OH-]^2}{Ksp(Zn(OH)_2)}\right) + \left(\mathcal{E}^{o}_{Hg} - \frac{0.02568}{2} \ln \frac{[OH-]^2}{Ksp(Hg(OH)_2)}\right)$$

$$= -\mathcal{E}^{o}_{Zn} + \mathcal{E}^{o}_{Hg} + \frac{0.02568}{2} \ln \frac{Ksp(Hg(OH)_2}{Ksp(Zn(OH)_2}$$

$$= 0.763 + 0.850 + \frac{0.02568}{2} \ln \frac{1.7 \times 10^{-26}}{4.5 \times 10^{-17}}$$

$$= 1.613 + \frac{0.02568}{2} \ln 3.8 \times 10^{-10}$$

$$\mathcal{E}_{cell} = 1.613 - 0.278 = 1.335 \text{ V}$$

This value compares favorably with the actual cell voltage of about 1.35 V. Thus, the cell potential is independent of the concentration of the electrolyte, [OH⁻]. The cell has a low internal resistance and has a very long shelf life when compared to the other two dry cells discussed previously, as shown in Fig. 9.8. For example, after a 3-year storage, a typical cell voltage changed from an initial value of 1.357 to 1.344 V; that is, there was about a 1% change. Thus, the use of the mercury dry cell as a reference voltage is widespread. A comparison of cell voltage of the three types of dry cells with time during constant current drain is shown in Fig. 9.9. The remarkable constancy in voltage of the mercury cell in contrast to the sharp voltage drop in the zinc–carbon and manganese–zinc alkaline cells is obvious. The mercury cell, although initially about three times more expensive than the zinc–carbon cell, has a lower operating cost per hour than either of the other two dry cells. Therefore, it is not too difficult to understand why the mercury cell is being used increasingly as a convenient source of power and reference voltage.

Also shown in Fig. 9.9 is the voltage curve for the zinc–air cell. The cell consists of an anode of amalgamated zinc powder in contact with the electrolyte, which is concentrated potassium hydroxide, and a cathode of metal mesh, which is a catalyst for the conversion of oxygen to the hydroxide ion. The half-reactions are

$$Zn \rightarrow Zn^{2+} + 2e^- \tag{9.31}$$

$$O_2 + 2H_2O + 4e^- \rightarrow 4OH^- \tag{9.32}$$

and the overall reaction is

$$2Zn + O_2 + 2H_2O + 4OH^- \rightarrow 2Zn(OH)_4^{2-} \tag{9.33}$$

The cell is encased in a porous polymer that allows oxygen from the air to diffuse to the cathode but does not allow the electrolyte to leak out. The shelf life is almost indefinite when the cell is stored in an airtight container. The cell is used to best advantage when continuous high currents are required for a short period of time, since it cannot be left in contact with air without losing capacity. Resealing the cell or cutting off the air supply to the cell when it is not in use extends the life during intermittent use. The catalytic cathode for the zinc–air cell is a direct development from work on fuel cells, which are discussed later.

In contrast to the zinc–air battery, the lithium–iodine solid LiI electrolyte battery will last for almost 15 years. A 120-mAh battery with an initial voltage of 2.8 V drops to 2.6 V when discharged continuously at about 1 μA. The cell is written as

$$Li(s)|LiI(s)|P2VP.nI_2(s)$$

where P2VP.nI$_2$ is a complex between poly-2-vinylpyridine (P2VP) and iodine. The reaction is given as

$$Li(s) + \tfrac{1}{2}I_2(s) \rightarrow LiI(s) \tag{9.34}$$

This type of cell is highly reliable, and it is commonly used in cardiac pacemaker batteries which are implanted.

The temperature coefficient of a cell's potential is determined by the change in free energy, ΔG, with temperature and is given by

$$\frac{d\varepsilon^\circ}{dT} = \frac{\Delta S^\circ}{n\mathcal{F}} \tag{9.35}$$

where ΔS° is the standard entropy change for the reaction.

9.6.2 Secondary Batteries

The most common secondary battery is the lead storage battery, which has as an essential feature and ability to be recharged. The cell consists of a lead plate for the negative electrode, separated by a porous spacer from the positive electrode, which is composed of porous lead dioxide. The electrolyte is sulfuric acid—about 32% by weight. The electrode reactions are as follows:

Negative electrode

$$
\begin{aligned}
Pb &\rightleftharpoons Pb^{2+} + 2e^- \\
\underline{Pb^{2-} + SO_4^{2-}} &\underline{\rightleftharpoons PbSO_4} \\
Pb + SO_4^{2-} &\rightleftharpoons PbSO_4 + 2e^-
\end{aligned}
\tag{9.36}
$$

Positive electrode

$$
\begin{aligned}
PbO_2 + 4H^+ + 2e^- &\rightleftharpoons Pb^{2+} + 2H_2O \\
\underline{Pb^{2+}SO_4^{2-}} &\underline{\rightleftharpoons PbSO_4} \\
PbO_2 + 4H^+ + SO_4^{2-} &= PbSO_4 + 2H_2O
\end{aligned}
\tag{9.37}
$$

The net overall reaction is as follows:

$$
PbO_2 + Pb + 2H_2SO_4 \underset{\longleftarrow}{\overset{\text{discharge}}{\longrightarrow}} 2\mathcal{F} \quad 2PbSO_4 + 2H_2O
\tag{9.38}
$$

Polarization by hydrogen is minimized by the PbO_2 electrode, which is also a depolarizer. The discharge of the battery consumes acid and forms insoluble lead sulfate and water; that is, the density of the solution decreases from about 1.28 g/cm^3 in the fully charged condition to about 1.1 g/cm^3 in the discharged state. The overall open cell voltage (when no current is being drawn) depends on the acid concentration (i.e., SO_4^- ion concentration, which in turn controls the concentration of the Pb^{2+} ion via the K_{sp} for $PbSO_4$). The voltage varies from 1.88 V at 5% H_2SO_4 by weight to 2.15 V at 40% acid by weight. The conductivity of aqueous H_2SO_4 is at a maximum when H_2SO_4 is about 31.4% by weight at 30°C (or 27% at −20°C); it is best to control concentration in this range since the internal resistance of the battery is at a minimum. Another factor influencing the choice of the acid concentration is the freezing point of the sulfuric acid solution; thus, in cold climates, a higher acid level (38% H_2SO_4 by weight, specific gravity 1.28) is required in order to minimize the possibility of the electrolyte freezing at the relatively common temperature of −40°C. The amount of lead and lead dioxide incorporated into the electrodes is three to four times the amount used in the discharging process because of the construction of the electrodes and the need for a conducting system that makes possible the recharging of a "dead" battery.

In 1988, a collection of 8,256 lead–acid batteries was used by a California electric power plant to store energy and to deliver it during peak power demands, that is, load leveling. The batteries contained over 1,800 tonnes of lead and could supply 10 MWe for 4 h, enough to meet the electrical demands of 4,000 homes. The efficiency of the system was rated at 75%.

The capacity of a battery is rated in terms of amp-hours and depends on the rate of discharge and, even more significantly, on the temperature. For example, a battery with a rating of 90 amp-h at 25°C has a rating of about 45 amp-h at −12°C and about 36 amp-h at −18°C. The lead–acid battery in the fully discharged state slowly loses capacity since the lead sulfate recrystallizes, and some of the larger crystals are then not available for the reverse charge reaction. When this happens, the battery is said to be sulfonated. This can be remedied by the process of removing the "insoluble" sulfate, recharging the battery, and reconstituting the acid to the appropriate specific gravity.

In respect to this property as well as others, the nickel–alkaline battery is superior to—although about three times more costly than—the lead–acid battery. There are two types of nickel–alkaline storage batteries: the Edison nickel–iron battery and the nickel–cadmium battery.

In the Edison battery, the cell can be represented as follows:

$$\text{Steel} \mid Ni_2O_3; \;\; Ni(OH)_2 \mid KOH(aq \; 20\%) \mid Fe(OH)_2 Fe \mid Steel$$

The overall reaction is as follows:

$$Ni_2O_3 + 3H_2O + Fe \underset{\text{charge}}{\overset{\text{discharge}}{\rightleftarrows}} 2Ni(OH)_2 + Fe(OH)_2 \tag{9.39}$$

The cell potential has an average value of 1.25 V. Although the overall reaction does not apparently involve the electrolyte, the KOH does in fact participate in each of the half-cell reactions. Although the Edison battery is designed and suitable for regular cyclic service, the efficiency of charge is only 60%; thus, it has now been almost completely replaced by the more efficient (72%) nickel–cadmium battery, which is itself inferior in energy efficiency to the lead–acid battery with an efficiency of about 80%.

In the nickel–cadmium alkaline storage battery, the iron of the Edison cell is replaced by cadmium to give the following equivalent reaction:

$$Ni_2O_3 + 3H_2O + Cd \underset{\text{charge}}{\overset{\text{discharge}}{\rightleftarrows}} 2Ni(OH)_2 + Cd(OH)_2 \tag{9.40}$$

The average cell voltage of 1.2 V is slightly lower than that of the Edison cell. Cadmium is preferred to iron in the nickel–alkaline cell because cadmium hydroxide is more conductive than iron hydroxide. The absence of higher oxidation states for cadmium minimizes side reactions, which occur in the Edison cell. The nickel–cadmium cell can also be charged at a lower voltage since there is no overvoltage, as there is at the iron electrode.

One major disadvantage of the nickel–alkaline battery is the alkaline electrolyte, which picks up CO_2 from the atmosphere and must therefore be replaced periodically. However, the advantages of the nickel–cadmium cell over the lead–acid battery are numerous; some of these are as follows:

1. The freezing point of the KOH electrolyte is low (about $-30°C$) regardless of the state of charge.
2. The capacity does not drop as sharply with drop in temperature.
3. The cell can be charged and discharged more often and at higher rates (without gassing) and thus has a longer useful life.

The storage battery has become an accepted source of power in our modern technological world, and in an environment-conscious society, the storage battery will play an ever increasing role.

Table 9.5 Values of standard cell voltages of selected fuel cell reactions at 25°C

Reaction	$\mathcal{E}^o_{cell}$ (V)
$2C + O_2 \rightarrow 2CO$	0.70
$C + O_2 \rightarrow CO_2$	1.02
$CH_4 + 2O_2 \rightarrow CO_2 + 2H_2O$	1.04
$C_3H_8 + 5O_2 \rightarrow 3CO_2 + 4H_2O$	1.10
$4NH_3 + 3O_2 \rightarrow 2N_2 + 6H_2O$	1.13
$CH_3OH + 3/2O_2 \rightarrow CO_2 + 2H_2O$	1.21
$H_2 + \frac{1}{2}O_2 \rightarrow H_2O(l)$	1.23
$2CO + O_2 \rightarrow 2CO_2$	1.33
$N_2H_4 + O_2 \rightarrow N_2 + 2H_2O$	1.56
$2Na + H_2O + \frac{1}{2}O_2 \rightarrow 2NaOH$	3.14

9.7 Fuel Cells

The discovery of the fuel cell followed soon after Faraday developed his laws of electrolysis. In 1839, Grove showed that the electrolysis of water was partially reversible. Hydrogen and oxygen formed by the electrolysis of water were allowed to recombine at the platinum electrodes to produce a current or what appeared to be "reverse electrolysis." Using the same fundamental principles but somewhat more advanced technology, Bacon in 1959—after about 20 years of intensive effort—produced a 6 kW power unit that could drive a small truck.

It was recognized early that the overall thermodynamic efficiency of steam engines is only about 15%. The efficiency of modern electrical generators is about 20–50%, whereas the efficiency of the fuel cell (in which there is direct conversion of chemical energy into electrical energy) does not have any thermodynamic limitation. Theoretically, the efficiency of the fuel cell can approach 100%, and in practice, efficiency of over 80% can be achieved.

Interest in the fuel cell has increased remarkably in the last decade primarily because of (1) the high efficiency associated with the energy conversion, (2) the low weight requirement essential for satellite and spacecraft power sources that is readily satisfied with hydrogen as a fuel, and (3) the recent requirement of a pollution-free power source.

Any redox system with a continuous supply of reagents is potentially a fuel cell. Some reactions that have been studied are given in Table 9.5 with the corresponding theoretical $\mathcal{E}^\circ_{cell}$ values, which are calculated from thermodynamic data ($\Delta G^\circ = -n\mathscr{F}\mathcal{E}^\circ$). The temperature coefficients of the $\mathcal{E}^\circ_{cell}$ values of some of the reactions in Table 9.5 are shown in Fig. 9.10.

In practice, the suitability of a reaction system is determined by the kinetics of the reaction, which depends on temperature, pressure of gases, electrode polarization, surface area of electrodes, and presence of a catalyst. A fuel cell that is thermodynamically and kinetically feasible must be considered from an economic viewpoint before it is accepted. Thus, since hydrogen, hydrazine, and methanol are too expensive for general application, their use in fuel cells has been limited to special cases. Hydrogen has been used for fuel cells in satellites and space vehicles, in which reliability and lightness are more important than cost. Hydrazine fuel cells have been used in portable-radio power supplies for the United States Army because of their truly silent operation. Methanol fuel cells have been used to power navigation buoys and remote alpine television repeater stations because such power systems are comparatively free from maintenance problems over periods of a year or more. The polarization at the electrodes of a fuel cell is the most important single factor that limits the usefulness of the cell. The various polarization characteristics for a typical fuel cell are plotted separately as a function of current density in Fig. 9.11.

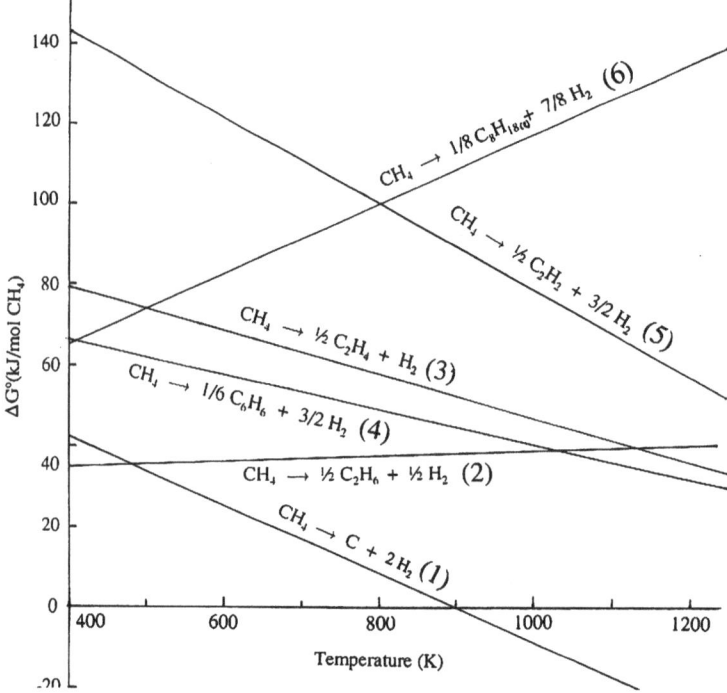

Fig. 9.10 Effect of temperature on the cell voltage, $\mathcal{E}°_{cell}$, for some fuel cell reactions. *Note:* Slope is related to the entropy change of the reaction

Fig. 9.11 Operating characteristics of a typical fuel cell. Net polarization is given by $\eta total = \eta anodic + \eta cathodic + \eta ohmic + \eta conc$

Fig. 9.12 Bacon hydrogen–oxygen fuel cell with gas-diffusion electrodes

The most successful fuel cell to date is the hydrogen–oxygen fuel cell, which deserves special attention since it has been used in the Apollo and Gemini space flights and moon landings. The reaction

$$H_2(gas) \rightleftharpoons 2H(solid) \rightleftharpoons 2H^+(solution) + 2e^- \qquad (9.41)$$

occurs at the gas $\rightleftharpoons$ solid interface.

$$\text{liquid}$$

To facilitate the rapid attainment of equilibrium, a liquid gas-diffusion electrode was developed whereby concentration polarization could be minimized. The ohmic polarization (the RI drop between the electrodes, which gives rise to an internal resistance) is also minimized when the anode-to-cathode separation is reduced. The apparatus of the hydrogen–oxygen fuel cell developed by Bacon with gas-diffusion electrodes is shown in Fig. 9.12. The operating temperature of 240°C is attained with an electrolyte concentration of about 80% KOH solution, which with the high pressures of about 600 psi for H_2 and O_2, allows high current densities to be drawn with relatively low polarization losses. Units such as these with power of 15 kW have been built and used successfully for long periods.

Today, fuel cells are still in the development stage, and much further work must be done before an efficient economical fuel cell is produced. The oxidation of coal or oil to CO_2 and H_2O has been achieved in a fuel cell; the system uses platinum as a catalyst and an acid electrolyte at high temperature, and thus the cost of materials for the cell construction is very high. The economic fuel cell-powered automobile, although a distinct possibility, is not to be expected in the immediate future.

9.8 Hybrid Cells

The hybrid cell is one which is not rechargeable by simply reversing the voltage. Some of these use oxygen in air as the cathode material

$$O_2 + 4H_2O + 4e^- \rightarrow 4OH^- \qquad (9.42)$$

Table 9.6 Some properties of selected metal–air batteries

Battery	Electrolytes	Cell volt (V)	Energy density (Wh/kg)		Peak power (W/kg)	Cycle life	Comments
			Theoretical	Actual			
Lithium–air $2Li + \frac{1}{2}O_2 \rightarrow Li_2O$	LiOH	2.9	11.148	290		Mechanical	Unlikely to be developed for commercial use. High Li costs
Aluminum–air $2Al + 3O_2 + 3Al_2O_3$	NaOH	2.71	8.081	440		Mechanical rechargeable	Prototype developed and tested. Good energy density, low cost
Magnesium–air $2Mg + O_2 \rightarrow 2MgO$	NaCl	3.09	6.813			Mechanical rechargeable	No advantage over Al, not being considered seriously at present

and a metal, for example, Al, as the anode material

$$Al \;\rightarrow\; Al^{3+} + 3e^- \tag{9.43}$$

Such systems are called *metal–air batteries* and are mechanically rechargeable (anode metal is replaced). Such batteries have only recently become practicable due to the developments of the O_2 – electrode in fuel cells. Some characteristics of selected metal–air batteries are given in Table 9.6.

The aluminum–air battery has recently received some attention as a result of work done by the Lawrence Livermore National Laboratory. It was estimated that a 60-cell system with 230 kg of aluminum can power a VW for 5,000 km before requiring mechanical recharging. Periodic refill with water and removal of $Al(OH)_3$ would be required after 400 km. The conversion of the $Al(OH)_3$ back to Al at an electrolytic refinery completes the recycling process. In 1986, an Al–air battery producing 1,680 W was shown to power an electric golf cart for 8 h.

A battery where the active components are flowed past electrodes in a cell with two compartments separated by an appropriate membrane is called a *flow battery*. One such battery is the Fe/Cr redox system

cathode

$$Cr^{2+} \rightarrow Cr^{3+} + e^- \qquad \mathcal{E}^\circ = -0.410 \text{ V.}$$

and

anode

$$\frac{Fe^{3+} + e^- \rightarrow Fe^{2+} \qquad \mathcal{E}^\circ = 0.771 \text{ V}}{Cr^{2+} + Fe^{3+} \rightarrow Cr^{3+} + Fe^{2+} \quad \mathcal{E}^\circ = 1.181 \text{ V}} \tag{9.44}$$

The overall voltage is given by

Fig. 9.13 Schematic diagram
of a redox cell (battery) using
Cr^{2+} and Fe^{3+} aqueous
solutions as reactants

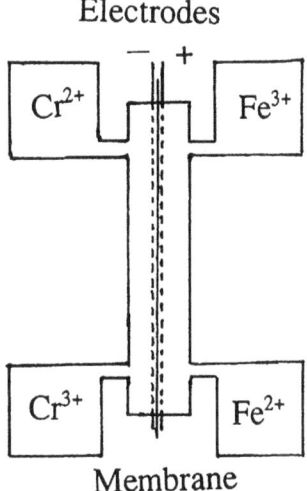

$$\mathcal{E}^{o} = \mathcal{E}^{o}_{cell} - 0.059 \log K \tag{9.45}$$

$$\mathcal{E}^{o}_{cell} = 1.181 - 0.059 \log \frac{[Fe^{2+}][Cr^{3+}]}{[Fe^{3+}][Cr^{2+}]} \tag{9.46}$$

The reactant solutions Cr^{2+} and Fe^{3+} are reacted as shown in Fig. 9.13.

The product solutions are kept separate, and the Fe^{2+} can be oxidized by air back to Fe^{3+}, whereas the Cr^{3+} can be electrolytically reduced back to Cr^{2+}. An Australian redox flow battery has been described which uses vanadium both as oxidant and reductant in the following reactions:

$$\text{Negative V(III)} + e \xrightleftharpoons[\text{discharge}]{\text{charge}} V(II) \tag{9.47}$$

$$\text{Positive V(1V)} \xrightleftharpoons[\text{discharge}]{\text{charge}} V(V) + e \tag{9.48}$$

The electrolyte is 2 M $VOSO_4$ in 2 M H_2SO_4 with graphite plates acting as electrodes to collect the current. The open circuit voltage (OCV) is 1.45 V with a 95% charging efficiency and little or no H_2 or O_2 evolution. One of the major advantages of this battery is that if the membrane leaks, then the separation of the two flow streams is not necessary as in the Fe/Cr system. Several such redox cells are available and being studied primarily as potential power sources for the electric vehicle (EV) which will most assuredly be a reality in the near future.

9.9 Electric Vehicle

The first EV was built in 1839 by Robert Anderson of Aberdeen, Scotland. The first practical one was a taxi introduced in England, 1886, which had 28 bulky batteries and a top speed of 12.8 km/h. By 1904, the electric vehicle was common throughout the world, but its production peaked at about 1910

Table 9.7 A comparison of performance and cost goals for a practical EV battery with those of deep-cycling industrial lead–acid battery

Parameter	Goal	Lead–acid
Cost ~ \$/kWh	45	90
Life (cycles)	1,000	700
Life (years)	10	5
Energy efficiency $\left(\dfrac{\text{discharge}}{\text{charge energy}}\right)$	0.80	0.65
Charge time (h)	1–6	6–8
Discharge time (h)	2–4	2–4
Energy density (Wh/kg)	140	35
Power density—peak (Wh/kg)	200	80
Power density—sustained (Wh/kg)	70	30
Volume density (Wh/L)	200	50
Typical size (kWh)	20–50	20–40

when the self-starting gasoline-powered internal combustion engine began to dominate. This was due to the availability of cheap gasoline and mass-produced cars.

However, the EV is due to make a comeback because of the rising cost of gasoline and diesel fuel, the pollution of the environment, and the prevalence of a second small car in most families. Some of the major American automobile producers have been planning to have an EV on the market for the past 30 years. The major stumbling block is the batteries which must be reliable, lightweight, take hundreds of full discharges and recharges, and be inexpensive as well. The desirable features of an ideal battery are compared in Table 9.7 with the lead–acid battery still used at present in EV. A list of possible batteries and some of their properties are given in Table 9.8. The use of the fuel cell and hybrid fuel cell type power sources must also be included.

The choice of batteries available for an EV is both expanding in number and narrowing in type. Several batteries listed in Table 9.8 are being given commercial pilot production tests. It must be recognized that winter restricts the choice or design of a suitable system for cold-climate regions. Recent tests in Winnipeg, Canada, of a US-made EV using lead–acid batteries showed it to be appropriate in summer (about 80 km/charge), but in winter, the lower capacity resulted in less than 8 km/charge. This could undoubtedly be corrected by an integrated design. Since the batteries are only about 70% efficient on charge, the excess energy (heat) could be stored by insulating the batteries or adding a heat-storing medium such as Glauber's salt (see Chap. 1) between the batteries and the insulation. This, however, adds both weight and volume to the system.

The modern design EV will be lightweight and have minimal aerodynamic drag and rolling resistance, efficient motor control system, and transmission as well as regenerative (battery charging) braking. The usual goal of EV is a range of about 100 km, a maximum speed of 90 km/h, and a cruise speed of 45 km/h, with a recharge time of 8–10 h. It would be interesting to speculate that as the EV becomes common and recharging is performed at night, the resulting power drain may invert the peak load, that is, the greater load would occur overnight. The low vehicle emissions set for California are readily met by the EV, and major automobile manufacturers are striving to meet the demand.

One example is the Mercedes-Benz 5-seater 190 Electro car which develops up to 32 kW (44 hp), has a maximum speed of 115 km/h and an operating range of 150 km. The sodium–nickel chloride batteries were chosen over nickel–cadmium and sodium–sulfur alternatives. The car is shown in Fig. 9.14.

A second example is the use of a hydrogen fuel cell to run a bus. Using a Proton Exchange Membrane Fuel Cell (PEMFC), Ballard of Vancouver has built a prototype bus for Chicago Transit Authority. The bus stores hydrogen at high pressure in cylinders on the roof of the bus—enough to give the bus a 560-km range (see Fig. 9.15). Designs have been developed for a more compact

Table 9.8 Evaluation and characteristics of electric vehicles batteries (values depend on source)

Battery and reaction electrolyte Discharge	Electrolyte	Temp (°C)	Cell voltage (V)	Energy density Mass Theoretical (Wh/kg)	Actual (Wh/kg)	Vol. Actual (Wh/L)	Power peak (W/kg)	Cycle life	Depth charge (%)	Charge efficient (%)	Cost initial ($/kWh)	Projected	Advantage	Disadvantage
Improved lead-acid °Pb + PbO$_2$+H$_2$SO$_4$ →2PbSO$_2$+H$_2$O	H$_2$SO$_4$	0–40	2.05	171/30		30/46	50–100	500	90–60	65	90	50	Now available	Low specific energy marginal peak power limited to about 100 miles (150 km)
Nickel–zinc °Zn+°2NiO (OH) →2H$_2$O+2Ni (OH)$_2$	KOH 2H$_2$O	−45+40	1.7	321/66		66/140	150	400	65	65	150	50	Excellent power and volume density. (UK prospect)	Solubility of Zn in KOH shortees shelf life. High Ni cost
Nickel–iron °Fe+2NiO (OH)+ 2H$_2$O→Zn (OH)+Fe(OH)$_2$	KOH	−40+40	1.37	267/55		55/100	75	1,500	90	60	120	65	Developed in 1901 by Edison, now mosure and rugged	Low power density H, garing during charge. Poor peak power at low temperature
Zinc–chlorine °Zn+Cl$_2$ 6H$_2$O- 2nCl$_2$+6H$_2$O	ZnCl$_2$	<9.0+40	2.12	465/100		100/60	80	400	–	65	100	35	Inexpensive lightweight materials, high power density	Complex system requiring refrigeration and heating, Cl$_3$ hazard
Sodium–sulfur xNa+YS→Na, Sy x = Z, y = 53	Al$_2$O$_3$	350	1.76–2.08	664/150		150/160	200	300	60	85	50	35	Inexpensive light material, high energy density	Short life due to seals and corrosion, Na hazard. High temperature requirement
Lithium–sulfur °2Li+S→Li$_2$S	LiCl/KC	450	22	2567									Improved with Li/ Al alloy (~5 at. % Li)	Too reactive, Li cost high

Fig. 9.14 Mercedes-Benz zero emission class A EV. The car uses a 40 kW (54 hp)—three-phase induction motor developing a rated torque of 155 Nm which can accelerate the car to 100 km/h in 17 s with a top speed of 120 km/h and a normal usage range of 150 km. Recharging can be made in 6–12 h using normal household sockets. The battery system is sodium/nickel chloride with an energy storage capacity of 100 Wh/kg and a life of over 100,000 km

(volume of 32 L) stack of fuel cells which delivers 32.3 kW and intended for a small passenger vehicle.

General Motors has announced (February 1996) the start of a mass-produced EV powered by lead–acid batteries and which will have a range of about 110 km per charge and a maximum speed of 100 km/h. The estimated cost is expected to be US$35,000.

It is obvious that much experimental work remains to be done before a reliable and economical EV will be on the market. The incentive—lower fuel costs and a cleaner environment for the EV in comparison to the ICE—will not diminish with time.

Exercises

1. What is a concentration cell?
2. What is a redox cell?
3. Distinguish between primary and secondary batteries with examples.
4. What factors determine the voltage of a cell and its temperature coefficient?
5. Explain how the capacity of a storage battery is measured.
6. Explain why the capacity of the lead–acid battery drops so rapidly with decrease in temperature.
7. What factors influence the drop in cell voltage of a battery when current is withdrawn?

Fig. 9.15 (**a**) The Ballard H_2 fuel cell-powered bus and (**b**) the FirstEnergy fuel cell unit (Ballard Power)

8. The voltage of the H_2–O_2 fuel cell is given by $\Delta G° = -n\mathscr{F}\mathscr{E}°$ where n is the number of electrons transferred in the reaction, $\mathscr{F}$ is the Faraday, and ΔG is the standard free energy change in the reaction. Show that $\varepsilon° = 1.23$ V from thermodynamic data.

9. Explain why in the electrolysis of water into H_2 and O_2, the voltage required is greater than 1.23 V.

10. (a) Calculate the voltage of the methanol–oxygen fuel cell.
    (b) Explain why the voltage will be lower if air was used instead of pure oxygen.
    (c) Calculate the standard voltage for the methanol–air fuel cell.

11. The approximate energy consumption for an EV is 0.16 kWh/km to 0.32 kWh/mile. If electrical energy is priced at 3 ¢/kWh, calculate the energy cost/km for an EV.

12. The minimum energy required to dissociate H_2O into H_2 and O_2 is 1.23 V. It is possible to use a carbon anode and produce CO_2 instead of O_2. Calculate the minimum cell potential for such a cell.

$$H_2O_{(1)} \rightarrow H_{2(g)} + \tfrac{1}{2}O_{2(g)} \qquad \Delta G^\circ = 237.2 \text{ kJ}$$

$$C_{(s)} + O_{2(g)} \rightarrow CO_{2(g)} \qquad \Delta G^\circ = -394.4 \text{ kJ}$$

Note: The actual voltage required is from 0.85 to 1.0 V. Give some reasons for the discrepancy between the calculated and actual values (Fuel 58 705 (1979)).

13. Consider an automobile with an internal combustion engine running on hydrogen which can somehow be stored for a modest run of 50 km. The hydrogen is prepared by the electrolysis of water (in one cell at about 2.5 V) during the night (from 10:00 p.m. to 7:00 a.m.). What would be the required current? Make the following assumptions: (1) heat of combustion of gasoline and hydrogen is given in Table 6.8; (2) the automobile has the efficiency of the new generation of cars, namely, 14 km/L gasoline; and (3) the efficiency of the hydrogen-driven vehicle is related by its comparable heat of combustion to the gasoline efficiency.
    Note: Do not be surprised by the large current that is necessary.

14. An ICE vehicle has been described running on H_2 produced on board the vehicle by the reaction of an Al wire with KOH/H_2O. Assume that the vehicle is the same one as in Exercise 9.14. Calculate the mass of Al required for the 50-km trip.

15. Calculate the standard cell potential for the Ti/Fe flow battery where

$$\mathcal{E}^\circ_{\text{Ti4+/Ti3+}} = 0.04 \text{ V}$$

$$\mathcal{E}^\circ_{\text{Fe3+/Fe2+}} = 0.771 \text{ V}$$

16. Calculate the RI drop of a lead–acid battery (in which the internal resistance is 0.01052 Ω) when it cranks an engine drawing 200A.

17. Estimate the amount of Glauber's salt needed to keep a lead–acid battery pack from cooling to 0°C if the outside temperature drops to −40°C. The excess heat for charging the 40-kW power supply will be insulated with 8 cm of Styrofoam. Assume that the volume of the battery pack is 0.3 m^3, its area is 2.5 m^2, and its heat capacity is 3 kJ/K. The thermal conduction of the polystyrene foam is 0.0003 J s^{-1} cm^{-1} K^{-1}. Assume that the EV is stored at work from 8:00 a.m. to 6:00 p.m. at −40°C.

18. From (9.6) and (9.7), calculate the ratio of CO_2/CO produced when TiC is electrolytically machined.
    Note: $Z = 6.6$.

Further Reading

1. Schmickler W (2010) Interfacial electrochemistry, 2nd edn. Springer, London
2. Zinola CF (ed) (2010) Electrocatalysis, vol 149, Computational experimental & industrial aspects. CRC Press, Boca Raton
3. Smil V (2010) Energy myths and realities. Bringing science to the energy policy debate. The AEI Press, Washington, DC
4. Bard AJ et al (eds) (2010) Electroanalytical chemistry, vol 23. CRC Press, Boca Raton
5. Ceroni P et al (eds) (2010) Electrochemistry of functional supramolecular systems. Wiley, Hoboken
6. Domenech-Carbo A (2010) Electrochemistry of porous materials. CRC Press, Boca Raton
7. Koper M (ed) (2009) Fuel cell catalysis: a surface science approach. Wiley, Hoboken
8. Plieth W (2008) Electrochemistry for material science, corrosion & protection. Elsevier, Amsterdam
9. Wendt H, Kreysa G (1999) Electrochemical engineering. Springer, New York
10. Society of Automotive Engineers (1999) Fuel cell power for transportation. SAE, Warrendale
11. Koppel T (1999) Powering the future, the Ballard fuel cell and the race to change the world. Wiley, New York
12. Bereny JA (ed) (1996) Battery performance, research and development. Business Technology Books, Orinda

13. Turrentine T, Kurani K (1996) The household market for electric vehicles. Business Technology Books, Orinda
14. Kordesch K, Simader G (1996) Fuel cells and their applications. VCH Pubs, New York
15. Linden D (1996) Handbook of batteries, 2nd edn. McGraw-Hill, New York
16. Saxman D, Grant S (eds) (1995) Batteries and EV industry review. BCC Press, Norwalk
17. Blomen LJ, Mugerwa MN (eds) (1994) Fuel cell systems. Plenum Press, New York
18. Staff OECD (1994) Electric vehicles: performance and potential. OECD, Washington, DC
19. Pletcher D, Walsh FC (1994) Industrial electrochemistry, 2nd edn. Chapman and Hall, London
20. Brant B, Leitman S (1993) Build your own electric vehicle. TAB, New York
21. Koryta J, Dvorak J (1987) Principles of electrochemistry. Wiley, New York
22. Rieger PH (1987) Electrochemistry. Prentice-Hall, Englewood Cliffs
23. Chowdari BVR, Radhakrishna S (eds) (1986) Materials for solid state batteries. World Scientific, Singapore
24. Rand DAJ, Bond AM (eds) (1984) Electrochemistry: the interfacing science. Elsevier, New York
25. Vincent CA (1982) Modern batteries. Edward Arnold, London
26. Unnewehr LE, Nasar SA (1982) Electric vehicle technology. Wiley, New York
27. Barak M (ed) (1980) Electrochemical power sources. Peter Peregrinus Ltd., New York
28. Graham RW (1978) Primary batteries, recent advances. Noyes Data Corp., Park Ridge
29. Antropor LI (1977) Theoretical electrochemistry. MIR Publishers, Moscow
30. Palin GR (1969) Electrochemistry for technologists. Pergamon Press, Oxford
31. Lyons EH Jr (1967) Introduction to electrochemistry. Heather and Co., Boston
32. Potter EC (1956) Electrochemistry—principles and application. Clever-Hume Press, London
33. E.V. Association of America http://www.evaa.org/Canadian Electric Vehicle Ltd., http://www.canev.com
34. Electrochemistry lecture notes,
35. http://www.chem.ualberta.ca/courses/plambeck/pl02/p0209x.htm
36. http://chemed.chem.purdue.edu/genchem/topicreview/bp/ch20/electroframe.html
37. http://edu.cprost.sfu.ca/rhIogan/electrochem.html
38. http://www.chem 1 .com/CB l.htm
39. Help files for Electrochemistry http://www.learn.chem.vt.edu/tutorials/electrochem/
40. Batteries http://www.panasonic.com/
41. Batteries http://www.varta.com/
42. All about batteries http://www.howstuffworks.com/battery.htm

Chapter 10
Corrosion

10.1 Introduction

Corrosion is the unwanted reaction or destruction of a metal component by the environment. The annual cost of corrosion to the US economy has been estimated to be over $70 billion. Similar costs are associated with other industrialized countries. Many of the problems can be avoided if basic precautions and design processes are followed.

The mechanism of corrosion is electrochemical and can be induced by the flow of current or will cause a current to flow. When a corroding metal is oxidized, the reaction

$$M \rightarrow M^{+n} + ne^-$$ (10.1)

must be accompanied by a reduction reaction which is usually the reduction of oxygen whether in the air or dissolved in water.

$$O_2 + 4H^+ + 4e^- \rightarrow 2H_2O$$ (10.2)

O.V. Roussak and H.D. Gesser, *Applied Chemistry: A Textbook for Engineers and Technologists*,
DOI 10.1007/978-1-4614-4262-2_10, © Springer Science+Business Media New York 2013

Or

$$O_2 + 2H_2O + 4e^- \rightarrow 4OH^- \tag{10.3}$$

In some cases, the reduction of hydrogen occurs.

$$2H^+ + 2e^- \rightarrow H_2 \tag{10.4}$$

The usual classification of corrosion is according to the environment to which the metal is exposed or the actual reactions which occur. We have seen that the concentration cell is a simple cell in which a metal can corrode as dissolution takes place.

10.2 Factors Affecting the Rate of Corrosion

It is convenient to classify the corrosion of metals in terms of (a) the metals and (b) the environment.

The reduction potential is the most important characteristic of a metal that determines its susceptibility to corrosion. This has been illustrated by Table 9.4. Thus, the noble metals, gold and platinum, are resistant to corrosion and will only dissolve in strong oxidizing solutions which also contain complexing halides or other ions, for example, (CN^-). For metals in seawater, the relative order of the reduction potential of metals and alloys has been established. This is illustrated in Table 10.1 where distinction is made between active and passive surfaces for some metals. Magnesium is a most active metal, whereas platinum and graphite are the least active materials. The voltages are given with respect to the saturated calomel electrode (SCE).[1] The oxidation reaction (10.1) represents corrosion which must be accompanied by a reduction reaction (10.2), (10.3), or (10.4) as well as reactions such as

$$Fe^{3+} + e^- \rightarrow Fe^{2+} \tag{10.5}$$

and

$$3H^+ + NO_3^- + 2e^- \rightarrow HNO_2 + H_2O \tag{10.6}$$

The reaction which occurs depends on the solution in which the metal corrodes, but in most cases the cathodic reaction involves O_2.

The corrosion rate will thus depend on the partial pressure of oxygen. This is shown in Table 10.2. Hence, the removal of oxygen from water in steam boilers is one method of reducing corrosion.

If hydrogen evolution is the cathodic reaction (10.4), then it can be reduced by increasing the overvoltage. The overvoltage of H_2 on mercury is very high (see Table 9.2), and reaction (10.4) can be inhibited if mercury is used to coat the metal surface and to form an amalgam (see the zinc–air cell, Sect. 9.6). The overvoltage is dependent on current density which is determined by the area of the metals. Hence, as the cathode area decreases, the polarization can be expected to increase resulting in a decrease in rate of corrosion. In the case of iron (anode) on a large copper sheet (cathode), the large cathode/anode ratio favors corrosion of the iron. This is shown in Fig. 10.1.

---

[1] The saturated calomel electrode is a convenient reference electrode often used instead of the standard hydrogen electrode: $\frac{1}{2}Hg_2Cl_2 + e^- \rightarrow Hg + Cl^-$, $\mathcal{E}° = 0.2224$ (25°C).

	−V (V)
Table 10.1 Galvanic metal and alloy potential V (vs. SCE) in seawater	
Mg	1.6 ± 0.02
Zn	1.00 ± 0.02
Be	0.99 ± 0.01
Al alloys	0.89 ± 0.11
Cd	0.71 ± 0.01
Mild steel	0.65 ± 0.05
Cast iron	0.61 ± 0.05
Low alloy steel	0.60 ± 0.02
Austenite Ni	0.50 ± 0.03
Bronze	0.36 ± 0.05
Brass	0.35 ± 0.04
Cu	0.34 ± 0.04
Sn	0.32 ± 0.03
Solder Pb–Sn	0.31 ± 0.03
Al brass	0.31 ± 0.03
Manganese bronze	0.31 ± 0.02
410,416 stainless steel	0.31 ± 0.03
Active potential	0.51 ± 0.04
Silicon bronze	0.29 ± 0.02
Tin bronze	0.29 ± 0.03
Nickel silver	0.28 ± 0.02
Cu/Ni 90/10	0.26 ± 0.04
Cu/Ni 80/20	0.26 ± 0.04
430 stainless steel	0.24 ± 0.04
Active potential	0.52 ± 0.06
Pb	0.23 ± 0.03
Cu/Ni, 70130	0.21 ± 0.02
Ni/Al bronze	0.20 ± 0.05
Ni/Co 600 alloy	0.17 ± 0.02
Active potential	0.41 ± 0.06
Ag bronze alloys	0.15 ± 0.05
Ni 200	0.15 ± 0.05
Ag	0.13 ± 0.03
302,304,321,347, SS	0.08 ± 0.02
Active	0.51 ± 0.05
Alloy 2C, stainless steel	0.00 ± 0.06
Ni/Fe/Cr/Alloy 825	−0.08 ± 0.04
Ni/Cr/Mo/Cu/Si alloy	−0.07 ± 0.03
Ta	−0.09 ± 0.06
Ni/Cr/Mo alloy C	−0.07 ± 0.07
Pt	−0.13 ± 0.10
Graphite	−0.14 ± 0.16

The type and amount of impurities in a metal will affect the rate of corrosion. For example, a zinc sample which is 99.99 % pure (referred to as 4n zinc) would corrode about 2,000 times faster than a 5n sample. Even improperly annealed metals will show excessive corrosion rates.

Another factor which controls the rate of corrosion is the relative volume of the corrosion product (oxide) to the metal as well as the porosity of the oxide layer. For example, the volume ratio of oxide/metal for Al, Ni, Cr, and W is 1.24, 1.6, 2.0, and 3.6, respectively. The oxide layer on a metal can

Table 10.2 Effect of O_2
pressure on corrosion of iron
in seawater

$P(O_2)$ (atm)	Rate of corrosion (mm/year)
0.2	2.2
1	9.3
10	86.4
61	300

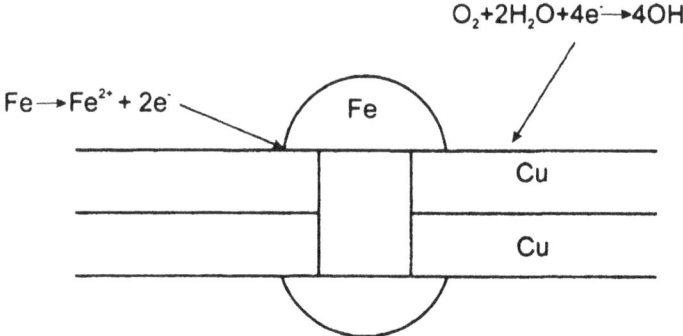

Fig. 10.1 The corrosion of an iron rivet in a copper plate. The large copper surface results in a low O_2 overvoltage, allowing the corrosion to proceed at a rate controlled by O_2 diffusion

convert a metal from one that corrodes to one that is inert. Aluminum can react with water to form hydrogen by the reaction

$$2Al + 6H_2O \rightarrow 2Al(OH)_3 + 3H_2 \tag{10.7}$$

followed by

$$2Al(OH)_3 \rightarrow Al_2O_3 + 3H_2O \tag{10.8}$$

However, the oxide layer which forms prevents the water from contacting the aluminum surface. Only in acid or alkali is the Al_2O_3 solubilized, and the aluminum reacts to liberate hydrogen.

An oxide layer is readily formed on many metals when they are made anodic in aqueous solutions. In the case of aluminum, this process is called anodization. It is also referred to as a passive film which reduces the corrosion rate. Such passive films can be thin, from 0.01 μm, and fragile and easily broken. Thus, when steel is immersed in nitric acid or chromic acid and then washed, the steel does not immediately tarnish nor will it displace copper from aqueous $CuSO_4$. The steel has become passive due to the formation of an adhering oxide film which can be readily destroyed by HCl which forms the strong acid $H^+ FeCl_4^-$.

The factors influencing the rusting of iron can be illustrated by the electrochemical treatment of the overall reaction.

$$2Fe + O_2 + 4H^+ \rightarrow 2Fe^{2+} + 2H_2O$$

$$Fe \rightarrow Fe^{2+} + 2e \quad \mathcal{E} = 0.440 \text{ V}$$

$$O_2 + 4H^+ + 4e^- \rightarrow 2H_2O \quad \mathcal{E} = 1.67 \text{ V}$$

From the Nernst equation (9.12)

$$\mathcal{E}_{\text{cell}} = \mathcal{E}^0{}_{\text{cell}} - (RT/4F) \ln \frac{[Fe^{2+}]^2}{Po_2[H^+]^4} \tag{10.9}$$

The corrosion reaction (10.9) ceases when $\mathcal{E}_{\text{cell}} \leq 0$

$$0 = 1.67 - (0.0591/4) \ln \frac{[Fe^{2+}]^2}{Po_2[H^+]^4} \tag{10.10}$$

Hence, $\mathcal{E}_{\text{cell}} \leq 0$ when

$$\log \frac{[Fe^{2+}]^2}{Po_2[H^+]^4} \geq 113 \tag{10.11}$$

Let us consider extreme conditions where

$$Po_2 = 10^{-6} \text{atm}, \quad [Fe^{2+}] = 10 \text{ M}, \quad [H^+] = 10^{-14} \text{ M}$$

$$\log 10^2/(10^{-6} \times 10^{-56}) = \log 10^{64} = 64$$

Since $64 < 113$, corrosion will continue to occur. In strong NaOH solution, rusting is reduced because the Fe_2O_3 forms a protective layer over the metal.

10.3 Types of Corrosion

The various forms of corrosion can be classified by their various causes. These are uniform corrosion attack (UC), bimetallic corrosion (BC), crevice corrosion (CC), pitting corrosion (PC), grain boundary corrosion (GBC), layer corrosion (LC), stress corrosion cracking (SCC), cavitation corrosion (CC), and hydrogen embrittlement (HE).

10.3.1 Uniform Corrosion

Such corrosion is usually easy to detect and rectify. The slow corrosion of a metal in aqueous acidic solution is an example of such corrosion. Impurities in a metal can result in local cells which, in the presence of electrolyte, will show corrosive action.

10.3.2 Bimetallic Corrosion

This type of corrosion, also called *galvanic corrosion*, is characterized by the rapid dissolution of a more reactive metal in contact with a less reactive more noble metal. For example, galvanized steel (Zn–Fe) in contact with copper (Cu) pipe is a common household error. A nonconducting plastic

Fig. 10.2 A gradient in O_2 concentration in the water drop makes the center portion of the iron anodic where Fe − Fe- + 2e~, while the edge is cathodic and oxygen is reduced

spacer would reduce the corrosion rate in the pipe. The rate of corrosion is partially determined by the difference in the standard cell potentials of the two metals in contact (see Table 9.4). The relative potential of metals in seawater is given in Table 10.1 and represents the driving force of the corrosion which includes the current, or more precisely, the current density, that is, A/cm^2.

An electrochemical cell is formed and the anodic metal dissolves. This can be corrected by applying a counter current or voltage or by introducing a more reactive, sacrificial anode, for example, adding magnesium alloy to the above Zn–Fe–Cu system, a procedure commonly used for hot-water pipes in renovated buildings.

10.3.3 Crevice Corrosion

A nonuniform environment or concentration gradient due to material structure or design leads to concentration cells and corrosion. Differential aeration is, for example, the cause of corrosion at the waterline or at the edges of holes or flange joints. The size of a crevice can range from 25 to 100 µm in width—small enough to create an oxygen concentration cell between the crevice solution and that on the outer surface. Oxygen can form a thin oxide layer on metals which acts as a protective passive film.

10.3.4 Pitting Corrosion

Like CC, PC is due to differential aeration or film formation (due to dust particles). The breakdown of a protective oxide layer at a lattice defects is another common cause of pits. The mechanism of pitting of iron under a water drop is shown in Fig. 10.2, and as in a CC, a differential concentration of oxygen in the drop creates a concentration cell. Rust has the composition of Fe_3O_4 and FeO(OH). Fe_3O_4 is a mixed oxide of FeO $\cdot$ Fe_2O_3 where iron is in the +2 and +3 oxidation state. The PC of various iron alloys induced by Cl⁻ in the presence of 0.5 M H_2SO_4 is given in Table 10.3. High chromium alloys are effective in reducing PC, but a limit is reached at about 25 % Cr, whereas nickel seems to have little effect on corrosion resistance. Other salts in solution also can affect the pitting rate as well as the depth of the pits.to the bulk alloy, and severe intergranular corrosion and pitting results. The corrosion rate of stainless steel (18/8) in aqueous HCl solutions depends on the concentration of acid,

Table 10.3 Minimum concentration of chloride ion necessary for starting pitting in 0.5 M H_2SO_4

Alloy	$[Cl^-]$ (M)
Fe	0.0003
5.6 Cr–Fe	0.017
11.6Cr–Fe	0.069
20 Cr–Fe	0.1
18.6 Cr, 9.9 Ni–Fe	0.1
24.5 Cr–Fe	1.0
29.4 Cr–Fe	1.0

Fig. 10.3 A comparison of the corrosion rates of metallic glasses (x,●) and crystalline stainless steel (O, A) as a function of HCl concentration at 30°C. No weight changes of the metallic glasses of $Fe-Cr_{10}P_{13}C_7$ were detected by a microbalance after immersion for 200 hr

temperature, and the oxygen pressure. In contrast, an equivalent metallic glass[2] ($Fe–Cr_{10}Ni_5P_{13}C_7$) showed no detectable corrosion. This is illustrated in Fig. 10.3 and clearly shows how important corrosion is along the grain boundaries in stainless steel. Similar results were obtained for immersion tests in 10 % wt. of $FeCl_3 \cdot 6H_2O$ at 60°C as an indication of PC. Again, the stainless steel (304, 136, 316) all showed significant pitting, whereas metallic glasses showed no detectable weight loss after 200 h. Not all metallic glasses are resistant to corrosion, and much more work is needed to understand these differences.

---

[2] Metallic glasses are amorphous noncrystalline solids which are usually prepared by rapidly cooling the molten metal. Such metals are devoid of grain boundaries.

10.3.5   Grain Boundary Corrosion

Coarse crystalline-rolled metals or alloys can corrode at the edge of the crystallites; thus, the iron impurity in aluminum is responsible for aluminum corrosion. Similarly, stainless steel (18/8 Cr/Ni) when heated (during welding) results in the precipitation of chromium carbide at the grain boundaries. This forms an enriched nickel layer anodic to the bulk alloy, and severe intergranular corrosion and pitting results. The corrosion of stainless steel (18/8) in aqueous HCl solutions depends on the concentration of acid, temperature, and the oxygen pressure. In contrast, an equivalent metallic glass* ($Fe-Cr_{10}P_{13}C_7$) showed the detectable corrosion. This is illustrated in Fig.10.3 and clearly showed important corrosion is along the grain boundaries in stainless steel. Similar results were obtained for immersion tests in 10% wt. of iron(III) chloride hexahydrate at 60° C as an indication of PC. Again the stainless steel (304, 136, 316) all showed significant pitting whereas metallic glasses showed no detectable weight loss after 200 hr. Not all metallic glasses are resistant to corrosion and much more work is needed to understand these differences.

10.3.6   Layer Corrosion

Like GBC, LC is caused by the dissolution of one element in an alloy and the formation of leaflike scale exfoliation. Some cast irons and brasses show flakelike corrosion products. The corrosion is due to microcells between varying compositions of an alloy.

10.3.7   Stress Corrosion Cracking

This is normally found only in alloys such as stainless steel and in specific environments. This type of corrosion is a result of the combined effects of mechanical, electrochemical, and metallurgical properties of the system.

The residual stress in a metal, or more commonly an alloy, will, in certain corrosive environments, result in mechanical failure by cracking. It first became apparent at the end of the nineteenth century in brass (but not copper) condenser tubing used in the electric power generating industry. It was then called *season cracking*. It is usually prevalent in cold-drawn or cold-rolled alloys which have residual stress. Heat treatments to relieve this stress were developed to solve the problem. It was soon realized that there were three important elements of the phenomenon: the mechanical, electrochemical, and metallurgical aspects.

The mechanical aspect is concerned with the tensile stress of the metal alloy. The mechanism of crack formation includes an induction period followed by a propagation period which ends in fracture. The kinetics of crack formation and propagation has been studied for high-strength alloys, and the overall process can be resolved into two or three stages depending on the alloy. The velocity of cracking is usually very slow, and rates of about 10^{-11} m/s have been measured. Activation energies for stages I and II are usually of about 100 kJ/mol and 15 kJ/mol, respectively. Stainless steel piping in nuclear reactors (BWR) often suffers such SCC and must be replaced before they leak. Zircaloy tubes used to contain uranium fuel in nuclear reactors are also subject to SCC.

An essential feature is the presence of tensile stress which may be introduced by loads (compression), cold work, or heat treatment. The first stage involves the initiation of the crack from a pit which forms after the passive oxide film is broken by Cl⁻ ions; the anodic dissolution reaction of metal produces oxide corrosion products with high levels of H^+ ions. Hydrogen evolution during the second stage contributes to the propagation of the crack. Stainless steel pipes used in nuclear power plants for cooling often suffer from SCC. This can be reduced by removing oxygen and chloride from the water, by using high purity components, and careful annealing with a minimum of weld joints.

The electrochemical aspect of the process is associated with anodic dissolution, accounting for high cracking velocities. The crack tip is free of the oxide protective coating in the alloy, and crack propagation proceeds as the alloy dissolves. Chloride ions present in solution tend to destroy this passivity in the crevice, which is depleted in oxygen. In stainless steels, the dissolution of chromium in the crevice occurs by the reactions:

$$Cr \rightarrow Cr^{3+} + 3e^- \tag{10.12}$$

$$Cr^{3+} + 3H_2O \rightarrow Cr(OH)_3 + 3H^+ \tag{10.13}$$

and accounts for the major cause of the autocatalytic process whereby the increased acidity in the crevice increases the rate of corrosion. Titanium is resistant to CC because its passive layer is not attached by chloride ions. This explains the specificity of the corrosive environment for a particular alloy since the reformation of the protective surface layer would stop the crack from propagating further.

The metallurgical aspect is exemplified by the effect of grain size—reducing grain size reduces SCC. SCC is increased by cold working and reduced by heat treatment annealing. Other metallurgical properties of an alloy can contribute to its susceptibility to SCC. Solutions to the problem include heat treatment, the use of corrosion-resistant cladding, and—in the case of nuclear power plants—the use of a nuclear-grade stainless steel.

10.3.8   Cavitation Corrosion

Cavitation is due to ultrasonics or hydrodynamic flow and is associated with the formation of microbubbles which collapse adiabatically to form thermal shocks and localized *hot spots* sufficient to decompose water and form hydrogen peroxide and nitric acid (from dissolved air). The resulting corrosion is thus due to a mechanical and chemical effect and can be reduced by cathodic protection or by the use of chemically resistant alloys.

Cavitation is normally associated with motion of metal through water which forms low-pressure bubbles. These microbubbles, upon collapsing adiabatically, heat the entrapped oxygen, nitrogen, and water to above decomposition temperatures with the resulting formation of a variety of compounds such as NO_x, HNO_3, H_2O_2, and at times O_3. Cavitation is thus produced in the turbulence formed by propeller blades of ships, water pumps and mixers, and in the steady vibrations of engines. Cavitation also has the effect of disrupting the protective surface coating on metals, and when pieces of the metal are actually removed by the flow of bubbles, the process is called *cavitation erosion* (CE).

Figure 10.4a shows the cylinder casing of a diesel engine which was water cooled. Vibrations caused cavitation resulting in pitting which penetrated the casing. The lower Fig. 10.4b shows the blades of the water pump in the diesel which had also corroded for the same reasons.

Cavitation corrosion can be reduced by the proper design and vibration damping of systems. It has also been shown that the addition of drag reducers (see Appendix B) to the water reduces CE and transient noise. High Reynolds number (Re = 124,000) can be achieved without cavitation. It would seem advantageous to add water-soluble drag reducers such as polyethylene oxide to recirculating water cooling systems to reduce CC.

Fig. 10.4 (**a**) Cavitation
corrosion of a water-cooled
cylinder casing of a diesel
engine. Corrosion holes
have penetrated the wall.
(**b**) The water pump
propeller in the same
diesel engine corroded
by cavitation

10.3.9   Hydrogen Embrittlement

The migration of hydrogen dissolved in a metal lattice usually occurs along grain boundaries where cracks occur during stress. The embrittlement of steels is due to hydrogen atoms which diffuse along grain boundaries. They then recombine to form H_2 and produce enormous pressures which result in cracking. The H-atoms are formed during the corrosion of the metal or a baser metal in contact with the steel.

10.4   Atmospheric Corrosion

The major cause of corrosion of metals in the air is due to oxygen and moisture. In the absence of moisture, the oxidation of a metal occurs at high temperatures with activation energies E_a, ranging from 100 to 250 kJ/mol which is determined by the work function ϕ, where

$$E_a(\text{kJ/mol}) = \phi - 289 \tag{10.14}$$

At ambient temperature, however, all metals except gold have a thin microscopic layer of oxide.

An example of a noncorroding steel structure is the Delhi Iron Pillar (India) which dates from about 400 A.D. It is a solid cylinder of wrought iron 40 cm in diameter, 7.2 m high. The iron contains 0.15 % C and 0.25 % P and has resisted extensive corrosion because of the dry and relatively unpolluted climate.

Table 10.4 Relationship between the resistivity of soil corrosion activity and estimated lifetime of buried steel pipe

Resistivities (H-cm)	Corrosion	Normal duration (years)
<800	Severe	<10
800–5,000	Moderate	15
5,000–10,000	Mild	20
>10,000	Unlikely	>25

The industrial corrosive effluents could include NO_x, SO, and H_2S, whereas natural occurring corrosive substances are H_2O, CO_2, and, in coastal areas, NaCl from sea sprays. These two sources of corrosive substances were enough to corrode the Statue of Liberty in New York Bay. (The statue, which is 46 m high, was a gift from France in 1886 and was erected on a pedestal 46 m above ground level to commemorate the centenary of the American Revolution.) It was constructed of 300 shaped copper panels (32 t), 2.4 mm thick, riveted together and held in place by 1,800 steel armatures which slipped through 1,500 copper saddles. Thus, though the iron touched the copper, there was no direct bonding of the two metals. This did not stop the electrochemical corrosion when rainwater and ocean spray penetrated the structure. More than one third of the 12,000 rivets had popped by 1975. To commemorate the second centennial of the USA, the rebuilt Statue of Liberty was unveiled after renovations costing about $60 million. The copper panels are now sealed on the inside of the structure by silicone sealant to prevent water from entering the statue. The iron armatures were replaced by stainless steel with a Teflon-coated tape to separate the two metals. Though the copper skin is expected to last over 1,000 years, the durability of the wrought iron structure is much shorter, and it will corrode quickly if not protected from the elements. This normally involves lead-based paints or silicone rubber sealants which are used for bridges.

10.5 Corrosion in Soil

The resistivity of soil is an important characteristic which often determines the rate of corrosion—low resistivity is usually associated with high rates of corrosion. This is shown in Table 10.4. Soluble salts and high moisture content account for low resistivity–high conductivity. The density and particle size can control the moisture level and permeability of the soil to water and oxygen.

The groundwater level determines the depth of *dry* soil. Oxygen depletion by decaying organic substances or living organisms tends to inhibit corrosion. Oxygen transport from air into soil is facilitated by water and leads to higher corrosion rates above groundwater than below.

A low pH of soil (pH 3.5–4.5)—high acid level—contributes to the corrosion rate. Soil of pH > 5 is much less corrosive. Alkaline soil, pH > 7, can be corrosive to aluminum, and if ammonia is formed by bacterial activity, then even copper will be attached. The weak organic acids present in humic acid can solubilize surface oxides and lead to corrosion of metals by complexation processes. Anaerobic bacterial action in soil can lead to H_2S (and CH_4 plus CO_2) which, though a weak acid, will form insoluble metallic sulfides, reducing the free metal ions in the soil and shifting the equilibrium toward metal dissolution.

10.6 Aqueous Corrosion

As indicated previously, the corrosion of metals in aqueous environments is determined by the Nernst equation in terms of the electrode potential and pH—called a *Pourbaix* diagram. This is shown in Fig. 10.5 for iron where the vertical axis is the redox potential of the corroding system and the pH

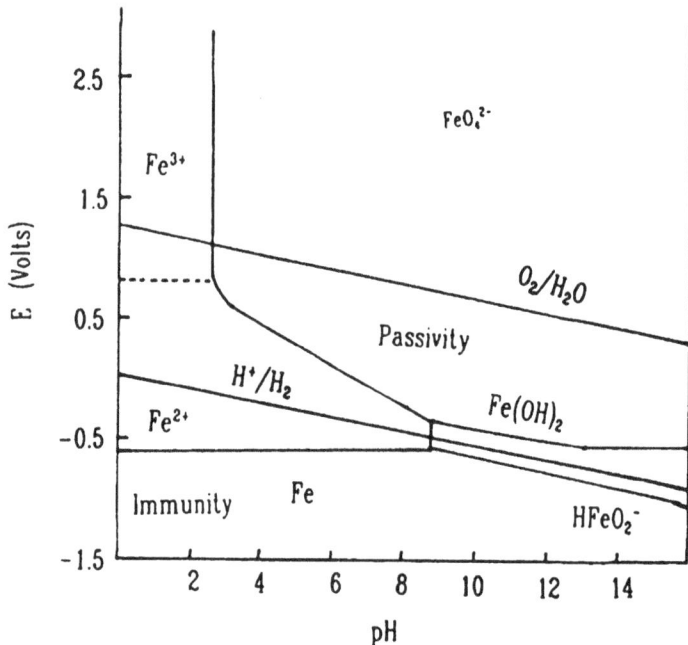

Fig. 10.5 A Pourbaix diagram for iron showing the general conditions under which the metal is passive, corrosive, or stable (immunity)

scale is the horizontal axis. The dashed lines show the H^+/H_2 and O_2/H_2O redox reactions which have slopes of 0.059 V/pH. Water is stable between the two lines. Sloping lines indicate that the redox potential is pH dependent. Horizontal lines reflect redox potential which is not pH dependent, whereas vertical lines refer to changes which do not involve a change in oxidation state. Above the O_2/H_2O line, oxygen is evolved, while below the H^+/H_2 line, hydrogen is liberated.

When a solid insoluble product is formed, it may protect the metal from further reaction. This corresponds to the passive region and assumes low concentration for metal ions in solution (e.g., 10^{-6} M). Under condition where the metal is stable, a state of immunity exists, and corrosion cannot occur. Iron corrodes, forming Fe^{2+} at low pH, but at high pH, the $Fe(OH)_2$ dissolves to form $HFeO_2^-$. This region is referred to as caustic cracking of steel (pH > 12) analogous to stress corrosion cracking.

Iron will corrode in acids except H_2CrO_4, conc. HNO_3, $H_2SO_4 > 70\%$, and HF > 90 %. Pourbaix diagrams are available for most metals and help define the corrosion-free conditions.

10.7   Corrosion Protection and Inhibition

The Royal Navy's first submarine, *Holland I*, sank in 1913 off the coast of England and for 70 years lay in 63 m of seawater. The wreck was recently located and raised. She was in remarkable condition considering that the hull contained a mass of dissimilar metals, steel, cast and malleable tin, brass, bronze, and lead. The doors opened, springs sprang, the engine turned, rivets were tight, and a battery when cleaned, refilled, and recharged, delivered its specified 30 amps. The explanation for the absence of the corrosion expected is due to the protection given to the surface by the rapid colonization of a cold-water coral and the deposition of a 3–4-mm layer of calcium carbonate. This prevented the

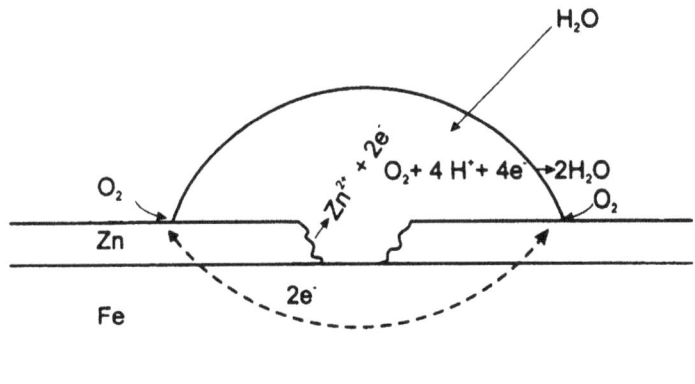

Fig. 10.6   The Fe/Zn system. A break in the zinc coating on iron (galvanized iron) will, to a limited extent, continue to protect the iron (cathodic) as the zinc (anodic) dissolves

diffusion of oxygen and electrolyte from reaching the metal surface. Coatings thus represent a simple and at times effective method of reducing corrosion.

Corrosion can be eliminated by covering metals with more noble ones by plating or cladding. This is impractical because of the expense involved. Protective metal coatings of chromium are familiar, being decorative as well as preventing corrosion. Plated jewelry with silver, gold, and rhodium are common. Steel coated with zinc is protected in both air and water. The standard potentials are $\mathcal{E}^0_{Zn+2/Zn} = -0.763$ V and $\mathcal{E}^0_{Fe+2/Fe} = -0.440$ V, means that any break in the zinc coating on iron will make Zn anodic and iron cathodic (see Fig. 10.6). In hot water, exposed iron can be protectively coated with $CaCO_3$ if the water is hard (i.e., contains $CaHCO_3$). The zinc is usually applied by hot-dipping and produces a continuous coat of 80–125 μm thick. Other coating processes include spray plating, electroplating, and for small items, tumbling. A galvanized surface can be repaired by painting with a zinc-rich paint consisting of metallic zinc powder bound in an epoxide or resin base.

Ordinary paints may be permeable to oxygen and water vapor (see Chap. 13), and though they may slow the rate of corrosion, they cannot prevent it completely. Hence, special paints with chromates or red lead (Pb_3O_4) have been used for many years as a protective coating for steel. Polymeric resins, though more expensive than the linseed oil-based paints, last longer and thus are more effective.

10.8   Corrosion in Boiler Steam and Condensate

Steam lines with air and CO_2 entrained can be very corrosive. To reduce corrosion, oxygen can be removed by the addition of hydrazine (N_2H_4).

$$N_2H_4 + O_2 \rightarrow N_2 + 2H_2O \tag{10.15}$$

or Na_2SO_3

$$2Na_2SO_3 + O_2 \rightarrow 2Na_2SO_4 \tag{10.16}$$

Other additives which are commonly added are basic amines which neutralize the acids (H_2CO_3) present in the water. One important property of the amine besides the pH of its solution is the distribution ratio (DR) which is the ratio of amine in steam to that dissolved in the condensate. A high

Table 10.5 Some characteristics of selected amines used in steam systems as corrosion inhibitors

	pH	DR
Morpholine	8–10	0.3–0.8
Cyclohexylamine	10–11	6–8
Diethylaminoethanol	11–12	2–4
Benzylamine	8–9	3–4

Values are dependent on concentration of the amine

DR value means that the metal is readily coated with a thin protective film of the amine. Some amines commonly used and their pH and DR are given in Table 10.5. The amine is slowly lost, and it must be replaced continuously. Steam lines invariably have these amines, and the use of brass, bronze, or copper results in the corrosive removal of copper.

Stored metallic equipment or parts are subject to corrosion. Sodium nitrite is an inhibitor which is often included in the enclosure or packaging. However, vapor phase corrosion inhibitors (VCI) such as dicyclohexylammonium nitrite and ammonium benzoate are superior corrosion inhibitors because *of* the film formed on the metal surfaces.

10.9 Cathodic Protection

It is possible to prevent the corrosion of a metal by connecting it to a more active metal. This active metal becomes anodic and tends to corrode, whereas the cathodic metal is preserved. Iron pipes in soil or water will not corrode if they are connected to a sacrificial anode such as aluminum, zinc, or magnesium. Steel pipes for water and gas are usually protected in this manner. Galvanized iron pipes for hot water lines have a limited life which can be extended by introducing a magnesium rod to act as a sacrificial anode

The potential needed to protect iron in seawater is −0.62 V with respect to the SHE or −0.86 V relative to SCE. Aluminum can provide this potential, −0.95 V relative to the SCE, and its use has been extended to offshore oil platforms, ship's hulls, ballast tanks, and jetty piles with life expectancies ranging from 3 to 10 years, depending on the mass of aluminum employed.

An alternate approach is to apply a potential onto the steel, making it cathodic relative to an inert anode such as Pb, C, or Ni. A potential of −0.86 V is suitable for the protection of iron.

Though more negative potentials, such as −1.0 V, can be used, it should be avoided in order to prevent hydrogen evolution and hydrogen embrittlement.

Exercises

1. Show how different oxygen concentrations in a cell for a single metal can result in corrosion.
2. What are the cathodic reactions which usually accompany the corrosive dissolution of a metal?
3. Explain why the standard reduction potential, $\mathcal{E}°$

$$Hg_2Cl_2 + 2e^- \rightarrow 2Hg + 2Cl^- \, (\mathcal{E}° = 0.2680 \text{ V})$$

   (Table 9.4) is different from that for the SCE ($\mathcal{E}° = 0.2415$ V).
4. Explain why drag reducers may decrease cavitation corrosion (see Appendix B).
5. Why is the corrosion rate of metallic glass orders of magnitudes lower than the crystalline metal?
6. When Ni and Cd are in contact, which metal will corrode?

7. Describe six types of corrosion and explain how the corrosion rates can be reduced.
8. How can a metal be made passive? Give three examples.
9. When two dissimilar metals are joined together, a potential is set up due to the Seebeck effect. This is the basis of the thermocouple and is due to differences in work function of the two metals. Explain how this applies to corrosion.
10. The tarnishing of silver by H_2S is a type of corrosion which requires the presence of O_2.

$$2Ag + H_2S + \frac{1}{2}O_2 \rightarrow Ag_2S + H_2O$$

Explain this in terms of a corrosion mechanism.
11. How does polarization affect the rate of corrosion?
12. Why is chloride ion (Cl^-) more corrosive to iron than nitrate (NO_3^-)?
13. Estimate the activation energy for the oxidation of the following metals in dry air. The values of the respective work functions are given in eV units. Cd (4.22), Cr (4.5), Fe (4.5), Mo (4.6), Ni (5.15), Ti (4.33), Zr (4.05).

Further Reading

1. Revie RW, Winston R (2011) Uhlig's corrosion handbook, 3rd edn. Wiley, Hoboken
2. Volkan C (2011) Corrosion chemistry. Wiley, Hoboken, N. J., Scrivener, Salem
3. McCafferty E (2010) Introduction to corrosion science. Springer, New York
4. Roberge PR (1999) Corrosion engineering handbook. McGraw-Hill, New York
5. Becker JR (1998) Corrosion and scale handbook. Penn Well Books, Tulsa
6. Schweitzer PA (ed) (1996) Corrosion engineering handbook. M. Dekker, New York
7. Craig BD, Anderson HD (1995) Handbook of corrosion data, 2nd edn. ASM International Materials Park, Ohio
8. Marcus P, Oudar J (eds) (1995) Corrosion mechanisms in theory and practice. M. Dekker, New York
9. Bogaerts WF, Agena KS (1994) Active library on corrosion CD-ROM and ALC network. Elsevier, Amsterdam
10. Flick EW (1993) Corrosion inhibitors: an industrial guide, 2nd edn. Noyes, Westwood
11. Bradford S (1992) Corrosion control. Chapman and Hall, New York
12. Scully JC (1990) The fundamentals of corrosion, 3rd edn. Pergamon Press, New York
13. Trethewey KR, Chamberlain J (1988) Corrosion for students of science and engineering. Longman Scientific, New York
14. Wrangler G (1985) An introduction to corrosion and protection of metals. Chapman and Hall, London
15. West JM (1980) Basic corrosion and oxidation. Ellis Harwood Ltd., Chichester
16. Chilton JP (1973) Principles of metallic corrosion, 2nd edn. The Chemical Society, London
17. Evans UR (1972) The rusting of iron: causes and control. Edward Arnold, London
18. Corrosion notes. http://www.cp.umist.ac.uk/
19. National Association of Corrosion Engineers. http://www.nace.org/index.asp
20. Corrosion Research Centre. http://www.cems.umn.edu/research/crc
21. Protective Coatings. http://www.corrosion.com/
22. Corrosion sources. http://www.corrosionsource.com/
23. Corrosion. http://www.intercorr.com/
24. Corrosion experts past and present. http://www.corrosion-doctors.org
25. Corrosion Prevention Association. http://www.corrosionpreventiorLorg.uk.cpu/defauiLhtm
26. Corrosion solution. http://www.importexporthelp.com/electroshield.htm
27. Correct corrosion. http://www.repairrustedoutcars.com/ http://www.metalogic.be/projects

Chapter 11
Polymers and Plastics

11.1 Introduction

A *polymer* is a large chain molecule of high molecular weight which is composed of a single molecule (monomer) that is repeated many times in the chain. In contrast, a macromolecule is a large molecule composed of many small molecules bound together with chemical bonds, e.g., a protein or DNA. An oligomer is a small polymer of only several monomer units.

Plastics are prepared by the melting, molding, extruding, or the compression of polymers. The word "polymer" implies a molecule consisting of a long chain of units of smaller molecules or monomers. Thus, the polymer is also called a macromolecule. Such large molecules exist in nature and common examples of these are cellulose, rubber, cotton, silk, wool, starch, and keratin.

The annual world production of polymers has increased from 11.5 Mt in 1940 to about 27 Mt in 1960, after which time production almost doubled every decade to more than 150 Mt in 1990. Fiber production at about 36 Mt is almost equally divided into natural and synthetic. The production of elastomers (flexible plastics) represents about one-tenth of the total polymers, with production of synthetic elastomers being about twice that of natural rubber.

11.2 Molecular Weight

Normally the number of monomers in a polymer molecule varies considerably, but the interesting range for the fabricator of plastics is generally between 10^3 and 10^6 units.

Since the precise number cannot be controlled, the molecular weight (MW) of a polymer is not a unique value, and the distribution can vary as a result of the method of preparation. There are two important average molecular weights of a polymer: the number average MW, $\bar{M}n$, and the weight average MW, $\bar{M}w$.

If we let w represent the total mass of a sample of polymer and w_i the weight of the ith species of MW M_i, then

$$n_i = \frac{w_i}{M_i} w_i \qquad \text{where } n_i \text{ is the number of moles of } i\text{th species} \tag{11.1}$$

$$\sum_{i=1}^{\infty} n_i = n_T \qquad \text{the total number of moles in the sample} \tag{11.2}$$

O.V. Roussak and H.D. Gesser, *Applied Chemistry: A Textbook for Engineers and Technologists,*
DOI 10.1007/978-1-4614-4262-2_11, © Springer Science+Business Media New York 2013

Fig. 11.1 Fraction of weight having an average MW

the total weight, $\quad w = \sum_{i=1}^{\infty} w_i = \sum_{i=1}^{\infty} n_i M_i \quad$ (11.3)

The number average MW $\bar{M}n$ is given by

$$\bar{M}_n = \frac{w}{\sum\limits_{i=1}^{\infty} n_i} = \frac{\sum\limits_{i=1}^{\infty} w_i}{\sum\limits_{i=1}^{\infty} n_i} = \frac{\sum\limits_{i=1}^{\infty} n_i M_i}{\sum\limits_{i=1}^{\infty} n_i} \quad (11.4)$$

The weight average MW, $\bar{M}w$, is given by

$$\bar{M}_w = \frac{\sum\limits_{i=1}^{\infty} w_i M_i}{\sum\limits_{i=1}^{\infty} w_i} = \frac{\sum\limits_{i=1}^{\infty} n_i M_i^2}{\sum\limits_{i=1}^{\infty} n_i M_i} \quad (11.5)$$

A typical distribution of MW of a polymer is shown in Fig. 11.1.

The MW of a polymer is the single most important physical characteristic of the plastic since it determines its mechanical properties and even solubility among other properties.

Another related concept is the degree of polymerization (DP) which represents the number of monomer units in the polymer chain. Since the value of DP differs from one polymer chain to another, the value of the degree of polymerization is usually an average and is related to the MW by the relation

$$M_w = M(DP) \quad \text{and} \quad \overline{M}_w = M(\overline{DP}) \quad (11.6)$$

where M is the MW of the monomer.

The MW of a polymer can be determined by a variety of methods.

Fig. 11.2 Mechanical strength of a plastic as a function of degree of polymerization (DP)

The colligative properties of polymers in solution give rise to the number average MW, $\bar{M}n$. Thus, boiling point elevation and osmotic pressure measurement are commonly used though the latter method is much more sensitive, though restricted by the choice of suitable membranes. The weight average MW, $\bar{M}_w$, of a polymer in solution can be obtained by light scattering measurements.

The simplest and most commonly used method of measuring the MW of a polymer is by viscosity measurements of its solution. The relationship is

$$[\eta] = K\bar{M}_v^{\alpha} \tag{11.7}$$

where K and a are empirical constants dependent on the polymer, solvent, and temperature.

M_v is the average viscosity MW, and $[\eta]$ is the intrinsic viscosity defined as

$$[\eta] = \frac{\text{Limit}}{c \to 0} \frac{\eta sp}{c} \tag{11.8}$$

c is the concentration usually expressed as grams of polymer/100 g solvent and ηsp is the specific viscosity determined from the measurement of the viscosity of the pure solvent η_0, and viscosity of the solution η where $\eta/(\eta_0)$ is usually referred to as the viscosity ratio η_r and

$$\eta sp = \frac{\eta}{\eta_0} - 1 \tag{11.9}$$

Values of α and K are available from handbooks on polymers and range from 0.5 to 1 for α and 0.5 to 0.5×10^{-4} for K. The value of $\bar{M}_v$ is usually about 10 to 20% below the value of $\bar{M}_w$ (see Appendix B).

It is now well established that all important mechanical properties, such as tensile strength, elongation to break, impact strength, and reversible elasticity of polymers, depend on DP. When DP is relatively low, the polymer has little or no strength. As DP increases, the mechanical properties improve and tend toward a constant value. This is illustrated in Fig. 11.2 which shows the typical shape of the curve. The critical value DPc below which the polymer is essentially friable is different for each polymer, as is the bend over point β. However, plastics have little strength when DP $<$ 30 and approach limiting strength at DP $>$ 600.

11.3   Copolymers

When a polymer is formed from two or more monomers then the polymer is said to be a copolymer. The relative positions of the two monomers can be random or regular or in chunks. Figure 11.3 shows the different possible arrangements.

Blends of copolymers can be used to obtain specific properties of a plastic. Thus, polyethylene is brittle at temperatures below 0°C. However, when copolymers are formed with vinyl acetate (15 mol %), the resulting plastic is more flexible down to −40°C.

Another example of a copolymer is vinyl chloride with about 5% propylene. Polyvinyl chloride (PVC) is a hard brittle plastic which is made soft and flexible by dissolving a plasticizer into the PVC. Up to 30% by weight of plasticizers such as dioctyl phthalate is used to make plastic tubing. The propylene copolymer is soft without the plasticizer, or less plasticizer is required at lower concentrations in the propylene/PVC copolymer.

The loss of plasticizer from vinyl upholstery is the cause of cracking commonly observed in automobile seats and furniture.

11.4   Classification of Polymers

Many polymers occur naturally, e.g., cotton, wool, silk, gelatin, rubber, leather. Some are even inorganic such as sulfur, glass, and silicones. The thermal property of polymers is another important characterization. Thermoplastic polymers become soft and, without cross-linking, can be molded and shaped into various forms which are retained on cooling. The process is reversible, and the plastics can be reformed into other shapes when heated. Examples include polyethylene, PVC, nylon, and polystyrene. Thermosetting polymers cross-link on setting and once formed cannot be reshaped. Heating decomposes the plastic. Examples include Bakelite, melamine, phenol formaldehyde, and epoxy resins.

The manner in which polymers are formed is also a distinguishing feature. Two common methods are described.

Fig. 11.3 Schematic arrangements of copolymers made from two monomers O and ©

11.4.1 Addition Polymers

The addition process where the monomer is converted into a free radical[1] which adds to another monomer. The process continues until the two growing chains combine, or one combines with a free radical. The process is as follows:

$$\text{Initiation} \quad \begin{cases} \text{Catalyst} \to R^* \\ M + R^* \to R - M^* \end{cases} \tag{11.10}$$

$$\text{Propagation} \quad \begin{cases} R - M^* + M \to R - M - M^* \\ RM_n^* + M \to RM_{n+1}^* \end{cases} \tag{11.11}$$

$$\text{Termination} \quad RM_x^* + R^* \to RM_x R$$

$$\text{Recombination} \quad RM_x^* + R^* M_y^* \to RM_x M_y R \tag{11.12}$$

$$\text{Disproportionate} \quad RM + RM \to RM_x - H + RM_y H$$

where $RM_X = H$ is RM; which has lost a H-atom forming a C=C double bond.

The initiation process is usually by the thermal generation of free radicals from a peroxide such as benzoyl peroxide

$$C_6H_5\overset{O}{\overset{\parallel}{C}}-O-O-\overset{O}{\overset{\parallel}{C}}-C_6H_5 \to C_6H_5^{\cdot} + C_6H_5-O^{\cdot} + CO_2 \tag{11.13}$$

The peroxide or other azo initiators can also be decomposed when exposed to UV light. Such processes are used for the setting of polymers which function as fillings of tooth cavities.

11.4.2 Condensation Polymers

Condensation polymers are formed from the reaction of two different bifunctional monomers A and B which form AB by the reaction with the elimination of a product, usually water

$$A + B \to A - B$$
$$AB + A \to ABA \tag{11.14}$$
$$ABA + B \to BABA$$

and so on.

Thus, the polymer grows at both ends by condensing and stops when at least one of the reagents is fully consumed. Nylon 66 is a condensation polymer between a 6-carbon diamine and 6-carbon dicarboxylic acid.

---

[1] A free radical is a molecule or fragment which has one or more unpaired.

$$NH_2 - (CH_2)_6 - NH_2 + HOOC - (CH_2)_4COOH$$
$$\rightarrow NH_2 - (CH_2)_6 - NHOC(CH_2)_4COOH + H_2O \tag{11.15}$$

resulting in nylon 6,6 when the chain has grown sufficiently.

11.5 Vinyl Polymers

The vinyl radical is $CH_2=CH^\bullet$ and is the basis of a wide variety of monomers having the general formula $CH_2=CHX$. For example, when X=H, the molecule is ethylene and the polymer is polyethylene. The major vinyl polymers are listed in Table 11.1.

Table 11.1 Addition polymers for vinyl polymers $CH_2=CHX$

X	Monomer	Polymer	Uses
H	$CH_2=CH_2$ ethylene	$-CH_2-CH_2-$	Bottles, plastic tubing
CH_3	$CH_2=CHCH_3$ propylene	$\begin{array}{c} CH_3 \\ \mid \\ -CH_2-CH- \end{array}$	Carpeting, textiles, ropes
Cl	$CH_2=CHCl$ vinyl chloride	$\begin{array}{c} Cl \\ \mid \\ -CH_2-CH- \end{array}$	Pipes, floor tiles, tubing
⬡	$CH_2=CH(C_6H_5)$ styrene	⬡ $\begin{array}{c} \mid \\ -CH_2-CH- \end{array}$	Clear film, foam insulation, cups
CN	$CH_2=CHCN$ acrylonitrile	$\begin{array}{c} CN \\ \mid \\ -CH_2-CH- \end{array}$	Orion, ABS, carpet
$O=\overset{\mid}{C}-CH_3$	Methyl methacrylate	$\begin{array}{c} CH_3 \\ \mid \\ -CH_2-C- \\ \mid \\ O=C-OCH_3 \end{array}$	Windows, outdoor signs, lighting
$-O-\overset{O}{\underset{\parallel}{C}}-CH_3$	Vinyl acetate	$\begin{array}{c} -CH_2-CH- \\ \mid \\ O-C-CH_3 \\ \parallel \\ O \end{array}$	Paints, adhesives
$CF_2=CF_2$	Teflon	$-CF_2-CF_2-$	Electrical insulation, heat resistant, lubricant

11.5.1 Polyethylene

Also referred to as polythene, *polyethylene* is similar to polymethylene ($—CH_2—)_x$ which was prepared about 100 years ago by the decomposition of diazomethane (CH_2N_2), an explosive gas.

Polyethylene ($^{\bullet}CH_2—CH_2—)_x$ was first produced commercially in 1939. The early process was under high pressure (1,000–3,000 atm) and at temperatures from 80 to 300°C. The polymerization mechanism is via free radical initiators such as benzoyl peroxide

which are added to the reaction mixture.

This process results in low-density polyethylene (0.915–0.94 g/mL).

High-density polyethylene is prepared at low pressure at about 70°C in the presence of a special catalyst (usually a titanium complex). The density is approximately 0.95 g/mL because of the higher degree of crystallinity and order in the polymer.

The original high-pressure process gave some branched polymers; polyethylene formed at low pressure has a higher melting point, higher density, and higher tensile strength. It is a linear crystalline polymer which costs approx. 1.5 times that of the high-pressure-low-density material.

Polyethylene films are commonly used as vapor barriers in housing insulation. For greenhouse covering or window material, it is transparent enough but will slowly disintegrate due to the presence of residual carbon–carbon double bonds (C=C) which are split by ozone. Ultraviolet light will also degrade plastics unless a UV stabilizer is added which converts the absorbed UV light into heat. To make a plastic biodegradable, a substance is added which absorbs UV light from the sun and forms free radicals which attack the polymer chain.

11.5.2 Polypropylene

Polypropylene was first produced commercially in 1957. Early attempts resulted in very low MW polymers having poor plastic properties. The titanium complex used to prepare high-density polyethylene was found to be effective in polymerizing propylene. Because of the asymmetry of the propylene molecule, three different types of stereochemical arrangements can occur in the polymer chain.

1. Isotactic

(all methyl groups on one side)

2. Syndiotactic

$$-CH-CH_2-CH-CH_2-CH-CH_2-CH-CH_2-$$

(with CH_3 groups: first CH has CH_3 below and CH_3 above on the third CH...)

methyl groups alternate

3. Atactic

$$CH-CH_2-CH-CH_2-CH-CH_2-CH-CH_2-$$

random orientation of CH_3 groups

Atactic polypropylene is completely amorphous, somewhat rubbery, and of little value. The isotactic and syndiotactic polymers are stiff, crystalline, and have a high melting point. Increasing the degree of crystallinity increases the tensile strength, modulus, and hardness.

Polypropylene is the lightest nonfoamed plastic, with a density of 0.91 g/mL. It is more rigid than polyethylene and has exceptional flex life. Polypropylene has found use in a wide variety of products which include refrigerators, radios, and TVs as well as monofilaments, ropes, and pipes.

11.5.3 Polyvinyl Chloride

Polyvinyl chloride is one of the cheapest plastics in use today. It is prepared by the polymerization of vinyl chloride (VCM) ($CH_2=CHCl$, B.P.—14°C) as a suspension or emulsion in a pressure reactor. The polymer is unstable at high temperatures and liberates HCl at $T > 200°C$. It can be injection molded or formed into a hard and brittle material. It can be readily softened by the addition of plasticizers such as diethylhexylphthalate to the extent of 30%. Plasticized PVC is used as an upholstery substitute for leather. Since the plasticizer is volatile to a small extent, it slowly leaves the vinyl which eventually becomes hard, brittle, and then cracks. This can be restored by replacing the plasticizer by repeated conditioning of the vinyl surface.

11.5.4 Polyvinylidene Chloride

Polyvinylidene chloride (PVDC) is prepared by free radical polymerization of vinylidene chloride ($CH_2=CCl_2$). This polymer, unlike PVC, is insoluble in most solvents. It forms copolymers with fiber forming polymers. Its films, known as Saran, have a very low permeability for O_2 and CO_2 and thus are used in food packaging. When heated to high temperatures, in the absence of oxygen, it liberates HCl, leaving a very active carbon with pores of about 1.6 nm. This "Saran carbon" has been used to double the storage capacity of CH_4 in cylinders. This is presently being considered for use in CH_4-fueled vehicles.

11.5.5   Polystyrene

Polystyrene (PS) is prepared by the polymerization of styrene (C_6H_5—CH=CH$_2$), also known as vinylbenzene. Commercial PS is mostly of the atactic variety and is therefore amorphous. The polymer, on decomposition, unzips and forms the monomer with some benzene and toluene. Its major defects are poor stability to weather exposure, turning yellow and crazing in sunlight. In spite of these drawbacks and its brittleness it has found wide use as molded containers, lids, bottles, electronic cabinets. As a foamed plastic it is used in packaging and insulation. The thermal conductivity of the expanded PS foam is about 0.03 $Wm^{-1} K^{-1}$. The foam can absorb aromatic hydrocarbons usually found in the exhaust of automobiles and buses, causing the foam to disintegrate after long periods of normal exposure to a polluted environment.

The copolymerization of a small amount of divinylbenzene results in a cross-linked polymer which is less soluble and stronger. Cross-linking can sometimes be accomplished by γ-radiation which breaks some C—H and C—C bonds and on rearrangement form larger branched molecules. This is the case for polyethylene which, after cross-linking, will allow baby bottles to withstand steam sterilization.

11.5.6   Polyacrylonitrile

Polyacrylonitrile (PAN) is formed by the peroxide-initiated free-radical polymerization of acrylonitrile (CH$_2$=CH—CN). The major application of PAN is as the fiber Orion. When copolymerized with butadiene, it forms Buna N or nitrile rubber, which is resistant to hydrocarbons and oils. As a copolymer with styrene (SAN), it is a transparent plastic with very good impact strength used for machine components and for molding crockery. As a terpolymer of acrylonitrile–butadiene–styrene (ABS), the plastic is known for its toughness and good strength and finds applications in water lines and drains.

Polyacrylonitrile fibers are an excellent source for high-strength carbon fibers which are used in the reinforcement of composite (plastic) materials. The process was developed by the British Royal Aircraft Establishment and consists of oxidizing the atactic polymer at about 220°C while preventing it from shrinking. Further heating to 350°C results in the elimination of water and cross-linking of the chains which continues with loss of nitrogen. The fibers are finally heated to 1,000°C. The reactions are illustrated in Figs. 11.4, 11.5. The high tensile strength (3.2 GNm^{-2}) and Young's modulus (300 GNm^{-2}) are attributed to the alignment of the polymer chains and their cross-linking.

Carbon fibers have also been made from the pyrolysis of viscose (cellulose), rayon, and jute and from pitch. Though these methods produce slightly lower strength carbon fibers as compared to PAN, the lower cost ($\sim \frac{1}{5}$ to $\frac{1}{2}$) makes them excellent reinforcement materials for noncritical items such as golf clubs, tennis rackets, skis, and related sports goods.

11.5.7   Polymethyl Methacrylate

Polymethyl methacrylate (PMMA), also called *Plexiglas, Lucite,* or *Perspex,* is a colorless clear transparent plastic with excellent outdoor stability if UV absorbers are added to the polymer—otherwise, it yellows on exposure to sunlight. Like styrene, it also unzips on heating to reform the monomer. It has poor scratch resistance but was the plastic of choice for early contact lenses.

Fig. 11.4 Structure of
(**a**) PAN, (**b**) PAN ladder
polymer, (**c**) oxidized PAN
ladder polymer

11.5.8 Polyvinyl Acetate, Polyvinyl Alcohol

Vinyl acetate (CH_2=CH($OCOCH_3$)) is polymerized to polyvinyl acetate (PVAc) which is used in adhesives and lacquers. Its major use, however, is in the preparation of polyvinyl alcohol (PVAl) which cannot be prepared from vinyl alcohol (CH_2=CHOH) which isomerizes into acetaldehyde (CH_3CHO).

Polyvinyl alcohol is a water-soluble polymer which can be cross-linked into a gel by sodium borate ($Na_2B_4O_7$). This is shown in Fig. 11.6. Fibers made from PVAl can be made insoluble in water by cross-linking with formaldehyde, shown in Fig. 11.7. Such fibers are excellent substitutes for cotton because they absorb moisture (sweat) readily.

11.5.9 Polytetrafluoroethylene or Teflon

This polymer was discovered by accident. An old cylinder of gaseous tetrafluoroethylene (C_2F_4 B. P.—76°C) was found to have no gaseous pressure but still contained the original mass of material. When the cylinder was cut open, a white waxy hydrophobic powder was found. The polymerization process is highly exothermic, and it must be conducted with caution. The highly crystalline polymer is stable up to 330°C (its melting point) and is inert to strong acids, alkali, and organic solvents. It reacts with sodium leaving a carbon surface and NaF. This surface activation process allows Teflon to be bonded to other surfaces. The reaction of Teflon with hydroxyl free radicals (OH) can make the surface hydrophilic and bondable with ordinary adhesives (see Chap. 12).

Teflon tends to flow under pressure and is thus readily distorted. When filled with glass, the composite is stabilized and can be machined to precise dimensions.

Teflon cannot be injection molded because of the high viscosity of the melt and must therefore be formed by a compression of its powders. Another fluorinated polymer of comparable properties to

Fig. 11.5 (a) Cross-linking of PAN by intermolecular elimination of water, (b) cross-linking of dehydrated PAN by intermolecular elimination of nitrogen

$$Na_2B_4O_7 \rightarrow 2Na^+ + B_4O_7^{2-}$$

$$B_4O_7^{2-} + 9H_2O \rightleftharpoons 4B(OH)_4^- + 2H^+$$

Fig. 11.6 The cross-linking of polyvinyl alcohol by borax

Fig. 11.7 The cross-linking of polyvinyl alcohol with formaldehyde

Teflon is a blend of PTFE and polyhexafluoropropylene (FEP) made by polymerization of perfluoropropylene (C_3F_6). This plastic is not as thermally stable as Teflon (M.P.$=290°C$), but it is less opaque than Teflon and can be extruded, injection molded, or blow molded and thus presents some advantage over Teflon in particular applications.

A Teflon-like surface is made when polyethylene bottles are blown with nitrogen containing about 1% F_2. This makes the bottles less permeable to organic solvents and thus increases its usefulness.

11.6   Condensation Polymers

Some condensation polymers are listed in Table 11.2.

11.6.1   Nylon

Nylon is classed as a polyamide polymer prepared by the condensation of a dicarboxylic acid [HOOC—$(CH_2)_n$—COOH] and a diamine [H_2N—$(CH_2)_m NH_2$]. The plastic is characterized by the values of n and m, i.e., nylon m, n. Thus, nylon 6,6 (M.P.$=250°C$) has good tensile strength, elasticity, toughness, and abrasion resistance and has use as a fiber as well as a plastic. The melting temperature range of nylon is from 250° to 300°C. The aromatic polyamides have very high melting points ($>500°C$) and unusually high strength/weight ratio, of which the fiber Kevlar is an example.

11.6.2   Polyester

A condensation of a dicarboxylic acid and a diol results in a polyester.

$$HOOC - (CH_2)_x - COOH + HO - CH_2 - (CH_2)_y - CH_2 - OH \rightarrow -OCH_2 - (CH_2)_y$$
$$- CH_2O\Big(OC - (CH_2)_x - COOCH_2(CH_2)_y - CH_2 - O\Big)_n OC - (CH_2)_x - +nH_2O$$

$$(11.16)$$

The aliphatic polyester has a melting point of about 65°C, whereas the aromatic substituted dicarboxylic acid has a melting point of 265°C. Thus, the polyester polyethylene terephthalate (PETP) is commercially one of the most popular polymers marketed as Terylene or terene.

Table 11.2 Some condensation polymers

Nylon 66	$HOOC(CH_2)_4COOH + HN,—(CH,)_6—NH_2$
	↓
	$-[—NH–(CH_2)_6-NH-C-(CH_a)_4–C-]-$
	$\quad\quad\quad\quad\quad\quad\;\; \| \quad\quad\;\; \|$
	$\quad\quad\quad\quad\quad\quad O \quad\quad\; O$
Polyester	$HO—CH_2—CH,—OH + HOOC—(C_6H_4)—COOH$
	↓
	$HO—CH_2—CH_2—O—[—C—(C_6H_4)—C—0—CH_2—CH_2 \sim 0—]—C—$
	$\quad\quad\quad\quad\quad\quad\quad\quad\;\; \| \quad\quad\quad\quad\;\; \| \quad\quad\quad\quad\quad\quad\quad\quad\quad\quad \|$
	$\quad\quad\quad\quad\quad\quad\quad\quad\; O \quad\quad\quad\; O \quad\quad\quad\quad\quad\quad\quad\quad\quad\quad\; O$
Polyurethane	$H—[—O— CH_2— CH_2—]—OH + CH_3—(C_6H_3)—(NCO)_2$
	↓
	$(CH_3)(NCO)(C_6H_3) —NH—C—[—O— CH_2— CH_2—]—O—C—NH—(C_6H_3)(CH_3)(NCO)$

11.6.3   Polycarbonates and Epoxides

The condensation of a diphenol, (bisphenol-A), with dicarboxylic acid

$$(11.17)$$

forms the polycarbonate. These polymers melt at about 265°C and have very high impact strength, making them useful for helmets and safety shields.

The reaction of epichlorohydrin with bisphenol-A forms the diepoxy or diglycidyl ether.

$$(11.18)$$

The reaction of the diglycidyl bis-phenyl ether with a polydiol or polydiamine will result in a hardened thermosetting resin. Other thermosetting polymers are discussed in the next section.

11.7   Thermosetting Polymers

Polymers which form 3D network solids are thermosetting and decompose when heated and thus cannot be reshaped once they have set. A selection of the common polymers is listed in Table 11.3.

Table 11.3 Some thermosetting resins

Phenol-formaldehyde

$$C_6H_5-OH + 2HCHO \rightarrow HO-(C_6H_3)(CH_2OH)_2 \rightarrow HO-(C_6H_4)-CH_2-(C_6H_3)-OH$$

$$\begin{array}{ccc} & CH_2 & CH_2 \cdot \\ & | & | \\ & (C_6H_3)-CH_2-(C_6H_3) \\ & | & | \end{array}$$

Urea-Formaldehyde

$$O=C(NH_2)_2 + HCHO \rightarrow -NH-CO-N-CH_2-NH-CO-N-CH_2-NH-CO-N-CH_2-$$

$$\begin{array}{ccc} CH_2 & CH_2 & \\ | & | & -CH-NH-CO- \\ OH & -NH-CO-N-CH_2-NH-CO-N-CH_2- \end{array}$$

Melamine-formaldehyde

$$(C_3N_3)(NH_2)_3 + 3HCHO \rightarrow -NH-CH_2-NH-(C_3N_3)-NH-CH_2-NH-$$

$$NH-CH_2-NH-(C_3N_3)-NH-CH_2-$$

$$(C_3N_3)-NH-CH_2-NH$$

Epoxy polymer

$$CH_2-CH-CH_2-Cl + HO-(C_6H_4)-C(CH_3)_2-(C_6H_4)-OH$$

$$CH_2-CH-CH_2-[-O-(C_6H_4)-C(CH_3)_2-(C_6H_4)-O-CH_2-CH-CH_2-]-O$$

$$\begin{array}{c} | \\ OH \end{array}$$

11.7.1 *Phenol Formaldehyde (Bakelite)*

The first industrial plastic was developed by Baekeland in about 1907 and was called Bakelite. This was prepared by the reaction of phenol and formaldehyde in the presence of catalysts.

(11.19)

When heated in excess formaldehyde, cross-linking occurs, and the resin novolac is formed for P/F ~ 1.25.

11.7.2 Urea Formaldehyde

Urea

$$\left(\begin{array}{c} O \\ \| \\ NH_2-C-NH_2 \end{array} \right)$$

reacts with formaldehyde to form a cross-linked resin which is an inexpensive adhesive.

$$NH_2-\overset{\overset{\displaystyle O}{\|}}{C}-NH_2 + 2HCHO \rightarrow HOCH_2NH-\overset{\overset{\displaystyle O}{\|}}{C}-NHCH_2OH \tag{11.20}$$

On further addition of urea and HCHO, $H(NHCO-NH-CH_2)_n-OH$ is formed. With an acid catalyst, it is possible to produce a foam product known as urea formaldehyde foam insulation (UFFI) having a thermal conductivity, K, of about 0.022 Wm^{-1} K^{-1}.

In 1977 the Canadian government subsidized the introduction of UFFI in older homes to conserve energy.[2] The UFFI proved to be unstable in some cases due to improper installation, and as a result formaldehyde levels in some homes exceeded the threshold limit value (TLV) of 0.10 ppm (120 $\mu g/m^3$). Ammonia was able to neutralize the acid, and it was also shown that the water-soluble polymeric amine, polyethyleneimine, could remove the liberated formaldehyde. Nonetheless, the Canadian government then paid the homeowners an estimated $272 million ($5,000 to 57,700 homes) to remove the UFFI. The urea formaldehyde resin is commonly used as the adhesive resins in plywood and particle board and will initially release formaldehyde if not sealed. As more composite wood products find their way into buildings, greater concern about indoor air is warranted.

11.7.3 Polyurethane

This condensation polymer is unique insofar as it can be a coating and varnish, a soft or hard foam, a resilient or rigid elastomer (rubber) as well as an adhesive. It is prepared by the reaction of a diisocyanate (OCNRNCO) with a diol (HOR'OH) where R can be an aromatic radical such as toluene (TDI-2,4, toluene diisocyanate).

---

[2] Each of the 100,000 homeowners was given $500 toward the cost of adding UFFI.

and R' is an aliphatic radical $(CH_2)_n$ where the length n determines strength, toughness, and elasticity of the plastic. The reaction is

$$OCN - R - NCO \ + \ HO - R' - OH \ \rightarrow \ -R' - [OCO - NH - R - NH - CO - OR'-]_x O - CO$$

$$(11.21)$$

For the preparation of foams, the R component is a polyether or polyester with reactive end groups of hydroxyl and carboxyl. The reaction is

$$- RCOOH \ + \ OCN - R'- \rightarrow -R - CO - NH - R' + \ CO_2 \qquad (11.22)$$

where the liberated CO_2 foams the plastic into an open or closed cell sponge with densities of 25–50 g/dm^3 and which is often used in upholstery. The hard and rigid foams, having a density of 50–300 g/dm^3, are used as insulation and elastomers.

It is possible to replace the air in inflatable tires by *polyurethane* foam. This is feasible for low-speed vehicles used in road construction, service equipment, snowplows, street sweepers, as well as many other applications. The two components are blended together to produce the resilient foam in the tire which is then not susceptible to flats or punctures, a feature which reduces downtime and tire replacement costs. Polyurethane foam (closed cell) has a thermal conductivity of 0.022 Wm^{-1} K^{-1} and is usually covered with aluminum foil to reduce the heat loss due to the transmission of radiation.

11.8   Glass Transition Temperature

The melting point of a polymer is not a unique value unless it can be formed into a crystalline solid. The amorphous glassy solid is really a supercooled liquid. A polymer which does not have long-range order cannot exist in a crystalline state. As the temperature of an amorphous plastic is increased, the polymer chains begin to achieve segmental mobility. This is called the *glass transition temperature* (T_g), and the material is in a rubbery state. On further heating, the polymer chains begin to move and have molecular mobility— the plastic begins to flow. A graph showing the transition in terms of the variation of the specific volume (the reciprocal of density) as a function of temperature is shown in Fig. 11.8. The T_g and melting points of some polymers are listed in Table 11.4.

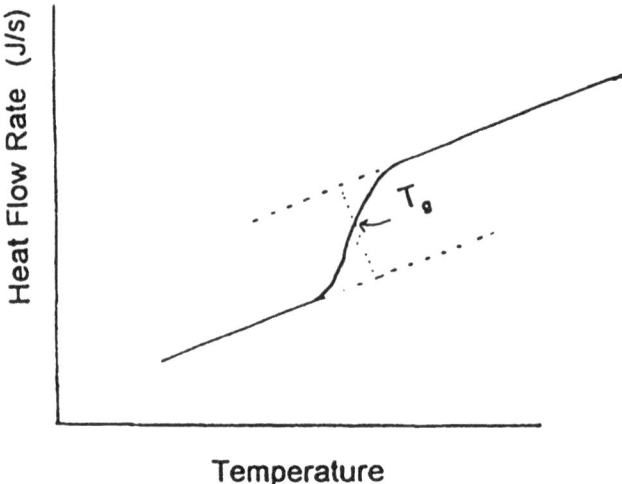

Fig. 11.8 The glass transition temperature is indicated by a change in heat flow of the material while the temperature increases linearly with time

Table 11.4 Theglass transition temperature (T_g) and melting temperature (T_m) of selected polymers

Polymer	Structural unit	Tg, °C	Tm, °C
1. Polybutadiene	$-CH_2-CH = CH-CH_2-$	Cis-102	6
		trans-58	100
2. Polyisoprene	$-CH_2-C=CH-CH_2-$ $\qquad\quad CH_3$	−75	65
3. Polystyrene	(benzene ring)$-CH-CH_3-$	100	240
4. Nylon 6,6	$-NH-(CH_2)_6-NH-O-CO-(CH_2)_4CO-O-$	50	270
5. Polyethylene	$-CH_2-CH_2-$	High density −125	140
6. Polypropylene	$-CH-CH_2-$ $\;\; CH_3$	Atactic-13	
		Isotactic-8	200
7. Poly methylacrylate	$-CH_2-CH-$ $\quad\;\; O=C-OCH_3$	5	
8. Polymethyl methacrylate	$\quad\;\; CH_3$ $-CH_2-C$ $\quad\;\; O=C-OCH_3$	72	
9. Polyvinylchloride	$-CH_2-CHCl-$	80	310
10. Polyethylene terphthalate	$-O-CH_2-CH_2-O-C$(benzene ring)$C-$ $\qquad\qquad\qquad O \qquad\qquad O$	70	260
11. Polycarbonate-basphenol A	$-O-$(ring)$-C-$(ring)$-O-C-$ $\quad CH_3 \qquad\qquad O$ $\quad CH_3$	150	225

Tg is dependent on degree of cross-linking area and molecular weight. The value of T_g increases as molecular weight of a polymer increases or as the branching or cross-linking increases. Thus, for PS #3, Table 11.4, $T_g = 100$°C which is much higher than for PE #5, $T_g = -125$°C. Similarly, the difference in T_g between polybutadiene (-102°C) #1 and #2 polyisoprene (-75°C) shows the effect of replacing a H by CH_3 in the side of a chain.

The T_g of a polymer can be reduced by the addition of a plasticizer to the solid plastic. This reduces the van der Waals interaction between the polymer chains and allows the molecules to move. The plasticizer may be considered as an internal lubricant. The plasticizer can also be considered to increase the free volume of the polymer by allowing increased motion of the chain ends, the side chains, or even the main chain. Another possible mechanism by which the plasticizer lowers the T_g is in terms of the solvent/solute system that forms when the plasticizer can be considered to solubilize the polymer. The plasticizer is usually a low volatile, low molecular weight organic compound which is compatible with the polymer.

11.9 Elastomers

Flexible plastics composed of polymers with T_g well below room temperature are classed as elastomers or rubbers. Natural rubber was known to the natives of South America for centuries, though it was not until Goodyear's discovery of vulcanization in 1839 that it became a practical product. Prior to this, rubber was used for waterproofing boots, clothing, and other weatherproofing surfaces. Goodyear showed that sulfur cross-linked the rubber and made it a manageable product. The polymer is based on the monomer isoprene

$$CH_2 = C - CH = CH_2$$
$$|$$
$$CH_3$$

which results in *cis* and *trans* forms (see Fig. 11.9). Of the two forms, the *trans* is less elastic because of the more ordered structure and more close packing of the molecules. In general, *elastomers* differ from plastics only because the elastomer is in a mobile "liquid" state whereas the plastic is in a glassy state. The transition between these two states occurs at the glass transition temperature T_g when the glassy state changes into the rubbery state. Below this temperature, the molecules are frozen into position and held in place by van der Waals forces.

Because of the residual carbon–carbon double bonds (C=C) in natural rubber, it is readily degraded by ozone, which adds to double bonds forming ozonides that eventually decompose, splitting the polymer chain.

Synthetic rubbers are made from chloroprene and butadiene which form neoprene and buna, respectively. The copolymer of acrylonitrile with butadiene (1,3) is known as nitrile rubber and styrene with butadiene (1,3) is Buna S. The combination of acrylonitrile, butadiene, and styrene in various formulations is used to form the thermoplastic ABS.

Some fluorinated polymers which show exceptional thermal stability and chemical inertness are Kel-F elastomers, made of a copolymer of chlorotrifluoroethylene-vinylidene fluoride $ClCF=CF_2/CH_2=CF_2$, and Viton, a copolymer of hexafluoropropylene and vinylidene fluoride $CF_2=CF—CF_3/CH_2=CF_2$.

Though stable at high temperature, these fluorocarbons show limited low-temperature flexibility. Silicone rubbers are made from dimethyldichlorosilane which under controlled hydrolysis form oils, gels, and rubbers.

Silicone rubber is more permeable to oxygen and carbon dioxide than most other polymers. A comparison of the permeability of these gases and water through various plastics is given in Table 11.5.

Fig. 11.9 Structures of natural rubber: *cis*, natural, herea and *trans*; gutta, percha, balata

Table 11.5 The gas permeability of various plastics at 25°C

Film	$P_r = \left(\dfrac{\text{mL gas (NTP) cm}}{\text{sec, cm}^2, \ \Delta P(\text{cm Hg})} \times 10^9 \right)$		
	O_2	CO_2	H_2O
1. Polyvinylidene chloride (Saran)	6.3×10^{-4}	4.8×10^{-4}	5.3
2. Monochlorotrifluoroethylene (Trithene A)	9.2×10^{-4}	0.010	0.8
3. Polyester (Mylar)	6.7×10^{-3}	0.012	32
4. Cellulose acetate	0.067	0.35	1,900
5. Opaque high density polyethylene	0.087	0.218	5.3
6. Polypropylene	0.115	0.400	15
7. Clear high density polyethylene	0.14	0.845	
8. Polystyrene	0.1	0.98	152
9. Low density polyethylene	0.35	1.09	25
10. Tetrafluoroethylene (Teflon)	0.67	1.88	67
11. Ethyl cellulose (Ethocel)	0.98	4.07	1,600
12. Natural rubber	2.5	13	
13. Fluorosilicone	1.1	64	
14. Nitrile silicone	8.5	67	
15. Silicone rubber (Silastic 372)	60	325	3,600
16. MEM-213 (G.E. Silicone Block Copolymer)	$16_{(N_2 7)}$	$97_{(H_2 21)}$	

Saran plastic shows the lowest permeability to O_2, CO_2, and very low for water. This feature makes Saran wrap an excellent packaging material for food in which freshness and flavor are to be preserved.

Silicone rubber, however, shows the highest permeability rates for these gases, and in fact, silicone rubber is used in blood oxygenators required for open heart surgery. It is also used in extended wear contact lenses since the transport of O_2 and CO_2 through the lens allows the cornea to respire. Thus, cloudiness and rainbows are not generally experienced, even after continuous wear for a month.

The permeability of various gases through silicone rubber is given in Table 11.6 and shows a broad variation. The permeability (P_r) of a gas through a plastic film is usually considered as equal to the product of the solubility (S) of the gas in the plastic and the rate of diffusion (D) of the gas in the plastic (actually the diffusion coefficient). Thus,

$$ P_r = \frac{D(\text{cm}^2/\text{sec}) \times 10^6 \times S\left(\dfrac{\text{mL NTP}}{\text{mL atm}}\right)}{76 \, \text{cm/atm}} \tag{11.23} $$

Some values of P_r, D, and S for O_2 and CO_2 at various temperatures are given in Table 11.7. When comparison is made for other gases and other plastic films, it becomes obvious that the solubility of gases in silicone rubber is not much different from other elastomers. Hence, the higher permeability of gases such as O_2 and CO_2 in silicone rubber is primarily due to higher diffusion coefficients due to more flexible O—Si—O bonds and to a much lower T_g (T_g, silicone rubber, is $-123°C$).

The high permeability of oxygen and CO_2 through silicone rubber suggests its possible use as an artificial gill. This is demonstrated in Fig. 11.10 in which a hamster lived in a (0.03 m^2) silicone rubber lined cage (30 L) submersed in air saturated water. When 35 L/min of the air saturated water is pumped around the cage, oxygen can be supplied to the 30 g hamster at a rate of 2.5 mL/min which is enough for its needs. The CO_2 is removed by the water flow and though the experiment could be continued for days, the molding of the food limited the duration of the experiment.

It may be noted in Table 11.6 that the permeability of O_2 is about twice that of nitrogen. Hence, it is possible to obtain air enriched in oxygen by collecting the gases which pass through several large

Table 11.6 The permeability of various gases through dimethyl-silicone rubber at 25°C

$$P_r = \frac{\text{mL gas (NTP) cm}}{\text{sec, cm}^2, (\Delta P) \text{ cm Hg}} \times 10^9$$

Gas	P_r	Gas	P_r
H_2	65	C_3H_8	410
He	35	n-C_4H_{10}	900
NH_3	590	n-C_5H_{12}	2,000
H_2O	3,600	n-C_6H_{14}	940
CO	34	n-C_8H_{18}	860
N_2	28	n-$C_{10}H_{22}$	430
NO	60	Freon 11	1,500
O_2	60	Freon 12	138
H_2S	850	H_2CO	1,110
Ar	60	CH_3OH	1,390
CO_2	325	Acetone	1,980
N_2O	435	Pyridine	1,910
NO_2	760	Benzene	1,080
SO_2	1,500	Toluene	913
CS_2	9,000	$COCl_2$	1,500
CH_4	95	Phenol	2,100
C_2H_6	250	Freon 22	382
C_2H_4	135	Freon 114	211
C_2H_2	2,640	Freon 115	51
CCl_4	5,835	Xe	171

Table 11.7 Effect of temperature on P_r, D, and S for O_2 and CO_2 in silicone rubber

Temp. (°C)		$P_r \times 10^9$	$D \times 10^6$	S
O_2	28	62	16	0.31
	−40	20	3.9	0.39
	−75	0.74	0.0012	47
CO_2	8	323	11	2.2
	−40	293	2.7	8.2
	−75	22	0.0022	770

Note: Unit for $P_r = \dfrac{\text{mL gas (NTP) cm}}{\text{sec, cm}^2(\Delta P \text{ cm Hg})} \times 100$

$D = \dfrac{\text{cm}^2}{\text{sec}} \times 10^6$

$S = \dfrac{\text{mL (NTP)}}{\text{mL atm}}$

membranes. The use of oxygen instead of air is advantageous in many processes such as combustion, steel manufacture, heating, welding and many others. For example, the removal of N_2 from the air used in the burning of natural gas results in a higher temperature and therefore more heat for the same amount of gas burnt. This is because when nitrogen is present, some of the heat of combustion is used to heat up the nitrogen which is both reactant and product.

This effect is even more pronounced if the nitrogen were to be removed from the air used in an internal combustion engine. The result would be a higher temperature of combustion and less work expended in the compression of the gases (see Exercise 11.8).

It is difficult to obtain continuous sheets of silicone rubber having a large area free of holes. It is, however, easier to draw capillary tubes and to stack these together giving very large areas. However, silicone rubber is not a thermoplastic and capillary tubing cannot be extruded. To get around this difficulty, General Electric prepared a block copolymer of silicone rubber and polycarbonate and

Fig. 11.10 Hamster in submerged cage fitted with silicone rubber membrane sides

because the polycarbonate is thermoplastic—the plastic (MEM 213) made from the copolymer can be molded and small bore capillary tubing can be readily fabricated. The polymer has most of the properties of pure silicone rubber with permeability rates of about 60% of the pure material.

Many modifications of the elastomers above have been developed chiefly by copolymerization with various synthetic resins. These modified elastomers furnish the engineer with a wide selection of materials suitable for special uses. An example of a relatively new hydrocarbon rubber is a copolymer of ethylene and propylene, EPR. This rubber has outstanding resistance to ozone weathering. It has lower tear resistence and higher air permeability than Buna S (SBR), but can be lined with butyl to correct the latter property.

11.10   Mechanical Strength of Plastics

The mechanical properties of materials are usually studied by means of tensile testing machines or dynamometers. The stress–strain curve obtained characterizes the plastic and determines its usefulness for specific applications.

Some design properties of common plastics are shown in Fig. 11.11. Shown also are relative costs as well as useful temperature range. Many of these properties can be improved by incorporating solid fillers into the plastics. Fillers which increase the mechanical strength are called *active fillers,* and include carbon black, titania, limestone, kaolin, silica, and mica. Their application to rubber and elastomers has been practiced for many years. The T_g of elastomers are usually increased by the addition of fillers. Good wetting of the filler by the polymer is essential for maximum effect. Thus, coupling agents are used (see Chap. 12) which bond to the solid and can react with the polymer. The size and shape of the filler particles also has an influence on its effectiveness. Thus, mica, which is a

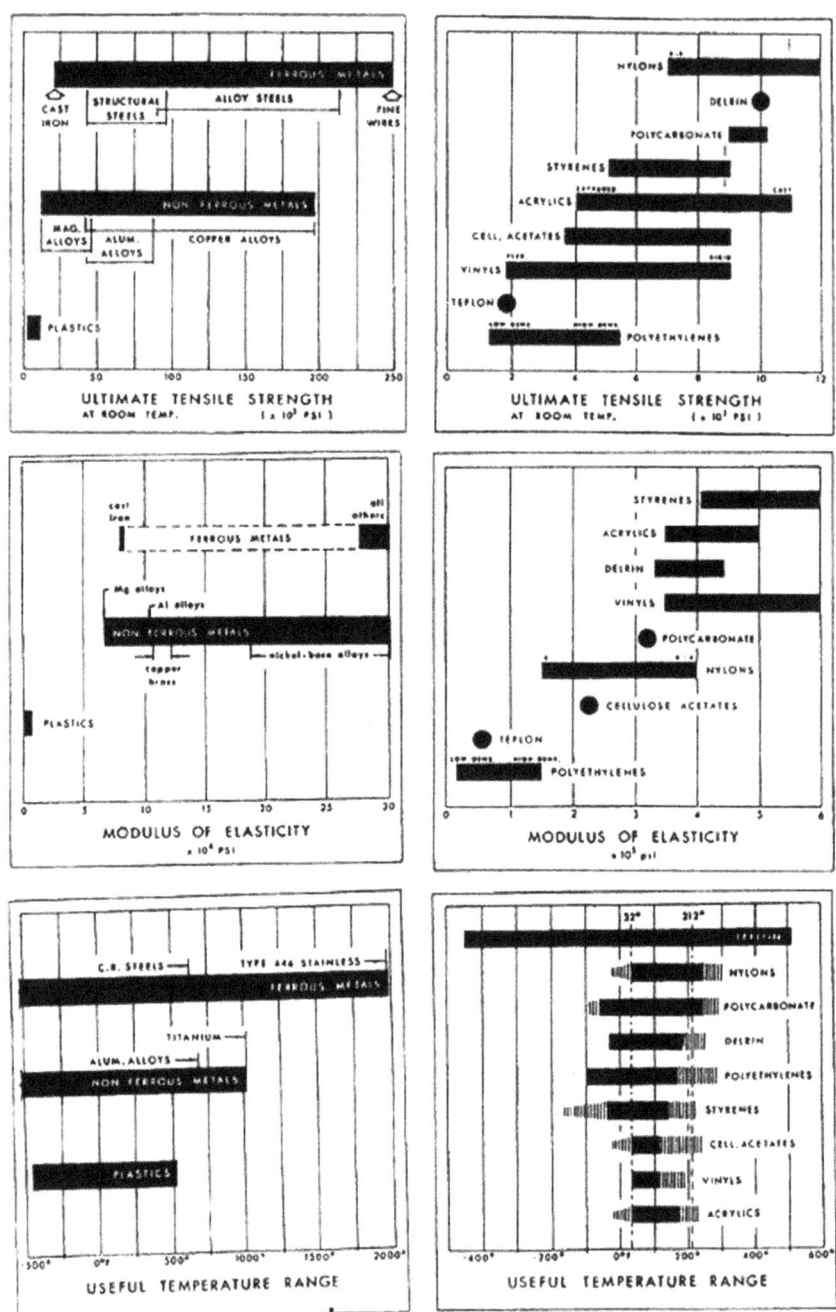

Fig. 11.11 Selected design properties or some plastics as compared to other common materials

layer lattice, is not spherical particles but thin platelets which can be split into thinner particles by ultrasonics. The length to thickness dimensions is called the *aspect ratio*. High aspect ratio (HAR) mica is much superior as a filler to ordinary mica or comparable amounts of silica or other fillers. The role of reinforcing fibers and binders in composite materials is discussed in Chap. 16.

11.11 Fire Retardants in Plastics

Plastics composed of polymer which have carbon and hydrogen are combustible. During the flammable process both thermal decomposition and combustion occur. A substance is classified as noncombustible if it does not produce flammable vapors when heated to 750°C. Few organic polymers can pass this test. Hence, most plastics burn, producing combustion products which can be toxic. Some of these gases are listed in Table 11.8. In the case of hydrogen cyanide (HCN), the amount produced is from 20 to 50% of the nitrogen present in the polymer. The major fire hazard is not the toxic gases but the smoke and lack of oxygen. Thus, smoke and fire retardants are essential ingredients in the formulation of plastics. The relative decrease in light transmission or degree of obscuration for some materials as determined in a specific apparatus is given in Table 11.9. The addition of flame retardants to materials may reduce fire but can at times increase the formation of smoke. The ease with which a substance will burn is determined by the minimum O_2 concentration (in N_2 as %) which will support combustion. This is called the limiting oxygen index (LOI) and some selected valves are given in Table 11.10. Thus, those materials with LOI $\leq$ 21 are combustible in air and must be treated to increase the LOI values

Fire retardants are additives to plastics and are usually based on some of the following elements: Al, B, Br, Cl, Mg, N, P, Sb, Sn, Zn. Halogen compounds (RX) produce halogen atoms (X) which act as chain terminators in the reacting vapors. Bromine compounds are often used but the formation of HBr in a fire makes it a corrosive retardant. A more inert fire retardant is alumina trihydrate ($Al_2O_3 \cdot 3H_2O$) which also acts as a smoke suppressant. It absorbs heat while liberating water at 230° to 300°C. Another inorganic fire retardant is zinc borate ($ZnO \cdot B_2O_3 \cdot H_2O$) which liberates water

Table 11.8 Toxic degradation products from combustible materials in air

CO and CO_2	From all substances containing carbon
NO_x	Polyurethanes
HCN	Wool (150), silk, nylon (100), polyurethanes (40), polyacrylonitrile (200)
SO_2	Rubber
Hydrohalides	PVC, fluorinated plastics, polyvinylidene chloride
Phosgene	Chlorohydrocarbons
NH_3	Melamine, nylon, urea formaldehyde
Benzene	Styrene
Phenol	Phenol formaldehyde
CO	PUF (600), nylon (450), wool (300)

Values in parentheses are average values in mg/g of material

Table 11.9 Relative obscurance due to smoke formation during the combustion of various materials

Material	Obscuration (%)
Oak	2
Pine	48
Acrylic	2
Polystyrene	100
PVC	100
Polyester	90
Plaster board	1
Flame-retardant acrylic	97
Polyester	99
Plywood	15

Table 11.10 Limiting oxygen index (LOI) values of selected materials

Material	LOI
Polyurethane foam	16.5
Polymethyl methacrylate	17.3
Polyethylene	17.4
Polypropylene	17.4
Polystyrene	18.0
Acrylic fiber	18.2
Cotton	18.4
Nylon fiber	20.1
Polyester fiber	20.6
Oak	23
Polycarbonate	25.0
Wool	25.2
Nylon 6,6	28.7
Kevlar	29
Polyvinylidene fluoride	43.7
Polyvinyl chloride	47
Teflon	95
Polyvinylidene chloride	60
Carbon	65

Table 11.11 Selected fire retardants

Al_2O_3–$3H_2O$	Alumina trihydrate
$ZnOB_2O_3$–H_2O	Zinc borate
Sb_2O_3	Antimony oxide
NH_2CONH_2	Urea
$(NH_4)_2HPO_4$	Diammonium hydrogen phosphate
$Mg(OH)_2$	Magnesium hydroxide
$Br_3C_6H_4$—$CH{=}CH_2$	Tribromostyrene
C_6Br_5—CH_2—O—CO—$CH{=}CH_2$	Pentabromobenzyl acrylate
MoO_3	Molybdenum trioxide
CuC_2O_4	Cupric oxalate

at 245°–380°C and is used to supplement halogen retardants. A list of selected fire retardants is given in Table 11.11. The increase in LOI of 10% (from 18% to 28%) represents an effective application of retardants.

The flame characteristics of a plastic and a measure of its density can often be used to identify the polymer. Polymers can conveniently be identified by a quick test of IQ infrared absorption spectra. A short list of tests of a few plastics is given in Table 11.12. In general, aromatic substances burn with smoky flames. Chloride can be tested with the plastic-coated copper wire which shows a green color in the colorless part of a Bunsen flame.

The growth of the polymers and plastics industries has meant that a major fraction of the workforce in developed countries is either directly or indirectly employed by plastic-related jobs. As more stable and less costly plastics are developed, more applications are found and growth continues. Because polymers and plastics are based on petroleum and since petroleum is a limited resource, it is essential that continued efforts be made to recycle our plastic wastes—something that is slowly being realized.

Table 11.12 Combustion and density tests for plastics identification

Polymer	Density[a] (g/cm^3)	Type	Odor	Color	Character
PE	0.91–0.98	TP	Candle	Yellow	Burns, melts, drips
PP	0.89–0.92	TP	Wax, candle, wax	Yellow	Burns, melts, drips
PS	1.05–1.07	TP	Styrene	Smoky	Burns
PMMA	1.24	TP	Methanol	Yellow	Burns slowly
PET	1.39	TP	Aromatic	Smoky	Slightly acidic fumes
PC	1.2	TP	Benzene	Smoky	Neutral vapors, difficult to ignite, SE
PAN	1.2	TP	HCN	Smoky	
PF	1.28	TS	Phenol	Yellow smoky	Basic fumes, SE, difficult to ignite
PU	1.2	TS	Acid	Yellow	Acidic or basic fumes
PVC	1.39	TP	Acid	Yellow	Acidic fumes, SE
N	1.14	TP	Ammonia	Bluish	Burns, basic fumes, SE
ABS	1.0	TP	Styrene	Yellow	Burns

[a]Values ±0.1 to 0.2, *SE*, self-extinguishing, *TP* thermoplastic, *TS* thermosetting, *PE* polyethylene, *PP* polypropylene, *PS* polystyrene, *PMMA* polymethyl methacrylate, *PET* polyethylene terephthalate, *PC* polycarbonate, *PAN* polyacrylonitrile, *PF* phenol formaldehyde, *PU* polyurethane, *PVC* polyvinyl chloride, *N* nylon, *ABS* acrylonitrile–butadiene–styrene

Exercises

1. From the data given in Sect. 11.1, calculate the annual production of rubber in 1990.
2. The table below gives the fraction of molecules of a polymer sample having a given average molecular weight. Calculate the number average molecular weight, $\overline{M}_n$ and the weight average MW, $\overline{M}_w$.

MW	10,000	14,000	16,000	19,000	22,000	24,000	28,000	32,000
Fraction	0.12	0.14	0.16	0.22	0.12	0.1	0.08	0.06

3. What conclusion can you reach if both $\overline{M}_w$ and $\overline{M}_n$ are determined to be identical?
4. The MW of a polymer dissolved in a solvent was determined from viscosity measurements at 25°C for various concentrations

C(g/100 mL)	0.25	0.50	1.00	2.00
η/η_Q	1.36	1.8	2.8	6.1

The ratio η_{sp}/C is to be plotted against C and extrapolated to zero concentration in order to obtain the intrinsic viscosity $[\eta]$. The constants for Eq. (11.7) are $K=3.8 \times 10^{-4}$ and $\alpha = 0.92$ for this polymer-solvent-temperature system when the concentration is in g/100 mL Calculate the MW of the polymer.

5. Polyvinyl alcohol (PVA) is soluble in water. What will be the freezing point of a 3% solution of PVA (MW 50,000 g/mol) in water? [*Note:* 1 molal solution depresses the F.P. by 1.86°C]
6. Write the chemical reaction showing the formation of the following polymers from initial reactants: (a) nylon, (b) lexan, (c) terylene, (d) polyurethane.
7. Floor tiles of plasticized PVC can be made more flexible, more scratch resistant, to have a longer lifespan and have increased color fastness when treated with γ-radiation. Explain.
8. The permeability equation of a gas through a membrane is

$$N = A(P_i X_i - P_0 X_0)P_r/l$$

where N = mL/s (NTP) gas flow; l = film thickness in cm; P_r = permeability; P_i = pressure of feed gas in cmHg; X_i = mole fraction of diffusing component in feed gas; P_0 = pressure of product gas in cm of Hg; X_0 = mole fraction of component in product gas; A = area in cm^2. For air P_0 = 76 cm, X_i=0.2.

    (a) If we assume l = 1 miL (0.025 mm), P_0—38 cm, X_0—0.3, calculate the area required for an internal combustion engine burning up to 150 mL/min of gasoline.

    (b) If P_i is increased to 3 atm, would the required area be reduced significantly?

9. A man at rest uses 300 mL O$_2$/min. Seawater contained 4 mL O$_2$/L. What area of membrane is required to act as an artificial gill if water is initially saturated at 160 mmHg of O$_2$ and can be depleted to 100 mm? What volume of water must be passed over the membrane each minute to achieve this oxygen Mow to support a man under water?

10. It has been suggested that polymeric membranes be used to separate alcohol from water. What advantage would this have over distillation?

11. Distinguish between thermosetting plastics, thermoplastics, and elastomers.

12. How is polyvinyl alcohol prepared?

13. How much dioxygen is used by an automobile engine:
    (a) idle; (b) moving 50 km/h?

Further Reading

1. Kacapio, Norton IT, Ubbink JB (2009) Modern biopolymer science, 5 edn.
2. Brydson JA (1999) Plastic materials, 7th edn. Butterworth-Heinemann, Newton
3. Strong AB (1999) Plastics, 2nd edn. Prentice-Hall, Upper Saddle River
4. Elias HG (1997) An introduction to polymer science. VCH, New York
5. Harper CA (ed) (1996) Handbook of plastics, elastomers, and composites, 3rd edn. McGraw-Hill, New York
6. Luise RR (1996) Applications of high temperature polymers. CRC Press, Boca Raton
7. Mark JE (ed) (1996) Physical properties of polymers handbook. American Institute of Physics, Cincinnati
8. Salamone JC (ed) (1996) Polymeric materials encyclopedia. CRC Press, Boca Raton
9. Fried JR (1995) Polymer science and technology. Prentice-Hall, Englewood Cliffs
10. Ravve A (1995) Principles of polymer chemistry. Plenum, New York
11. Osswald TA, Menges G (1995) Materials science of polymers for engineers. Hanser-Gardner, Cincinnati
12. Santappa M (1995) State of the art in polymer science and engineering. International Special Book, Portland
13. Zaikov GE (ed) (1995) New approaches to polymer materials. Nova Science, Commack
14. Zaikov GE (ed) (1995) Polymers in medicine. Nova Science, Commack
15. Zaikov GE (ed) (1995) Flammability of polymeric materials. Nova Science, Commack
16. Zaikov GE (ed) (1995) Kinetic and thermodynamic aspects of polymer stability. Nova Science, Commack
17. Zaikosv GE (ed) (1994) Degradation and stabilization of polymers: theory and practice. Nova Science, Commack
18. Ward IM, Hadley DW (1993) An introduction to the mechanical properties of solid polymers. Wiley, New York
19. Domininghaus H (1993) Plastics for engineers: materials, properties, applications. Hanser-Gardner, Cincinnati
20. Rosen SL (1993) Fundamental principles of polymeric materials, 2nd edn. Wiley, New York
21. Ulrich H (1993) Introduction to industrial polymers. Hanser-Gardner, Cincinnati
22. Sperling LH (1992) Introduction to physical polymer science, 2nd edn. Wiley, New York
23. Vieth WR (1991) Diffusion in and through polymers: principles and applications. Hanser-Gardner, Cincinnati
24. Brandrup J, Immergut EH (eds) (1989) Polymer handbook, 3rd edn. Wiley, New York
25. Gowariker VR, Viswanathan NV, Sreedkar J (1986) Polymer science. Wiling Eastern, New Delhi
26. Billmeyer FW (1984) Textbook of polymer science, 3rd edn. Wiley-Interscience, New York
27. Rudin A (1982) The elements of polymer science and engineering. Academic, Toronto
28. J Chem Educ (1981) Full issue devoted to polymer chemistry
29. Brydson JA (1980) Plastic materials, 2nd edn. Newnes-Butterworths, London
30. MacGregor EA, Greenwood CT (1980) Polymers in nature. Wiley, New York
31. Tager A (1978) Physical chemistry of polymers, 2nd edn. (English). MIR, Moscow
32. Lee H, Neville K (1971) Handbook of biomedical plastics. Pasadena Technology Press, Pasadena
33. Palin GR (1967) Plastics for engineers. Pergamon Press, Oxford

34. American Plastic Council, http://www.ameriplas.org/
35. History of plastics, http://www.polymerpIastics.com
36. About silicone rubber, http://www.chemcases.com/silicon/
37. Sources of polymers, http://www.polymer-search.com
38. Chemical searcher, http://www.chemconnect.com
39. Polymer processing, http://www.polymerprocessing.com/sites.html
40. http://www.plastics.com/
41. http://www.canplastics.com/
42. Polymers in industry, http://www.chemindustry.com/
43. www.wikipedia.org

Chapter 12
Adhesives and Adhesion

12.1 Introduction

Adhesives play an important part in modern technology. They are usually defined as any substance which holds two solids together—by forming bonds between the solids as distinct from bolting, riveting, or even brazing, welding, or soldering. One of the most common examples of *adhesion* and one not yet understood completely is the barnacles which strongly adhere to ships' hulls and which are supposed to have influenced the battle of Salamis in 480 B.C.

Early glues were materials found in nature and include resins from trees, gums such as gum arabic, pitch and tar, egg, cheese, fish extracts, and flour paste. The introduction of rubber improved the versatility of the glues, but it was not until aircrafts were produced for the World Wars (I and II) that the demand for strong lightweight adhesives forced the development of special glues. The Mosquito aircraft of World War II was an example of the application of modern resin adhesives.

Adhesives offer many advantages over normal or conventional methods such as bolting, riveting, welding, stitching, clamping, or nailing. It offers the following:

1. The ability to join different materials such as plastic to metal, glass, or rubber.
2. The ability to join thin sheets together effectively.
3. Improved stress distribution in the joint which imparts a greater resistance to fatigue in the bonded components.
4. Increase in the flexibility of design.
5. Convenient and cost-effective method in production.

These advantages have led to an ever increasing application of adhesives to a wide variety of industries which include automobiles and aircrafts, garments, furniture, appliances, buildings, and floor coverings.

It is interesting that the theory of adhesion is less developed than its application, and it is only as a result of increasing demands by industry that the science of adhesion is developing. One difficulty is that the study of adhesion is a multidisciplinary subject and involves aspects of surface chemistry and physics, organic and inorganic chemistry, polymer chemistry and physics, rheology, stress analysis, and fracture phenomena. Thus, it is easier at present to practice the art of adhesion than the science of adhesion. Nevertheless, some simple principles have been shown to be effective in designing a strong adhesive joint, and these will be presented and discussed.

O.V. Roussak and H.D. Gesser, *Applied Chemistry: A Textbook for Engineers and Technologists*,
DOI 10.1007/978-1-4614-4262-2_12, © Springer Science+Business Media New York 2013

12.2   Classification and Types of Adhesives

Adhesives can be classified in many different ways, e.g., by application or setting, chemical composition, cost, materials to be bonded, and end products or use.

Adhesives can be applied as (a) a melted solid which sets when cooled, (b) a solution or suspension in a solvent which evaporates, and (c) unpolymerized or partly polymerized blend which sets when polymerization is completed.

(a) Hot melt adhesives are usually thermoplastic polymers or waxes which are heated to a temperature sufficiently high to allow the melt to wet the substrate and readily flow in the joint. Examples of such adhesives are asphalt, ethyl cellulose, cellulose acetate butyrate, polyethylene copolymers, and waxes. Safety glass is made by softening polyvinyl butyral by heat and bonding the two glass pieces under pressure.
(b) These adhesives are usually slow setting if the solvent must diffuse through the joint or the material to be bonded. To circumvent this aspect, contact (adhesive) cements have been developed where the solvent is allowed to evaporate from the two surfaces before they are brought together. The cement consists of low molecular weight elastomers, or for aqueous systems, suspended latexes are used usually at over 50% solids, but still fluid enough to flow and wet the surfaces. The elastomers used are natural and synthetic rubbers, vinyl resins, and acrylics.
(c) This type of reaction adhesive includes all thermosetting resins, as well as elastomers, which can be cross-linked, and some vinyl thermoplastics. The adhesive is usually applied as a low viscosity wetting blend which polymerizes to a strong hard adherent glue. Examples of these types include epoxy adhesives, cyanoacrylate esters (crazy glue), urea–formaldehyde resins, and urethanes.

The further classification of adhesives can be made by referring to the nature of the material, organic, inorganic, or hybrid. Thus, organic adhesives are either natural (e.g., starch, gelatin, shellac), semisynthetic (e.g., cellulose nitrate, castor oil–based polyurethanes), or synthetic (e.g., all vinyl polymers, plastics). The inorganic adhesives include cement, silicates (water glass), and sulfur and ceramic cements, whereas hybrid adhesives include litharge (PbO + glycerol 3:1), $AlPO_4$ in kerosene, and silicone rubbers.

Some of the adhesives can be used in strong acid, at high temperatures, in high vacuum, or other highly specialized applications. A detailed comprehensive treatment of adhesives can devote a full volume to classification. We shall hereafter concern ourselves with the form of the adhesive, e.g., some adhesives have been encapsulated (NCR), others are pressure sensitive (adhesive tapes), and some are flexible (rubbers). The electrical or thermal conductivity of adhesives (usually achieved by the addition of metal powders) is required in special applications. Adhesives are slowly penetrating into the garment industry where their replacement of stitching can speed up production and reduce costs. Similar changes are being made in the automobile where plastic is replacing metal at ever increasing rates.

12.3   The Adhesive Joint

The adhesive joint (AJ) is primarily composed of five parts (see Fig. 12.1): the two solid materials bonded (A and A'), the two interfaces or boundary layers between the materials and the adhesive (B and B'), and the bulk adhesive (C). A strong AJ implies (1) strong boundary layers, (2) strong interfacial bonds, and (3) a strong or hard-set adhesive. Weak boundary layers have been attributed by Bikerman to be responsible for most weak joints. In the case of elastomers or plastics, it is possible to show that even if interfacial or surface bonding of the adhesive is strong, a weak boundary layer results in a weak joint as the materials are separated and the surface molecules are pulled out of the

Fig. 12.1   The adhesive joint: a representation of the five component parts on an adhesive joint. The adherend A and A′, the interfaces or boundary layers B and B′, and the adhesive C

bulk of the material. Thus, though strong bonds may be formed between the materials and the adhesive layers, the surface bound molecules are not held very strongly by the material bulk. This can be corrected by cross-linking the surface and creating one giant surface molecule which is anchored to the bulk by numerous bonds thereby making its withdrawal from the surface energetically too high, resulting in strong boundary layers. This process of cross-linking the surfaces of a plastic or elastomer is called CASING (cross-linking by activated species of inert gases) and was developed in 1966 by Hansen and Schonhorn at the Bell Telephone Laboratories. CASING can be effected by subjecting the surface to hydrogen atom reactions in a vacuum discharge through He–H_2 mixtures.

The surface of a solid is usually contaminated by adsorbed gases and vapors, and as a result, the adhesives may form a poor or weak contact with the actual surface. The surface adsorbed vapors can be displaced by a liquid adhesive which "wets" the solid. In some cases, a surface-active agent may be added to an adhesive to facilitate the wetting process (Fig. 12.2).

An alternate method of increasing the strength of an AJ is to add coupling agents to the solid surface, thereby introducing an intermediate molecule which has reactive functional groups that can bond to the adhesive and thereby strengthening the bond. This is done to the surface of glass fibers used to reinforce radial tires and in fiberglass plastics. Some examples of coupling agents and their use are listed in Table 12.1.

These compounds which have alkoxide groups will hydrolyze on a hydroxylated surface to form an alcohol and the surface-bonded organic residue. For example,

$$\begin{aligned}
&Si\!\!-\!\!OH \quad + NH_2\!\!-\!\!CH_2CH_2CH_2Si(OC_2H_5)_3 \\
&Si\!\!-\!\!OH \qquad Si\!\!-\!\!O \\
&Si\!\!-\!\!OH \rightarrow \quad Si\!\!-\!\!O\!\!-\!\!Si\!\!-\!\!CH_2CH_2CH_2NH_2 \quad +3C_2H_5OH \\
&\qquad\qquad\qquad Si\!\!-\!\!O
\end{aligned}$$

(12.1)

The organic amino group can react further with the adhesive and contribute to the bond strength of the joint.

Another method used to improve the strength of a bond is to introduce the chemical reactive functional groups directly on the surface of the material. For example, the bonding of Teflon or polyethylene can be improved by introducing OH groups onto the plastic surface by sparking moist air next to the surface to be joined. This is best done by means of a Tesla coil or a Corona discharge,

Fig. 12.2 The effect of coupling agents on the binding of glass fibers and glass beads in adhesives and plastics. The example is that of a reaction injection molded polyurethane composite which has been fractured to expose the reinforcing glass fibers (**a** and **b**) and glass beads (**c** and **d**). (**a**) Untreated glass fibers; the photomicrograph shows holes where the poorly bonded fibers have been clearly removed. The extreme close-up shows a typical glass fiber which is not coated with polymer. (**b**) Glass fibers treated with coupling agent: The chemically coupled system shows fewer fibers pulled out in comparison with (**a**). The extreme close-up shows an excellent coat and adhesion of polymer attainable when silanes produce a chemical bond between the organic matrix and the inorganic reinforcement. (**c**) Untreated glass beads: The filler is clean and free of adhering polymer. (**d**) Treated glass beads: the glass shows strongly adhering polymer

Table 12.1 Some selected commercial silane coupling agents

Name	Formula	Application
Vinyltriethoxysilane	$CH_2=CHSi(OC_2H_5)_5$	Unsaturated polymers
Vinyl-tris(b-methoxyethoxy)silane	$CH_2=CHSi(OCH_2CH_2OCH_3)_3$	Unsaturated polymers
Vinyltriacetoxysilane	$CH_2=CHSi(OOCCH_3)_3$	Unsaturated polymers
γ-Methacryloxypropyltrimethoxysilane	$CH_2=C(CH_3)COO(CH_2)_3Si(OCH_3)_3$	Unsaturated polymers
γ-Aminopropyltriethoxysilane	$H_2NCH_2CH_2CH_2Si(OC_2H_5)_3$	Epoxies, phenolics, nylon
γ-(2-Aminoethyl)aminopropyltrimethoxysilane	$H_2NCH_2CH_2NH(CH_2)_3Si(OCH_3)_3$	Epoxies, phenolics, nylon
γ-Glycidoxypropyltrimethoxysilane	$\overset{O}{\overset{\displaystyle\diagup\!\diagdown}{CH_2CHCH_2O(CH_2)_3Si(OCH_3)_3}}$	Almost all resins
γ-Mercaptopropyltrimethoxysilane	$HSCH_2CH_2CH_2Si(OCH_3)_3$	Almost all resins
β-(3,4-Epoxycyclohexyl)ethyl-trimethoxysilane	$OC_6H_9\text{—}CH_2CH_2Si(OCH_3)_3$	Epoxies
γ-Chloropropyltrimethoxysilane	$ClCH_2CH_2CH_2Si(OCH_3)_3$	Epoxies

but a vacuum high voltage ac discharge through water vapor is most effective. Such surfaces become wet with respect to water and can form strong bonds with common glues such as epoxy adhesive.

More examples of the AJ will be given after a discussion of the theory of the adhesive bond.

12.4 The Theory of the Adhesive Bond

The theory of the adhesive bond (AB) is basically an attempt to generalize the mechanistic account of adhesion into a comprehensive explanation. One of the major difficulties in realizing a uniform consistent theory is due to the many variable factors which determine the strength of an AJ and the methods used to evaluate them, e.g., geometrical factors, loading factors, rheological energy losses, and interfacial interactions, to mention only a few. There are four principal mechanisms used to account for adhesion. These are the following:

1. Mechanical interlocking (E_M)
2. Diffusion theory (E_D)
3. Electrostatic theory (E_E)
4. Adsorption theory (E_A)

12.4.1 Mechanical Interlocking

This early theory was successfully applied to joints of porous or rough adherends such as wood. The liquid adhesive penetrates the porous and irregular surface and when hardened adds strength to the adhesive joint. Thus,

1. The greater the surface irregularity and porosity, the greater is the strength of the joint.
2. The joint strength will be proportional to the film strength of the adhesive when the adherend is stronger than the adhesive.

Some examples in which the mechanical mechanism is important are in the adhesion of polymers (elastomers and rubbers) to textiles. Another example, though somewhat contentious, is the metal plating of a plastic which usually requires a pretreatment to modify the surface topography of the polymer. Usually the increase in adhesion is also attributed to an improved surface force component due to the increased rugosity.

The bonding of maple wood samples with urea–formaldehyde resins at 5 psi gluing pressure was tested in shear as a result of surface treatment.

Surface treatment	Increasing roughness → Planed	Sanded	Sawn	Combed
Shear strength (psi)	3,120	2,360	2,690	2,400

This indicates that the smoother surface forms the stronger joint. This is due to the wood fibers becoming damaged by treatment and are easily removed from the bulk when the joint is stressed and separated. The trapping of air within the pores prevents the anchoring of the glue. Thus, mechanical interlocking has limited application, and usually strength is associated with other forces.

12.4.2 Diffusion Theory

This theory, put forth by Voyutskii, accounts for the autoadhesion of polymers by the interdiffusion of polymers across the interface and their mutual solubility. Direct radiometric and luminescence experiments have confirmed that the diffusion boundary may be as deep as 10 μm. Interdiffusion is also important in solvent welding of plastics where the solvent essentially plasticizes the surfaces and promotes diffusion between the two materials. Thus, though diffusion does occur in special cases, the theory has very little general application because many adhesives show no solubility in the adherends, e.g., glass, metal, and wood.

12.4.3   Electrostatic Theory

This theory, proposed by Deryaguin, arose from the crackling noises and flashes of light observed when adhesive films are peeled off rapidly in the dark. Likewise, film strengths (of the order of 1–100 mJ/cm^2) depend on the speed at which it is peeled. The Russians proposed that an electrical double layer is formed at the junction of two different materials due to the difference in the energy bands of the materials. The separation of the adhesive–adherend interface is then analogous to the plate separation of a capacitor. This should depend on the dielectric, but such peel strengths show no difference when measured in air or vacuum. Calculated values of the work of adhesion for gold, copper, and silver on glass show that the electrostatic contribution is 5, 80, and 115 mJ/m^2, respectively, whereas the van der Waals contribution is 950, 400, and 800 mJ/m^2, respectively. Measured values for the three metals were 1,400, 800, and 1,000 mJ/m^2, respectively, with errors of about ± 200 mJ/m^2. Thus, though electrostatic forces contribute to the strength of an adhesive joint, the major contributing factor appears to be the ever present van der Waals force.

12.4.4   Adsorption Theory

This theory relies on the orientation of surface forces resulting in van der Waals interaction. There are three general types of van der Waals forces: (a) Keesom, (b) Debye, and (c) London.

(a) Keesom forces are due to the dipole–dipole interaction. Since the dipoles are assumed to be aligned, this is also called the *orientation effect* and is given by the potential energy of attraction V_K.

$$V_K = \frac{-2\mu^4}{3kTr^6} \quad \text{or} \quad V_K = \frac{-2\mu_1^2\mu_2^2}{3kTr^6} \tag{12.2}$$

for molecules with different dipole moments μ_1 and μ_2, and where k is the Boltzmann constant, T is the absolute temperature, and r is the distance of separation.

(b) Debye forces are due to an induced dipole effect or polarization. The attractive potential energy V_D of two different molecules is

$$V_D = -\frac{\alpha_1\mu_2^2 + \alpha_2\mu_1^2}{r^6} \tag{12.3}$$

where α is the polarizability and μ is dipole moment. This force is temperature independent and therefore important at high temperatures.

(c) London forces are due to induced dipole–induced dipole attraction. Thus, asymmetrical molecule will exhibit oscillations inducing dipoles in neighboring molecules. The London attractive potential energy V_L is

$$V_L = \frac{-3h v_1 v_2}{2(v_1 + v_2)} \cdot \frac{\alpha_1 \alpha_2}{r^6} \tag{12.4}$$

where α_1 and α_2 are the polarizabilities, h is Planck's constant, and v_1 and v_2 are the characteristic frequencies of the molecules.

Since $hv = J$, the ionization energy of the molecule, then

$$V_L = -\frac{3}{2}\frac{I_1 I_2}{(I_1 + I_2)} \cdot \frac{\alpha_1 \alpha_2}{r^6} \tag{12.5}$$

The London forces are also called *dispersive forces* and contribute the major proportion of the bonding for nonpolar or weakly polar molecules.

The adsorption theory requires intimate molecular contact between surfaces for a strong adhesive bond. Van der Waals forces when operative are considered as secondary bonds, whereas ionic, covalent, and metallic forces contribute to primary adhesive bonds.

The hydrogen bond is intermediate in strength between primary and secondary forces. The results of corona discharge on polymers lead to the formation of hydroxyl groups on the surface, which improves the adhesive strength by hydrogen bonds or by covalent bonds (as in the case of coupling agents).

Intimate interfacial contact is promoted by the wetting of surfaces by a fluid adhesive which also displaces air from the surface if it has low viscosity prior to setting. This is the major characteristic of crazy glue that makes it so effective.

Wetting alone may not always give rise to a strong adhesive joint. A weak boundary layer can cause failure at what appears to be the interfacial boundary of adhesive–substrate but which is claimed to be at the surface layer of the substrate. CASING is supposed to strengthen this layer and thereby improve the strength of the joint. An alternate explanation for the observed effects of glow discharge treatment is that the surface free energy of the polymer substrate may be increased or chemically active species may be formed on the surface—both of which could account for the observed increase in adhesion.

Insofar as it is possible to cite examples which could fit each of the above mechanism, it is best to describe the overall adhesive strength to result from a combination of each such that

$$E_{\text{total}} = \alpha E_M + \beta E_D + \gamma E_E + \delta E_A \tag{12.6}$$

where the coefficients vary depending on the adhesives and adherends. It is not possible, at present, to assign values to these coefficients except in special cases. This is of little consequence—especially if the adhesive does the job and the bonded joint holds together.

12.5 Chemistry of Selected Adhesives

One of the most common and versatile adhesives is the epoxy resin, usually used as a two-part formulation which sets into a hard adherent mass from 5 min to 24 h, depending on the length of the prepolymer and the choice of the hardener.

The epoxy resins were discovered in 1938 during the search for a denture resin. The epoxy group,

$$-\overset{|}{\underset{|}{C}}-\overset{|}{\underset{|}{C}}-$$
$$\diagdown\!\!\diagup$$
$$O$$

is highly reactive, and the ring is readily opened by acidic or basic catalysts. Thus, an amine (primary or secondary) will react with an epoxide to form an amine alcohol

$$RNH_2 + H_2C\!\!-\!\!CH\!\!-\!\!CH_2\!\!-\!\!O\ldots \rightarrow RNH\!\!-\!\!CH_2\!\!-\!\!CH\!\!-\!\!CH_2\!\!-\!\!O\ldots \tag{12.7}$$

Other catalysts which react with the epoxide ring are anhydrides (which form ester alcohols).

The epoxide intermediates (short-chain nonviscous polymers) commonly used are made by the reaction of epichlorohydrin

$$(CH_2\text{---}CHCH_2Cl)$$
$$\diagdown O \diagup$$

with bisphenol A

e.g.,

$$(12.8)$$

where n varies from 0 to 20 or the molecular mass is between 900 and 5,000. When n is greater than 2, the resin is a solid with a melting point that increases as n increases. Since it is desirable for adhesives to be liquid, and thus flow in the joint, the prepolymer is small, i.e., $n \leq 2$. However, each polymer molecular has n hydroxyl groups (—OH) which increase the curing rate of the adhesive and because of its polarity increases the adhesion of the polymer to polar surfaces such as metal and glass. Thus, a balance between viscosity and strength results in a choice of n which is a compromise.

Catalysts are called reactive hardeners or curing agents and become bound to the polymer by opening the epoxide linkage. When the hardener is an amine, the reaction is

$$(12.9)$$

followed by

$$(12.10)$$

resulting in cross-linking which can be extended if the hardener is a diamine such as ethylene diamine NH_2—CH_2CH_2—NH_2. The concentration of the hardener is 3–100 phr (parts per hundred resin), depending on the system and curing time. Fillers commonly used include TiO_2, PbO, Fe_2O_3, Al_2O_3, and SiO_2 which are added to reduce cost, decrease shrinkage on setting, lower the coefficient of thermal expansion of the joint, and improve the heat resistance of the joint.

The high strength of epoxy adhesives has led to their application over a wide variety of conditions which include the fabrication of helicopter rotor blades, fiberglass-lined boats, aircraft, and in the building trades.

Another class of adhesives are those based on isocyanate ($—N = C = O$) of which the most common is the urethanes. The isocyanate group reacts with active hydrogen such as an alcohol (ROH) or an amine (R_2NH) as follows:

$$ROH + \phi NCO \rightarrow R - O - C - \overset{\overset{\textstyle H}{\textstyle |}}{\underset{\underset{\textstyle O}{\textstyle ||}}{N}} - \phi \tag{12.11}$$

$$R_2NH + \phi NCO \rightarrow R_2 - N - C - \overset{\overset{\textstyle H}{\textstyle |}}{\underset{\underset{\textstyle O}{\textstyle ||}}{N}} - \phi \tag{12.12}$$

where $\phi = C_6H_5$. Thus, by reacting a diisocyanate with a polymeric diol, a condensation reaction occurs, resulting in bonding to a hydroxylated surface such as glass or metal. The diisocyanates commonly used are toluene diisocyanate TDI and hexamethylene diisocyanate HDI.

$$\tag{12.13}$$

The high reactivity of the NCO group makes the diisocyanates toxic, and great care must be taken in using these adhesives. The isocyanates react with water to form CO_2 and an amine which can also react with the NCO group

$$R - NCO + H_2O \rightarrow RNH_2 + CO_2 \tag{12.14}$$

$$RNH_2 + R' - NCO \rightarrow RNH - C - \overset{\overset{\textstyle H}{\textstyle |}}{\underset{\underset{\textstyle O}{\textstyle ||}}{N}} - R' \tag{12.15}$$

The evolved CO_2 can lead to a highly cross-linked foam which is soft or hard depending on the prepolymers used. Similarly, the hardness of the set adhesive is determined by the hydrocarbon chain length in the polyol or polyamine used to react with the diisocyanate. Cross-linking by short-chain triols leads to a hard adhesive, whereas long diols result in elastomeric material, often used in textiles, or where a flexible joint is required.

A new urethane-based adhesive has recently been developed which combines the properties of "hot melts" and chemical reactivity. Usually a hot melt adhesive is applied at 150–200 °C. The heat weakens the van der Waals forces and allows the polymer to flow. When the material has cooled, the bonds reform and set within minutes. Chemically reactive adhesives rely on the reaction of polymerization or cross-linking to set a monomeric or prepolymeric adhesive, and setting may take hours during which the joint must be held together with clamps or in a jig. The new adhesive is a reactive polymer of short chains held together by weak forces which break upon heating to only 100 °C and sets again within minutes upon cooling. However, unlike previous material, the bond continues to strengthen with time because of further reaction with water vapor in the air which cross-links the polymer.

Another class of thermosetting resins used as adhesives involves the polymerization of formaldehyde with urea or phenols. These materials are very cheap and find extensive use in binding wood, e.g., plywood, chipboard, and particle board. Foamed urea–formaldehyde has also been used as insulation (UFFI) in homes because of its ability to be injected into the walls of older homes. Unfortunately, the foam is not very stable and shrinks and hydrolyzes to liberate formaldehyde and possibly other toxic vapors. Recent work has, however, shown that the addition (and subsequent removal) of ammonia (NH_3) to the foam can reduce the level of formaldehyde released by the polymer. This release of formaldehyde also occurs in plywood and particle board, and these materials must be sealed to avoid the formation of toxic levels (TLV $= 0.10$ ppm or 120 $\mu g/m^3$) of formaldehyde.

The urea–formaldehyde polymer is formed as follows:

$$(12.16)$$

When the methylols eliminate water, a methylene urea is formed which can condense to form a cross-linked polymer (under acid conditions)

$$(12.17)$$

Excessive residual acid in the resin is partly responsible for the subsequent release of formaldehyde from the set polymer.

The reaction of formaldehyde (H_2CO) with phenol,

has been studied for over 100 years. In 1909, Baekeland patented the resin which has become known as *Bakelite*. The prepolymers of phenol–formaldehyde are of two types: (1) resitols in which ϕ OH/H_2CO < 1 and (2) novolacs in which ϕ OH/H_2CO > 1.

1. Resitols are formed (under alkaline catalysis) by the reaction

$$(12.18)$$

The phenol alcohols condense to form resitols

which can further condense to form ether linkages

$$2R - CH_2OH \rightarrow R - CH_2 - O - CH_2R + H_2O$$

The cross-linked resin does not melt when heated, but carbonizes. This characteristic is used in efficient heat shields where the resins are used in the nose cone of missiles and space vehicles.

2. Novolacs are formed (under acid catalyzed with ϕ OH/H_2CO > 1):

$$(12.19)$$

The novolac resins are usually applied as solid adhesives incorporating some phenol and formaldehyde (from hexamethylene tetramine) to cross-link the prepolymer when heat is applied. Resitol resins are applied as a dispersion in dilute aqueous NaOH with fillers, surfactants, and extenders.

Phenol–formaldehyde resins have recently been applied to the fabrication of reconstituted lumber called *Parallam*. Small pieces of wood, approx. 0.3×20 cm, are laid end to end and staggered side by side, are impregnated with about 5 % of the resin, and are cured by microwave energy to give a continuous length of lumber with a width and thickness which can be readily controlled and varied. The strength of this lumber is superior to the equivalent natural lumber which it soon will be displacing. Other applications of phenol–formaldehyde resins include bonding abrasive grits in grinding wheels, wood and veneers, paper laminates, and many others.

A popular and highly advertised adhesive is based on cyanoacrylate—the most common being methyl-2-cyanoacrylate

also known as *crazy glue* (Eastman 910). This was first introduced commercially in 1959 and because of its unique properties has gained widespread use in industry, medicine, and by hobbyists. The monomer has a low viscosity and so can readily penetrate surfaces. Its rapid polymerization is initiated by water, alcohol, or other weak bases. Hence, when the monomer is placed on a dry grease-free joint, the air usually has sufficient moisture to initiate the polymerization while the two surfaces are held together. Though very expensive, its use without solvent, heat cures, or excessive pressure makes it an ideal adhesive. One important disadvantage is that it tends to glue things indiscriminately, e.g., hand, fingers, eyelids, etc. Since the body is seldom free of surface oils, it has been possible to loosen a crazy glue joint (of an eyelid) by applying mineral oil (Nujol) to dissolve the underlying oil film. In some cases, surgery is required.

Special adhesives are required for high temperature use. Silicone rubber is an excellent flexible adhesive with a temperature limit of 250 °C. For higher operating temperatures, one must resort to inorganic adhesives and cements. These are usually based on silicates (water glass), phosphates (phosphoric acid), and Portland cement.

Silicate adhesives using water-soluble sodium silicate (Na_2SiO_3, water glass) have fillers of silica (SiO_2), salt (NaCl), clay, talc, metals, asbestos, or limestone ($CaCO_3$), depending on its ultimate use.

These silicate adhesives are fire-resistant and can withstand temperatures up to 1,000 °C if the adhesives cure to a ceramic (homogeneous phase). Since water-soluble silicates are basic, it is essential that metals such as aluminum and zinc which form amphoteric oxides (the metal dissolves in basic solutions) be avoided since they would dissolve in the joint and form hydrogen.

Phosphate adhesives are made by reacting zinc oxide (ZnO) with phosphoric acid (H_3PO_4), forming $ZnHPO_43H_2O$ which sets to a crystalline material and is used as a dental cement. More elaborate dental materials are constantly being introduced and tested as we all verify on our trips to the dentist's office. However, progress is slow because the demands and requirements are high.

An old and still useful cement for joining glass to metal is litharge cement made by mixing litharge (PbO) with glycerol ($CH_2OHCHOHCH_2OH$) in the ratio PbO: glycerol: water of 6:2:1. The cement takes a day to set and will resist the action of dilute acid, ammonia, and hydrocarbons.

There are many different adhesives, but to all the common requirements of a good and lasting joint is the surface preparation. The following factors must be observed:

1. Dust, scale, grease, oil, plasticizer, and other surface contaminants must be removed.
2. The surfaces should be etched or roughened to give a larger surface area for bonding.
3. The surfaces may require planing or smoothing to allow intimate contact.
4. A priming coat may be applied to wet the surface and to displace air from the crevices.
5. The surfaces can be sealed to prevent the adhesive from penetrating the porous surface.
6. The surface can be treated chemically (e.g., CASING or coupling agents) to increase the surface energy and facilitate bonding.
7. The surface should be dried, and any adsorbed liquids must be removed.
8. A metal may be electroplated with another metal which forms stronger bonds with the adhesive (e.g., Ni on Au).
9. Use ultrasonics where appropriate to clean the surface to the true surface interface.
10. Remove residual cleaning substances from the surface.

The main problem in the selection of a proper adhesive is to match the adhesive to the adherend. Only experience and the ability to test the strength of trial sample joints can solve the problem. Since adhesives are here to stay, it is always worthwhile to follow the developments in new and interesting adhesive products.

Exercises

1. Airplanes today can have as much as a tonne of adhesives. Give some reason for this.
2. Classify adhesives according to their application and setting, giving examples of each type.
3. In what way does the chemical composition of a surface affect the strength of an adhesive joint?
4. What are coupling agents? How do they work? Give two examples.
5. Give a short account of the four different theories of the adhesive bond.
6. What is an epoxy group and how do epoxy adhesives work? (Write chemical reactions.)
7. Which hardener will result in a more cross-linked adhesive:
   ethylene diamine (H_2N—CH_2—CH_2—NH_2) or
   diethylene triamine (H_2N—CH_2—CH_2—NH—CH_2—CH_2—NH_2)?
8. Of what use are fillers in an adhesive joint? List five fillers.
9. (a) What is a urethane adhesive? (b) Why must care be taken when using them?
10. What surface feature of glass and metals makes urethane adhesive ideal for these materials?
11. How do the formaldehyde adhesives work? (Write chemical reactions.)
12. What is crazy glue and why is it an exceptional adhesive?
13. Write a short note on inorganic adhesives. What conditions are they best suited to?
14. List the primary surface treatments for a good adhesive joint.
15. (a) What is CASING? (b) To what type of surfaces can it be applied?
16. Draw typical strength time curves for (a) hot melt adhesive, (b) reactive adhesion, and (c) reactive melt adhesion.

Further Reading

1. Wolf RA (2010) Plastic surface modification: treatment and adhesion. Hanser Publications, Munich/Cincinnati
2. Wong C (2009) Electrical conductive adhesives with nanotechnologies. Springer, New York/London
3. Benedek I (2009) Technology of pressure—sensitive adhesives and products. CRC Press, Boca Raton
4. Ebnesajjad S (2008) Adhesives technology handbook, 2nd edn. William Andrew Pub, Norwich
5. Cognard P (2006) Handbook of adhesives and sealants, vol 2. Elsevier, Amsterdam/San Diego
6. Cognard P (2006) Handbook of adhesives and sealants, vol 2. Elsevier, Amsterdam/San Diego
7. Cognard P (2005) Handbook of adhesives and sealants, vol 1. Elsevier, Amsterdam/San Diego
8. Rimai DS et al (eds) (1995) Fundamentals of adhesion and interfaces. Coronet Books, Philadelphia
9. Young T (1995) The crafter guide to glues. Chilton, Radnor
10. Pizzi A, Mittal K (eds) (1994) Handbook of adhesive technology. Dekker, New York
11. Hodd J (1991) Epoxy resins. Pergamon, New York
12. Minford JD (ed) (1991) Treatise on adhesion and adhesives, vol 7. Dekker, New York
13. Lee LH (ed) (1991) Fundamentals of adhesion. Plenum, New York
14. Skeist I (1989) Handbook of adhesives. Chapman and Hall, New York
15. Kinloch AJ (1987) Adhesion and adhesives: science and technology. Chapman and Hall, New York
16. Bikerman JJ (1961) The science of adhesive joints. Academic, New York
17. The Adhesion Society. http://www.adhesionsociety.org/
18. Glossary of Terms. http://www.pprc.org/pprc/
19. Sources and Newsletter. http://www.adhesivesandsealants.com/
20. Epoxy Adhesives. http://www.airproducts.com/
21. Theory of Adhesive Bonding. http://joinplastics.eng.ohio-state.edu/kkwan/main.html http://www.umist.ac.uk/lecturenotes/JDS

Chapter 13
Paint and Coatings

13.1 Introduction

The purpose of *paints and coating* is primarily to prevent corrosion and wear. The aesthetic aspect is usually of secondary importance. The nature of the protective coating is usually determined by the surface and the duration of the desired protection. Needless to say, the cost of the coating is another factor which can vary considerably from place to place, as well as with time. The concern for the environment is another factor which must be taken into consideration since the drying process may involve the release of volatile organic compounds (VOC), which can contribute to smog or other undesirable atmospheric conditions.

Special paints are available for ship hulls, hot mufflers, luminous surfaces, waterproofing, etc.

13.2 Constituents of Paint and Coatings

There are four normal constituents of a paint or coating. These are the following:

1. *Binder*: Designed to hold the film together and provides the adhesive forces required to bond it to the surface.
2. *Pigment*: A fine powdered material that provides the color as well as the hiding characteristics and weathering resistance of the film. Pigment without binder, as in whitewash, has little permanence. Binder, such as linseed oil, without pigment can be an excellent sealer and preservative for wood. Varnish is resin and solvent without pigment.
3. *Solvent*: A volatile liquid which dissolves the binder and acts as a thinner to dilute the coating, allowing it to spread easily on the surface.
4. *Additives*: Small amounts of wetting agents, flattening agents, driers, plasticizers, emulsifiers, stabilizers, cross-linking agents.

13.3 Binder

The *binder* is also the paint vehicle and for oil-based paints, consists of a drying oil which contains double bonds. A list of various oils and their average compositions in terms of the fatty acids is given in Table 13.1. These oils are called triglycerides because they are fatty acid esters of glycerol.

O.V. Roussak and H.D. Gesser, *Applied Chemistry: A Textbook for Engineers and Technologists*,
DOI 10.1007/978-1-4614-4262-2_13, © Springer Science+Business Media New York 2013

Table 13.1 Compositions and characteristics of vegetable oils

	% In seed	Saturated (%)	Oleic (%)	Ricinoleic (%)	Linoleic (%)	Linolenic (%)	Eleostearic (%)	Licanic (%)	I_2 (no.)	Saponification (no.)
No. of double bonds		0	1	1	2	3	3	3		
Linseed	32–40	10	20		20	50			180	190
Perilla	30–50	8	14		14	64			200	
Tung	40–50	4	7		8		81		165	
Oiticica	52–62	12	7					81		
Soybean	16–22	12	28		54	6			132	
Tall		7	45		48					
Safflower	23–36	10	14		76					
Castor		3	7	88	2				110	190
Olive		12	83		5				85	
Coconut		90	8						10	
I_2 no.		0	89.9		181.0	273.5				

Stearic acid ($G_8H_{36}O_2$), CH3$(CH_2)_{16}$–COOH

Oleic Acid ($C_{18}H_{34}O_2$), CH_3–$(CH_2)_7$–CH=CH–$(CH_2)_7$–COOH

Ricinoleic ($C_{18}H_{34}O_3$), $CH_3(CH_2)_5$—CH—CH_2—CH=CH(CH_2)—COOH.

$\qquad\qquad\qquad\qquad\qquad$ |
$\qquad\qquad\qquad\qquad\quad$ OH

Linoleic ($C_{18}H_{32}O_2$), $CH_3(CH_2)_4$CH=CH–CH_2_CH=CH$(CH_2)_7$COOH

Linoleic ($C_{18}H_{30}O_2$), CH_3–CH_2–CH=CH–CH_2–CH=CH–CH_2–CH=CH–$(CH_2)_7$–COOH

Eleostearic($C_{18}H_{32}O_2$) CH_3–$(CH_2)_3$CH=CH–CH=CH–CH=CH$(CH_2)_7$COOH

Licanoic ($C_{18}H_{24}O_3$), $CH_3(CH_2)_3$CH=CH—CH=CH—CH=CH—$(CH_2)_4$—C—$(CH_2)_2$—COOH.

$\qquad\qquad\qquad\qquad\qquad\qquad\qquad\qquad\qquad\qquad\qquad\qquad$ ||
$\qquad\qquad\qquad\qquad\qquad\qquad\qquad\qquad\qquad\qquad\qquad\qquad$ O

$$\begin{array}{c}
\quad\quad\quad\quad O \\
\quad\quad\quad\quad || \\
H_2C-O-C\text{-}\text{-}R' \\
| \\
\quad\quad\quad\quad O \\
\quad\quad\quad\quad || \\
HC-O-C\text{-}\text{-}R'' \\
| \\
H_2C-O-C\text{-}\text{-}R''' \\
\quad\quad\quad || \\
\quad\quad\quad O
\end{array}$$

where R′, R″, and R‴ are the fatty acid chains which may or may not be identical radicals.

The oils with fatty acids having three double bonds are classed as fast drying oils. The rate of drying can be increased by boiling the oil in the presence of driers (metal oxides or salts of organic acids) such as those containing cobalt, manganese, etc. Lead as a drier or pigment is being phased out because of its toxic nature.

Boiled oils essentially contain the catalyst necessary to initiate the drying process. A raw oil which takes 4–6 days to dry can be made to dry in 12–15 h when "boiled."

Blowing air through hot oil can form peroxides which can also activate the drying process.

The iodine number is the number of centigrams of I_2 absorbed by 1 g of oil. The reaction is

$$— CH = CH — \; + I_2 \rightarrow —CHI — CHI — \qquad\qquad (13.1)$$

Thus, the iodine number is a measure of the degree of unsaturation of the oil and an indication of its drying rate.

The saponification value is the mass in mg of KOH needed to completely neutralize 1 g of oil:

$$
\text{RC} \overset{\displaystyle O}{\underset{\displaystyle OR'}{\diagdown}} + \text{KOH} \rightarrow \text{RC} \overset{\displaystyle O}{\underset{\displaystyle OK}{\diagdown}} \quad \text{R'OH} \tag{13.2}
$$

Usually an excess of base is used, and the excess is back-titrated with standard acid.

Other characteristics of the oil are the melting point, the peroxide value, the free fatty acid content, and color.

Linseed oil is the most commonly used fast drying oil for paint. It has one major disadvantage because it tends to yellow when used indoors. When exposed to sunlight, the yellowing tendency is inhibited or reversed. Hence, linseed oil-based paints are generally restricted to outdoor use or dark-colored indoor paints.

The natural vegetable oils can be improved by cross-linking the glycerides. This is done by first converting the triglycerides into monoglycerides by the reaction

$$
\begin{array}{c}
\text{CH}_2\text{—OH} \\
| \\
2\text{CH—OH} \\
| \\
\text{CH}_2\text{—OH}
\end{array}
+
\begin{array}{c}
\text{H}_2\text{C—C—OR} \\
\overset{\displaystyle O}{|} \\
O \\
| \\
\text{HC—C—OR} \\
| \\
\text{H}_2\text{C—C—OR} \\
\underset{\displaystyle O}{||}
\end{array}
\rightarrow 3
\left[
\begin{array}{c}
\text{CH}_2\text{OH} \\
| \\
\text{CHOH} \\
| \\
\text{H}_2\text{C—C—OR} \\
\underset{\displaystyle O}{||}
\end{array}
\right]
\tag{13.3}
$$

It is possible to control the degree of cross-linking by the use of trifunctional acids and the molecular weight of the resin. The alkyl resin, called *alkyd paint*, dries faster than the corresponding oil which is one of its major advantages.

13.4 Driers

Driers are additives which accelerate the drying process of a paint film. This is effected by (a) catalyzing the uptake of oxygen and (b) sensitizing the decomposition of the peroxides forming free radicals which add to the double bonds, causing polymerization to occur. The driers are generally metallic salts of the naphthenic acids. The mechanism by which they catalyze the decomposition of peroxides is shown for cobalt:

$$
\text{Co}^{2+} + \text{ROOH} \rightarrow \text{Co}^{3+} + \text{RO} + \text{OH}^- \tag{13.4}
$$

$$
\text{Co}^{3+} + \text{ROOH} \rightarrow \text{Co}^{2+} + \text{RO}_2 + \text{H}^+ \tag{13.5}
$$

$$
\text{Co}^{3+} + \text{OH}^- \rightarrow \text{Co}^{2+} + \text{OH} \tag{13.6}
$$

The overall reaction—the sum of reaction (13.4) and (13.5) is

$$
2\text{ROOH} \rightarrow \text{RO}_2 + \text{RO} + \text{H}_2\text{O} \tag{13.7}
$$

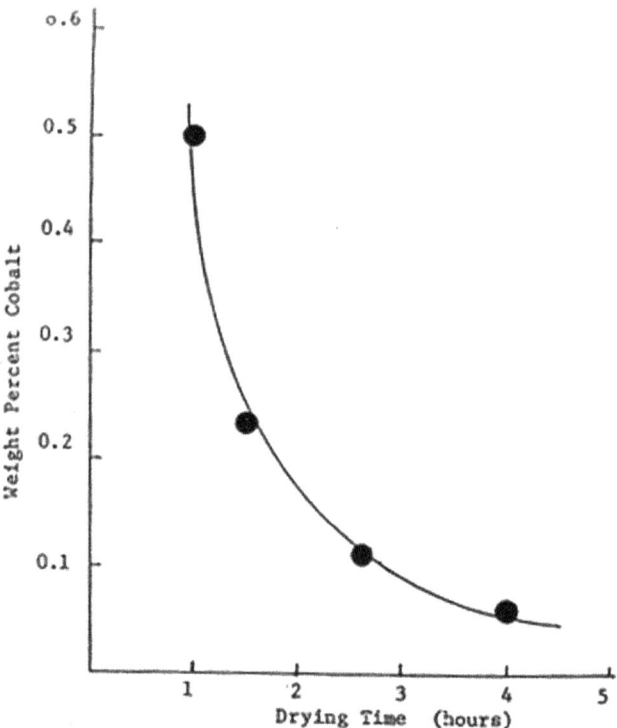

Fig. 13.1   Effect of increasing concentration of cobalt drier on the drying time of a coating film

The effect of various amounts of cobalt in a paint film on the time required to reach tack free dryness is shown in Fig. 13.1.

Lead has been a common drier in the past, and its use is limited to less than 0.5% in the dry film. This quantity may still be too large for use where children may ingest the paint.

13.5   Pigment

The *pigment* in a paint is usually fine powder of 0.01 μm in diameter and is meant to cover the surface, to provide color, to improve the strength of the film, to improve the adhesion of the paint to the surface, to improve the abrasion and weathering of the film, to reduce gloss, and to control the flow and application of the paint. An extender is cheaper than pigment and serves the same purpose except it does not provide the required color to the film. The hiding power of a pigment is determined by four properties of the interaction of light with a solid: absorption, transmission, reflection, and diffraction. Particle size, particle density, and refractive index of the pigment contribute to the hiding power. The optimum particle size (diameter in μm) of a pigment has been shown empirically to be given by

$$d = \lambda(\mu m)/1.414\pi n_b M \tag{13.8}$$

where M is the measure of scattering given by the Lorentz–Lorenz equation (13.9)

Table 13.2 Physical characteristics and hiding power of paint components

	Refractive index[a] $(Np)_D$	Hiding power (m^2/kg)	Tinting[b] strength	Optimum particle diameter[c] (μm)
Pigments				
TiO$_2$ rutile	2.76	32	1850	0.19
TiO$_2$ anatase	2.55	24	1750	0.22
ZnS	2.37	12	640	0.29
Sb$_2$O$_3$	2.09	4.5	300	
ZnO	2.02	4.0	210	0.4
Pb(OH)$_2$PbCO$_3$	2.00	3.7	160	0.25
Pb(OH)$_2$PbSO$_4$	1.93	2.8	120	
Extenders				
BaSO$_4$	1.64			1.65
CaSO$_4$	1.59	Very low		
CaCO$_3$	1.57			1.75
SiO$_2$	1.55			
Binders				
Tung oil	1.52			
Linseed oil	1.45			
Soybean oil	1.48			
Alkyl resins	1.5–1.6			

[a] Sodium D line (589 nm)
[b] Tinting strength is the relative ability of a pigment to impart color to a white base mixture
[c] At 560 nm

$$M = \frac{\left(\frac{n_p}{n_b}\right)^2 - 1}{\left(\frac{n_p}{n_b}\right)^2 + 2} \tag{13.9}$$

Some of the properties of the major pigments used today are given in Table 13.2. The hiding power of a pigment is a function of the difference in refractive index of the pigment (n_p) and that of the binder (n_b). It should be noted that extenders have a very low hiding power, whereas pigments such as TiO$_2$ are very effective, especially when used at the optimum particle diameter where scattering is at a maximum.

Organic pigments are used for colored paints. Carbon blacks are classed as organic and come in a variety of properties which depend on the method of production. Metallic powders and flakes are often used especially for hot surfaces. Aluminum, copper, and bronze powders also serve as decorative coatings.

13.6 Solvents: Thinners

Oil- and resin-based paints require a *solvent* to disperse the paint, reduce the viscosity, and allow for easy application. The essential requirement of the paint *thinner* is that it not be highly volatile, that is, have a boiling point within the range of 120–175 °C. The earliest paint thinners were the terpenes, cyclic hydrocarbons (C$_{10}$H$_{16}$) which come from wood and can range in boiling points of 150–180 °C.

Some of these are shown in Fig. 13.2 with a few of the oxygenated compounds which are important flavor and perfume additives. These terpenes are formed by the dimerization of isoprene.

α-pinene
Canphere
limonen
(dipentene)

B-pinene
α-terpineol
citronellol

Fig. 13.2 Structure and name of selected terpenes and oxygenated derivatives

$$\left(CH_2{=}\underset{\underset{}{\overset{\overset{CH_3}{|}}{C}}}{-}CH{=}CH_2 \right)$$

isoprene

Other solvents commonly used include the alkyl benzenes and xylenes.

The trend to reduce VOC in our environment has resulted in the rapid development of water-based latex paints which have little or no volatile solvent.

13.7 Water-Based Paints

Water-based paints make use of an emulsion of a polymer composed of vinyl acetate, vinyl chloride, acrylics, acrylonitrile, ethylene, styrene, butadiene, and isoprene. The paint consists of an emulsion of polymer known as latex because their appearance is similar to rubber latex or sap from the rubber tree. The oil-in-water emulsion has plasticizers to smooth out the drying surface and to lower the glass transition temperature of the polymer to below room temperature. As drying occurs and water is lost, the polymer particles coalesce, forming a continuous impermeable film.

Emulsifiers are usually bifunctional compounds, like soaps, in which one end is water soluble and the other end is oil or polymer miscible. Other additives include surface active agents, thickening agents, preservatives, and coalescing aids. Other additives such as freeze–thaw stabilizers and wet-edge extenders can be satisfied by the addition of high boiling glycols which evaporate slowly from the surface and allow the latex particles to merge in the film.

13.8 Protective Coatings

The protective nature of coatings with regard to the corrosion of steel is best illustrated by the reduced transport of water and oxygen to the steel coating interface. We have shown in Chap. 10 that the overall corrosion reaction is

Table 13.3 The permeability of coating films to water vapor from air at 95%

Binder	Pigment	Permeability H_2O (mg/cm^2/ year/100/μm)
Asphalt	–	190
Phenolic resin	–	718
Phenolic resin	Af leaf powder	191
Alkyd	–	825
Alkyd	Al leaf powder	200
Polystyrene	Lithopone 30% ZnS, 70% BaSO$_4$	485
Linseed oil	Zn	1,125

Table 13.4 Permeability of coating films to O_2 from air

Binder	Permeability (mg O_2/cm^2/year/100/μm)
Cellulose acetate	4–9
Polystyrene	13
Poly (vinyl butyral)	27
Asphalt + talc	39
Ethyl cellulose	51

$$2Fe + 3H_2O + 1.5O_2 \rightarrow 2Fe(OH)_3 \tag{13.10}$$

or for an intermediate step

$$Fe + H_2O + 1/2O_2 \rightarrow Fe(OH)_2 \tag{13.11}$$

When ordinary unpainted steel is exposed to an industrial atmosphere, the corrosion rate is approximately 70 mg/cm^2/year. Based on (13.10), the water requirements are 34 mg/cm^2/year and the oxygen needed is 30 mg/cm^2/year. Thus, any protective coating must have low permeability for H_2O and O_2 if corrosion is to be reduced.

Some permeability values for H_2O and O_2 by various coatings are given in Tables 13.3 and 13.4 and show that some coatings are not suitable as rust inhibitors.

A Mossbauer spectroscopic study on rust formed under coated steel panels exposed to a marine environment for 18 months showed the presence of γ-Fe$_2$O$_3$, αFeOOH, and γFeOOH next to the paint surface. The rust at the steel interface consisted primarily of γFeOOH. The coatings used for the different tests were the following:

(a) Three coats of red mud zinc chromate alkyd-based primer and two coats of alkyd-based olive green paint
(b) Two coats of zinc chromate epoxy-based primer and two coats of epoxy gray paint
(c) Two coats of red oxide zinc phosphate alkyd-based primer and two coats of chlorinated rubber-based gray paint
(d) Two coats of manganese phosphate barium chromate alkyd-based primer and two coats of vinyl-alkyd-based paint
(e) Two coats of basic lead silica chromate alkyd-based primer and two coats of chlorinated rubber gray paint

In all cases, there was sufficient permeability of water and oxygen to corrode the steel resulting in a delamination of the paint film which detaches the corrosion products from the metal surface.

13.9   Surface Preparation

The inhibition of rusting can be achieved by passivating the iron. This can be accomplished by including chromates, nitrites, or red lead in the paint. Proper surface preparation is an important aspect of effective protection against corrosion.

Metal surface treatment is usually divided into three classes: (1) the metal surface is not altered, (2) the metal surface is etched either chemically or mechanically, and (3) metallic salts are produced at the metal surface.

1. Solvent degreasing and oil removal treatment of a metallic surface is a common requirement. This can be done by dipping, hand wiping, or ultrasonic degreasing. Basic alkali solutions for surface cleaning are often used, though etching of the metal surface will occur unless inhibitors are present in the cleaner. Removal of the alkali requires a second rinse-wash if the last trace of a residue film of oil is to be removed.

   Vapor degreasing is another form of surface cleaning which is superior to solvent dipping because the tendency to leave an oil film on the surface is greatly reduced.

2. Sand blasting is an effective cleaning process since it can remove a thin layer of material which would include surface contaminants. Weathering, wire brushing, and flame cleaning are also used to clean surfaces. The most general surface treatment is, however, acid pickling where sulfuric acid at 1–10% by weight and at temperatures of 20–80 °C is used to clean the metal by removing the oxide layers. Some etching, which usually occurs, can aid in the adhesion of the new paint.

3. When phosphoric acid is used for surface cleaning, it can leave an inert metallic phosphate coat on the surface which can then act as a corrosion inhibitor. Chromic acid either by itself or with phosphoric acid will form insoluble metallic chromates which remain on the metal surface and inhibit corrosion while binding the coat film. This treatment is used for Al, Mg, Zn, Sn, and alloys of these metals.

The treatment of nonmetallic surfaces is generally determined by the porosity of the material, its moisture content, and roughness. Such surfaces are often treated with a primer or undercoat to seal the surface and act as an intermediate bond between substrate and coating.

13.10   Specialized Coatings

13.10.1   Formaldehyde Resins

Formaldehyde polymerizes by condensation with phenol and urea to form phenol—and urea—formaldehyde resins which are popular and inexpensive adhesives. They are paint and coatings often used in coatings when combined with alkyds, epoxides, polyesters, or acrylic to give strong flexible films.

Epoxy coatings are two-part systems which, similar to the adhesive, polymerize by condensation to form a thermosetting film. There are basically three types of epoxy resins used in coatings:

(a) Epoxy novolac resins

$$\text{(epichlorhydrin)} \qquad (13.12)$$

$$\underset{3HCl +}{CH_2\!-\!CH\!-\!CH_2\!-\!O}\quad CH_2\!-\!CH\!-\!CH_2\!-\!O \qquad\qquad O\!-\!CH_2\!-\!CH_2\!-\!CH_2 \tag{13.13}$$

(b) Cycloaliphatic epoxy resins where the epoxy group is part of the cyclic ring

$$O \diagdown \diagup \cdots C\!-\!O\!-\!CH_2 \cdots \diagup \diagdown O \tag{13.14}$$

(c) Acrylic epoxide resins where the epichlorohydrin is reacted with methacrylic acid to form glycidyl methacrylate

$$CH_2\!-\!CH\!-\!CH_2\!-\!Cl + CH_2\!\!=\!\!\underset{\substack{C=O \\ OH}}{\overset{CH_3}{C}} \rightarrow CH_2\!\!=\!\!\underset{\substack{C=O \\ O\!-\!CH_2\!-\!CH\!-\!CH_2}}{\overset{CH_3}{C}} \quad + HCl \tag{13.15}$$

The resin is prepared by mixing the epoxide with a diol to form the polymer. Some triols are usually added to effect cross-linking. The curing agent used is normally an amine which is also a reactive hardener, so each N—H group can react with the epoxy group by

$$\diagup\!\!NH + CH_2\!-\!CH\!-\!\! \rightarrow\ \diagup\!\!N\!-\!CH_2\!-\!\overset{OH}{\underset{}{CH}}\!-\! \tag{13.16}$$

Hence, each primary amine (R—NH$_2$) can also make cross-links of the epoxy polymer. Silicone resins make excellent durable coating which, except for cost, provide very satisfactory protection in corrosive environments. However, silicone rubbers and elastomeric coatings are highly permeable to oxygen and water and thus the need to include corrosion inhibitors when used on steel surfaces. Their high temperature stability (250 °C) makes them ideal for heat resistant coatings. They are nonyellowing, nonchalking, and ozone and UV resistant.

Polyurethane coatings are based on the reaction of a monoglyceride with a polyol and a diisocyanate to form a urethane oil or urethane alkyd. All the isocyanates react, leaving no toxic groups. These resins are alkali and water resistant and superior to ordinary alkyds as a weather-resistant coating and corrosion-inhibiting surface coating.

Though water-based paints meet most of the requirements of reducing the release of volatile organic compounds (VOC), other recent coatings are being developed as more restrictive measures are introduced. Three solvent-free systems based on thermoplastic powders applied to preheated surfaces are the following:

1. Dipping heated substrate into a heated fluidized bed of the plastic powders.
2. Electrostatic spraying.
3. Flame spraying of the powders which as molten particles coat the heated substrate. This method can be applied in the field. In the first two methods the coating must be baked on to the base metal to allow the polymer to melt, cross-link, and flow to create a uniform continuous coating.

The ability of such coatings to prevent or reduce corrosion of the metal substrate can be evaluated by the electrochemical methods such as the electrochemical noise method (ENM) and the electrochemical impedance spectroscopy (EIS). These methods rely on the permeation of electrolyte (usually a mixture of NaCl and $(NH_4)_2SO_4$) through the coating at elevated temperatures, changing the resistance noise or decreasing the impedance modulus when the T_g of the coating has been exceeded. This has been interpreted to indicate that electrolyte has penetrated the coating.

The common method of testing the durability of a paint or coating and its effectiveness in inhibiting or reducing corrosion of the metal substrate is to expose coupons or samples to an accelerated weathering conditions: that is, by exposure to harsh climate conditions that includes intense sunlight and rain or in a weatherometer which is an instrument that cycles these conditions rapidly and thus attempts to predict long-term exposure effects in very short time intervals.

13.11   Fire Retardant Paints

Paint can have an important role to play in the control of fire. If a coating releases a nonflammable gas or vapor, it can prevent the oxygen from reaching the surface.

Effective additives include urea, ammonium phosphate, magnesium carbonate, antimony oxide, alumina hydrate, silicates, and borates. Though brominated and chlorinated organic compounds are fire retardants, they decompose at high temperatures to produce HBr and HCl which, as corrosive gases, can often do more harm than good.

Another mechanism of flame retardance is the intumescent coating which forms an expanded carbon layer at the burning surface and so reduces heat transfer to the coated material. The additives include mixtures of paraformaldehyde, melamine, ammonium phosphate, and chlorinated hydrocarbons. Most intumescent paints are water sensitive and thus have limited applications.

13.12   Antifouling Paints

The problem of finding a suitable coating for ship hulls is twofold—first, the need to prevent corrosion of the steel; second, the need to keep the hull from fouling due to the growth of organisms such as barnacles, mussels, tube worms, and algal spores on the surface. The fouling results in a surface roughness which increases friction and therefore fuel consumption. A typical 10,000-tonne cruiser after 6 months at sea in temperate waters required almost 50% more fuel to maintain a speed of 20 knots.

Antifouling paints usually incorporate substances such as mercury or butyl tin which are toxic to the organisms but which must slowly leach out of the coatings. However, the prevalent use of such material has contaminated the coastal waters, and such materials are being replaced by less-dangerous toxins such as Cu_2O. The rate at which the copper is leached from the surface must be more than 10 mg Cu/cm^2/day in order to be effective. Hence, the binder pigment and other paint components must be controlled to permit such regular release of the copper.

The best type of surface coating for a ship's hull that will give the least resistance for the movement of the ship through water, that is, the fastest speed, has at times been proposed to be hydrophobic (Teflon-like surface) or hydrophilic (glasslike).

Teflon sprays are available as well as thin Teflon sheeting which have adhesive backing so it can be laminated to a ship's hull.

Hydrophilic paints have been claimed to be superior to hydrophobic surfaces for ship hulls if speed is desired. One such coating is the following:

Paint	
Hydroxyethyl methacrylate	10%
Ethyleneglycol monomethyl ether	40%
Ethanol	30%
Water	20%

Catalyst Ammonium dichromate 1.75% in water. The ratio is about 10:1 for paint: catalyst by volume.

If the polymer coating is slightly soluble in water, then it may be able to act as a drag reducer, reduce turbulence, and so increase the speed of the boat. A hydrophilic water insoluble coating of polyhydroxystyrene has been shown to reduce and conserve fuel when applied to the surface of boats.

The continuous development in polymers and coatings makes the optimum choice difficult for any specific application. However, adequate protection is usually readily available.

Exercises

1. Name the constituents of a paint and lacquer.
2. What type of coating would you use for (a) steel, (b) concrete, and (c) diesel exhaust pipes?
3. What are the components of a water-based paint?
4. What are the characteristic components of a fire retardant paint?
5. What are the active components of an antifouling paint?
6. What are the characteristics of the ideal coating?
7. Calculate the iodine number of (a) ricinoleic acid, (b) eleostearic acid, and (c) licanic acid.
8. A ship's hull can be treated with a coating which can be hydrophobic or hydrophilic. Which type would you recommend and why?
9. (a) Calculate the thickness of the corrosion layer of iron corresponding to 70 mg/cm^2/year. (b) Calculate the O_2 and H_2O requirements for iron rusting according to (13.11).
10. A white pigment is needed for a coating. Explain why TiO_2 is a good choice.
11. What precoating surface treatment would you consider for (a) a nonmetal surface, and (b) a metallic surface?

Further Reading

1. Gulrajani ML (2010) Colour measurement: principles, advances and industrial applications. Woodhead, Cambridge/Philadelphia
2. Weldon DG (2009) Failure analysis of paints and coatings. Wiley, Chichester
3. Hellio C (2009) Advances in marine antifolding coatings and technologies. Woodhead, CRC, Oxford/Boca Raton
4. Talbert R (2008) Paint technology handbook. CRC Press, Boca Raton
5. Marrion A (2004) The chemistry and physics of coatings. Royal Society of Chemistry, Cambridge, UK
6. Lambourne R, Strivens TR (1999) Paint and surface coatings: theory and practice, 2nd edn. Woodhead, New York

7. Ash M, Ash I (1996) Handbook of paint and coating raw materials, vol 1 and 2. Gower (UK), Ashgate Brookfield, Vermont

8. Innes H (1995) Environmental acceptable coatings. BBC, New York

9. LeSota S (ed) (1995) Coatings encyclopedic dictionary. Federation of Societies for Coatings Technology, Blue Bell

10. Bouquet FL (1994) Paint and surface coatings for space, 2nd edn. Systems Co., Carlsborg

11. Boxbaum G (1993) Industrial inorganic pigments. VCH, New York

12. Lambourne R (1993) Paint and surface coatings. Prentice Hall, Paramus

13. Stage D (ed) (1993) Paints, coating and solvents. VCH, New York

14. Skerry BS, Eden DA (1991) Characterisation of coatings performance using electrochemical noise analysis. Prog Organ Coat 19:379

15. Woodbridge R (ed) (1991) Principles of paint formulation. Chapman and Hall, New York

16. Labana S (1991) Advanced coatings technology. ESD, The Engineering Society, Southfield

17. Doorgeest T et al (1990) Elsevier's paint dictionary. Elsevier, New York

18. Morgan WM (1990) Outline of paint technology, 3rd edn. Wiley, New York

19. Schuerman G, Bruzan R (1989) Chemistry of paint. J Chem Edu 66:327

20. DeRenzo DJ (ed) (1987) Handbook of corrosion resistant coatings. Noyes, Westwood, New Jersey

21. Skerry BS, Eden DA (1987) Electrochemical testing to assess corrosion protective coatings. Prog Organ Coat 15:269

22. Weesmetal GE (ed) (1981) Paint handbook. McGraw-Hill, New York

23. Turner GPA (1980) Introduction to paint chemistry and principles of paint technology, 2nd edn. Chapman and Hall, London

24. (1974) *Surface coatings—oil and color chemists' association of Australia*, vol 2., TAFE Educational ... 41–42, Chapman and Hall, London

25. Toropov NA (1967) Heat resistant coatings (trans: Russian). Consultants Bureau, New York

26. Parker DH (1966) Principles of surface coating technology. Interscience, New York

27. Nylin P, Sunderland E (1965) Modern surface coatings. Interscience, New York

28. Taylor CJA, Marks S (eds) (1962) Paint technology manuals: I, nonconvertible coatings. II, solvents, oils, resin and driers. III, convertible coatings, oil and colour chemists association. Chapman and Hall, London

29. National Paint and Coating Association. http://www.paint.org/index.htm

30. International Centre for Coating Technology, PRA. http://www.pra.org.uk

31. Paint and Coating Mfg. Resources. http://www.pprc.org/pprc/sbap/painting.html

32. Paint and Coating Industry. http://www.pomag.com

33. Canadian Paint and Coating Association. http://www.cdnpaint.org

34. Finishing Info. http://www.paintcoatings.net/

35. Resource center, PCRC. http://www.paintcenter.org/

36. History of coatings. http://www.paint.org/history.htm

37. Papers on coatings. http://www.corrosion.com/techpage.html

38. Marine coatings. http://www.boatingreview.com

Chapter 14
Explosives

14.1 Introduction

The first recorded account of an explosive was the description of a crude form of gunpowder by Cheng Yin in China about 850 A.D. He cautioned about the dangers of burning the experimenter and the house. These warnings still apply today. The first European to refer to gunpowder was Roger Bacon who concealed the formula in a code which was not revealed for another century. By 1346, gunpowder (or black powder) was used to fire a cannon in battle. In the seventeenth century, it was used in mines as a blasting agent. This reaction is very practical at the present time and shown below:

$$4KNO_3 + 2S + 7C \rightarrow K_2CO_3 + K_2S_2 + 2N_2 + 3CO_2 + 3CO \tag{14.1}$$

During the Napoleonic Wars, England produced over 20,000 barrels of gunpowder a year with each barrel weighing 100 lb. The modern era of explosives began in 1846 when Sobrero discovered nitroglycerin.

An explosive substance is one which undergoes a very rapid chemical reaction that is highly exothermic and which is accompanied by high pressures at the reaction site and the evolution of a large quantity of gaseous products. When noise is also produced due to a shock wave, the process is called a *detonation*, and the substance is called a *high explosive*. The distinguishing characteristics of an explosive relative to fuels and propellants are given in Table 14.1, where it can be seen that the heat of reaction is not important. The most significant property of an explosive is its high rate of reaction, or the power generated, and the pressure produced. When the linear speed of reaction of a substance (the propagation of the flame front) is in the range of meters/sec, the substance is classed as a low explosive and the process is called a *deflagration*.

Explosives are further divided into primary and secondary explosives. Primary explosives are detonated by heat, spark, flame, or mechanical impact, whereas secondary explosives can only be detonated by an externally applied shock wave such as commonly produced by a primary explosive.

Secondary explosives when ignited by a flame will normally burn without detonating. However, even low explosives can be made to detonate under suitable conditions depending on the material.

14.2 Primary Explosives

Some selected *primary explosives* and their properties are given in Table 14.2. Most of these initiators are toxic as well as unstable. They are usually combined with other initiators or other substances such as $KClO_3$, KNO_3, $Ba(NO_3)_2$, PbO_2, and Sb_2S_3.

O.V. Roussak and H.D. Gesser, *Applied Chemistry: A Textbook for Engineers and Technologists*, DOI 10.1007/978-1-4614-4262-2_14, © Springer Science+Business Media New York 2013

Table 14.1 The distinguishing characteristics of burning fuel, propellants, and explosives

Property	Burning fuel	Propellant	Explosive
Typical material	Coal, oil, gas	Hydrazine	TNT
Linear reaction	10^{-6}	10^{-2}	10^3
$t^{1/2}$ (s)	10^{-1}	10^{-3}	10^{-6}
Factors controlling reaction rate	Heat transfer	Heat transfer	Shock transfer
Energy output (J/g)	10^4	10^3	10^3
Power output (W/cm^2)	10	10^3	10^9
Common initiation mode	Heat	Hot gases and particulate	High temperature and high pressure shock waves
Pressure developed (atm)	1–10	10–1,000	10^4–10^6

Table 14.2 Some primary explosives and their properties

	Mercury fulminate	Lead azide	Silver azide	Lead styphnate	Tetrazene[a]	Diazodinitro-phenol
Formula	Hg(ONC)$_2$	Pb(N$_3$)$_2$	AgN$_3$	C$_5$HN$_3$O$_8$Pb	C$_2$H$_8$N$_{10}$O	C$_6$H$_2$N$_4$O$_5$
Formula wt (g/mol)	285	291	150	468	188	210
Density (g/mL)	4.43	4.71	5.1	3.08	1.7	1.63
Color	Grey	White	White	Tan	Yellow	Yellow
Melting point (°C)	100 exp.	Dec.	252	206 explodes	Dec.	150 ± 10 exp.
Ignition temp. (°C)	215	330	300	280	130	170
Minimum value for igniting TNT (mg)	360	90	70			
Heat of formation (kJ/g)	−0.941	−1.48	−2.07	1.83	1.13	−1.59
Heat of explosion (kJ/g) to (H$_2$O)g	1.78	1.53	1.90	1.91	2.75	3.43
Specific gas volume ml (STP)/g	316	310		368	1190	865
Activation energy (kJ/g)	30	170	150	260		230
Detonation (mm/µsec)	4.5	4.5	6.8	5.0		6.9
Velocity (km/s) at density (g/mL)	3.3	3.8	5.1	2.7		1.6
Sensitivity to impact (kgm)	0.18	0.41		0.15	5	16
Sensitivity to friction (kgf)	0.43	0.12		0.08		
Drop hammer, 2 kg height (cm)	4	10		35		
Sensitivity to static electrical energy max. for non-ignition (J)	0.07	0.01	0.007	0.001	0.036	0.25

[a]

$$\text{N---N} \diagdown$$
$$\text{C---N}{=}\text{N---NH---NH---C}{=}\text{NH}\cdot\text{H}_2\text{O}$$
$$\text{N---N} \diagup \quad\quad \text{NH}_2$$
$$\text{H}$$

The azides are usually unstable and are formed by the reaction of hydrazoic acid (HN$_3$) with the metal oxide. Sodium azide (NaN$_3$) is used to explosively fill air bags in car crashes. Silver azide (AgN$_3$) is more expensive than the lead azide, Pb(N$_3$)$_2$, which is the common detonator. Care must be taken during its production since large crystals readily explode. Contact with copper must be avoided since copper azide (CuN$_3$) is extremely sensitive. The high thermal stability of Pb(N$_3$)$_2$ makes it suitable for long-term storage.

The amounts of primary initiator needed to detonate a secondary explosive vary from 10 to 400 mg and depend on both components. The major requirements of a good initiator are that it must be sufficiently stable for safe manufacturing, compatible with metal casing, easily loaded into detonators, and not too expensive. Its storage under adverse conditions must not alter its properties or stability.

The initiators are sensitive to friction and shock (blow). The percussion sensitivity is measured by the drop-hammer method where a 2 kg steel ball is dropped from increasing heights until detonation occurs. The amount of electrical energy needed to initiate the detonation depends on the initiator.

Diazodinitrophenol, though sensitive to impact, is not as sensitive to friction or electrostatic energy but somewhat less stable to heat than lead azide.

Lead styphnate ($C_5HN_3O_8Pb$), or lead trinitroresorcinate, is very sensitive to electrostatic discharge and often used to sensitize lead azide.

Tetrazene ($C_2H_8N_{10}O$) is readily decomposed in boiling water and is usually used as an ignition agent for lead azide. It is very sensitive to percussion and friction.

Mercury fulminate ($Hg(CNO)_2$) was the first initiating explosive which has found extensive use. However, in comparison with other primary explosives, it is relatively weak and does not store well under adverse conditions and hence is not used much at present.

14.3 Secondary Explosives

The substances which have high detonation velocities but are relatively insensitive to shock can act as *secondary explosives*. Some common secondary explosions are listed in Table 14.3. There are about 120 different chemical compounds which have been used or can be used as explosives. Many are too unstable to be used industrially or for military applications. A practical explosive is one that is not too difficult to produce on a large scale; is stable when stored under various conditions of temperature, humidity, and vibrations; and is not too expensive.

Explosives are generally characterized by five features: (1) chemical stability, (2) sensitivity to ignition, (3) sensitivity to detonation, (4) velocity of detonation, and (5) explosive strength. These characteristics can have quantitative values with statistical variations which must meet stringent requirements, especially for military use. However, the occurrence of an unexpected explosion or dud may be statistically predictable when thousands or millions of events have been tried. Perfect reliability is something which can seldom be achieved.

14.3.1 Chemical Stability

The *chemical stability* of an explosive is determined by its ability to maintain its reactive characteristics and to remain chemically unchanged while stored or aged under specific conditions.

Thus, extreme temperatures of $-30°C$ or $+45°C$ often encountered during military operations can destabilize the materials of an explosive. Chemical instability may thus be due to the nonexplosive degradation or decomposition of the explosive substance with the result that the reliability and strength is decreased.

14.3.2 Sensitivity to Ignition

Explosive solids can be detonated by heat, mechanical impact, friction, or electrical spark or discharge. The sensitivity of an explosive is the effect of the stimulus on its spontaneous detonation:

Table 14.3 Properties of some selected secondary explosives of importance

Name	Composition	Density (g/mL)	Ignition (°C)	Det velocity (km/s)	Gap test related to TNT	ΔH^0 expl. (kJ/g)	Spec. gas vol. (mL(STP)/g)	Sensitivity to friction (kg f)	Height of drop hammer 2 kg (cm)	Comments
Black power (BP)	KNO_3, 15 C, 10 S	1.2	310	0.4		2.8	280		60	Cheap, smoky, sensitive
Nitrocellulose (NC)	Nitrated cotton	1.5	180	6.3		4.3	765	10–36	20	Smokeless, grain size controls rate
Nitroglycerin (NG)	$C_3H_5(ONO_2)_3$	1.6	200	7.8		6.2	715	0–36	4	Volatile, very high sensitivity
Blasting gelatin	92/8: NG/NC(BG)	1.5	190	7.5		6.4	712		12	Strongest high brisance
Gelatin/ dynamite	BG+Dope (20–90 %)[a]	1.5	190	6.3		5.3	600		17	Jelly; waterproof, powerful
Trinitrotoluene	TNT, $CH_3C_6H_2(NO_2)_3$	1.57	225	6.9	100	4.0	690	29.5	60	Easily melted, powerful
Tetryl	$C_7H_5N_5O_8$	1.7	195	7.2	50	4.6	710	27–36	40	Cast. waterproof
Amatol	50/50: NH_4NO_3+TNT1:1	1.5	220	6.5	45	4.0	815		55	Low sensitivity (like TNT) hygroscopic
	80/20: NH_4NO_3+TNT4:1	1.4	220	5.4	150	3.9	860		60	High thermal stability, 1.5 times stronger than TNT
Hexogen, RDX cyclonite	$C_3H_6N_6O_6$	1.6	260	7.5	40	4.3	910	1.5	20	Plastic easily molded (vaseline)
Plastic explosive	RDX+elastomer	1.65		8.0		5.0				

[a] Dope is oxygen balanced explosive of either $NaNO_3$, NH_4NO_3 or explosive oils such as ethylene glycol dinitrate

(a) The thermal stability is determined by immersing a 5-mg sample in a thin-walled metal container in a heated bath. The temperature, ranging from 100 to 300°C, at which the sample detonates after 5 s, is the ignition temperature. The thermal stability is also determined by increasing the temperature of the sample by 5°C/min until detonation occurs.
(b) Sensitivity to percussion and impact is illustrated by the percussion cap in ammunition, and the impact fuses in bombs and shells. The test involves a given mass of steel which is dropped from increasing heights onto a given mass of sample of particle size <1 mm and spread over a given area. The height from which the sample detonates 50% of the time is an indication of the hazards involved in handling the explosive.
(c) The frictional force (kgf) required to detonate the sample 50% of the time is determined by the sliding torpedo test or the pendulum scoring test. This value determines the ease of handling the material and precautions which must be taken.
(d) The sensitivity of an explosive to a spark or electric discharge is determined by the minimum energy required to detonate the sample. The electrostatic charge on clothing has often accidentally detonated explosives while being handled. Nonsparking tools (some made of beryllium-copper alloys) are normally used in factories where explosives are handled. The maximum static energy which will not detonate a secondary explosive is usually much higher than for a primary explosive.

14.3.3   Sensitivity to Detonation

The sensitivity to detonation of an explosive is determined by the ease with which the explosive can be ignited by the primary detonation or another explosive in the vicinity which generates a shock wave. A maximum gap between the initiated detonator and the test explosive determines the sensitivity of the explosive to detonation. This gap test is an important characteristic which is usually determined relative to trinitrotoluene (TNT).

14.3.4   Velocity of Detonation

The *velocity of detonation* (VOD) of an explosive is the rate at which the detonation wave passes through the explosive. The greater the VOD, the larger is the power of the explosive. The VOD is determined by several methods: an optical, an electrical, and a comparative method.

In the optical method, a high-speed streak camera is used to follow the flame front at about 10 million frames/s, with exposures of as little as 0.01 µs.

In the electrical method, a resistance element is embedded along the explosive axis, and as the detonation proceeds, the resistance changes and can be followed as a function of time on an oscilloscope.

A comparative method uses a standard detonating fuse with known VOD. A fixed known length of the standard fuse is placed in a parallel loop with the test explosive. The two ends of the standard fuse are in contact with the test explosive and when detonated will ignite both standard and test explosive. The larger path of the standard fuse means that the start and end parts both ignite, causing the two flame fronts to meet at a point which is identified by an indentation on a lead plate adjacent to the standard fuse. The VOD of the sample can be calculated from the distances and the VOD of the standard explosive (see Exercise 14.10).

14.3.5   Explosive Strength

The strength of an explosive is its most important characteristic and is a measure of the conversion of its exothermic energy of combustion into mechanical disruption or power. The strength is usually

Table 14.4 Properties of some common explosives

Explosive	Formula	Mol. wt. (g/mol)	Blasting strength	Oxygen balance (g/atom/kg)
Trinitrophenol	$C_6H_3N_3O_7$	229.1	106	−28.4
Trinitrotoluene	$C_7H_5N_2O_6$	227.1	100	−46.2
Tetryl	$C_7H_5N_5O_8$	287.2	126	−29.6
RDX (hexogen)	$C_3H_6N_6O_6$	222.1	150	−13.5
PETN	$C_3H_8N_4O_{12}$	316.2	146	−6.3
Nitroglycerin	$C_3H_5N_3O_9$	227.1	140	+2.2
Ammonium nitrate	NH_4NO_3	80.1		+12.5

determined on the basis of unit weight, but the more practical basis is in terms of unit volume. Weight strength of an explosive is readily determined by comparison with a standard explosive such as blasting gelatin which is classified at 100%. For comparison, dynamite with 40% nitroglycerin has a weight strength of 40. The shattering power or brisance (B) of an explosive is usually compared with TNT which is set at 100. The strength values of some explosives are listed in Table 14.4. Most of the explosives are about the same strength as TNT. The B is difficult to determine experimentally, and a proposed calculated value is

$$B = d(VOD)^2 \qquad (14.2)$$

where d is the density. Other expressions used to calculate B include a force factor, but experimental values are still the most reliable. The experiment called the *sand-bomb* test consists of 80 g of Ottawa sand—20 + 30 mesh placed in a heavy-walled cylinder onto which is placed 0.400 g of the test explosive. An additional 120 g of sand is placed on top, and the explosive is detonated. The mass of sand that passes through the +30 mesh screen after the detonation is the B.

14.4 Oxygen Balance

The detonation of an explosive is an oxidation reaction in which it may be assumed that all the carbon forms CO_2, all the hydrogen forms H_2O, and all the nitrogen forms N_2. On this basis, an explosive with the composition $C_xH_vO_zN_D$ will have an *oxygen balance* (OB) of

$$OB = z - 2x - y/2 \qquad (14.3)$$

A positive value indicates a surplus of oxygen within the explosive whereas a negative value indicates that oxygen must be supplied, usually from the surrounding air. The values are expressed in terms of moles of oxygen (O) per kg of explosive. Explosives are often blended to give an oxygen balance of zero or slightly positive value. Hence, Table 14.3 shows an amatol 80/20 with 4:1 NH_4NO_3:TNT which would give an OB of −46.2 + (4 × 12.5) or a total value of +3.8. The detonation of mixture of explosives is, however, not only dependent on the OB but also on the combustion reactions as well as other physical properties.

14.5 Modern Explosives

The secondary explosives listed in Tables 14.3 and 14.4 represent the more common compounds. Some comments on a few of these will illustrate their relative characteristics.

14.5.1   Nitroglycerin

$$\begin{array}{ccc} NO_2 & NO_2 & NO_2 \\ | & | & | \\ O & O & O \\ | & | & | \\ CH_2\!-\!CH\!-\!CH_2 \end{array}$$

This colorless oily liquid has a density of about 1.6 g/mL, a freezing point of 13.2°C, and decomposes before it boils at 1 atm. It is slightly soluble in water but much more soluble in acetone, ether, benzene, and chloroform. It is very sensitive to shock, especially when the liquid contains air bubbles. It is made by reacting glycerol with a mixture of HNO_3 and H_2SO_4 at -20°C.

When *nitroglycerin* is adsorbed onto kieselguhr (25% diatomaceous earth, SiO_2), a plastic cheesy mass is formed. Present stabilizers use combustible material such as sawdust, flour, starch, or cereal products. When mixed with about 8% collodion cotton, it forms blasting gelatin. Plasticizers such as ethylene glycol dinitrate $(CH_2ONO_2)_2$ are added to reduce the freezing point of the dynamite and to increase its OB.

14.5.2   Trinitrotoluene (TNT)

Trinitrotoluene is easily made by the reaction of HNO_3 and H_2SO_4 with toluene. The order in which the reagents are mixed influences the output and safety of the process. It is thermally stable for over 40 h at 150°C and can be stored for 20 years at ambient temperatures. It is insensitive to shock and friction. It is slightly toxic at concentrations greater than 1.5 mg/m^3. Its use is extensive in the military as bombs, grenades, shells, torpedoes, and depth charges.

14.5.3   Tetryl

Tetryl, also called *tetranitromethylaniline*, has a melting point of about 130°C with some decomposition. It is nonhygroscopic and practically insoluble in water but highly soluble in acetone and benzene. It can be stored for over 20 years at ambient temperatures with no noticeable change in properties.

Tetryl is prepared by the nitration of dimethyl aniline followed by the oxidation of one of the *n*-methyl groups to CO_2 and more nitration.

Tetryl is used as a booster where it is mixed with TNT (called tetrytol) and some graphite to help pressing the mixture.

14.5.4 Ammonium Nitrate, NH₄NO₃

Ammonium nitrate is easily made by reacting nitric acid with ammonium hydroxide:

$$HNO_3 + NH_4OH \rightarrow NH_4NO_3 + H_2O \qquad (14.4)$$

The reaction is exothermic and can be carried out in a borehole prior to its detonation. The thermal decomposition of NH_4NO_3 yields nitrous oxide, N_2O:

$$NH_4NO_3 \xrightarrow{220°C} N_2O + 2H_2O \qquad (14.5)$$

The ignition temperature of NH_4NO_3 is about 465°C with heat of combustion of 2.62 kJ/g and an oxygen balance of +12.5 g/atom/kg. It is difficult to detonate, but accidental explosions of stored tonnes of the material on land or in ships have caused thousands of deaths and injuries during the last 100 years.

Its VOD is between 1.1 and 2.7 km/s. When fuel oil is added to the salt, it is known as ammonium nitrate fuel oil (ANFO). It is usually added to other explosives to increase the overall oxygen balance of the mixture.

14.5.5 Hexogen

Cyclonite or *hexogen* (RDX) was first prepared in 1899 and was initially recommended for medical use. Later applications (1916) included its use as a smokeless propellant. During World War II, Germany manufactured 7,000 t of RDX per month, whereas the US production was more than twice this rate.

Table 14.5 Some propellants and selected properties

Fuel	Oxidant	Storage temp. (°C)	ΔH^0_{comb} (kJ/mol)	I_s (s)
Cryogenics				
H_2	O_2	$-253/-183$	-241.8	362
H_2	F_2	$-253/-188$	-271.1	352
Storables				
N_2H_4	H_2O_2	20/20	2686/N_2H_4	300
Ethylene oxide[a]		20	5237	170
Nitromethane[a]		20	675	220

[a]Monopropellants

This substance was the explosive of choice in the last world war. It is prepared by the reaction of ammonia with formaldehyde to form hexamethylenetetramine that when nitrated forms RDX with tetramethylene tetranitramine (HMX) as a by-product:

$$(14.6)$$

The fuels listed in Table 14.5 are practical propellants. More exotic choices such as diborane, B_2H_6; acetylene, C_2H_2; ozone, O_3; oxygen difluoride, OF_2; and nitrogen trifluoride, NF_3 are a few examples.

Solid propellants usually consist of ammonium nitrate (NH_4NO_3), ammonium perchlorate (NH_4ClO_4), and aluminum powder bound in a resin of polystyrene, polyurethane, or polybutadiene. Specific impulse values of 250–300 s are achieved depending on the choice of fuel and oxidizer.

14.6 Applications

Explosives and deflagrators are used in a variety of *applications* which include propellants, fireworks or pyrotechnics, welding, riveting, and cladding.

14.6.1   Propellants

In an age of rockets and space travel, there is a constant demand for new and stronger *propellants*. These are classified into cryogenics and storable liquids and characterized by the specific impulse *(I_s)* which is the thrust per unit mass flow of fuel and by convention reported in units of seconds (see Table 14.5). The thrust depends on the combustion temperature and the exhaust and combustion chamber pressures, P_E and P_c, respectively.

RDX is often mixed with TNT (60 RDX/40 TNT) to achieve a higher VOD and thus greater strengths. When mixed with lubricating oil (12%), it became the common plastic explosive of World War II. The plastic which does not lose its flexibility, C_3, consists of 77% RDX and 23% gel composed of dinitrotoluene, nitrocellulose, and dibutyl phthalate. The putty- or doughlike material has an advantage in demolition work by forming space charges which can slice through a steel pipe or bridge support.

14.6.2   Pyrotechnics

Fireworks usually use low explosives which are reliable, stable, and of low toxicity and reasonable cost. The original oxidizer (in black powder) potassium nitrate (KNO_3) is still in use today. Potassium perchlorate ($KClO_4$), ignition temperature 560°C, is much preferred over potassium chlorate ($KClO_3$) which ignites at 220°C when mixed with sulfur. The colors are obtained from the various salts: yellow is readily obtained from sodium salts (such as Na_3AlF_6 or $Na_2C_2O_4$) due to the 589-nm line. Red is due to the strontium salts such as $Sr(NO_3)_2$ and $SrCO_3$ which emits the 606-nm band from SrO whereas SrCl gives the 636–688-nm red bands. Green is produced from barium as the nitrate or chloride. The 505–535-nm band is due to emission by BaCl. Blue is due to cuprous chloride (CuCl) which has a flame emission in the 420–460-nm range. Since $KClO_3$ is unstable in the presence of copper salts, it is customary to use NH_4ClO_4 with $CuCO_3$.

The fuels consist of resins, charcoal, and sulfur. A typical aerial display shell used in fireworks is shown in Fig. 14.1. The black powder is the propellant which fires the shell to suitable heights. A fuse ignites and fires the various compartments to provide the colors and flashes in sequence.

14.6.3   Metalworking

The use of explosives in *metalworking* was initially employed in 1880 to make spittoons. Today, the missile and rocket industry use explosives to shape bulkheads, nosecones, and even large rocket sections. A small 50-g charge can do the work of a 1,000-t press shaping a thick metal plate 2–3 m in diameter. An example of a metal forming system is shown in Fig. 14.2. The dye is usually cheap material of concrete or plaster which can be evacuated. The sheet metal is held in place by a bed of water in which the explosive charge is detonated. The pressure and shock wave forces the metal into the evacuated mold in microseconds.

A less hazardous method is the use of an exploding wire to generate the shock wave. A thin resistive wire 0.03 mm in diameter is connected to a condenser (1–10 μF) and is charged to high voltage. When the condenser discharges through the wire, it explodes in microseconds creating the shock wave through the water bed. The method has been used in cladding and welding. For example, it is possible to apply a thin layer of platinum on a steel crucible which will have most of the properties of a 100% Pt crucible but which contains less than 5% Pt. Normally, the metal oxides on the surface prevent the two metals from bonding. However, the shock of the explosion forces the

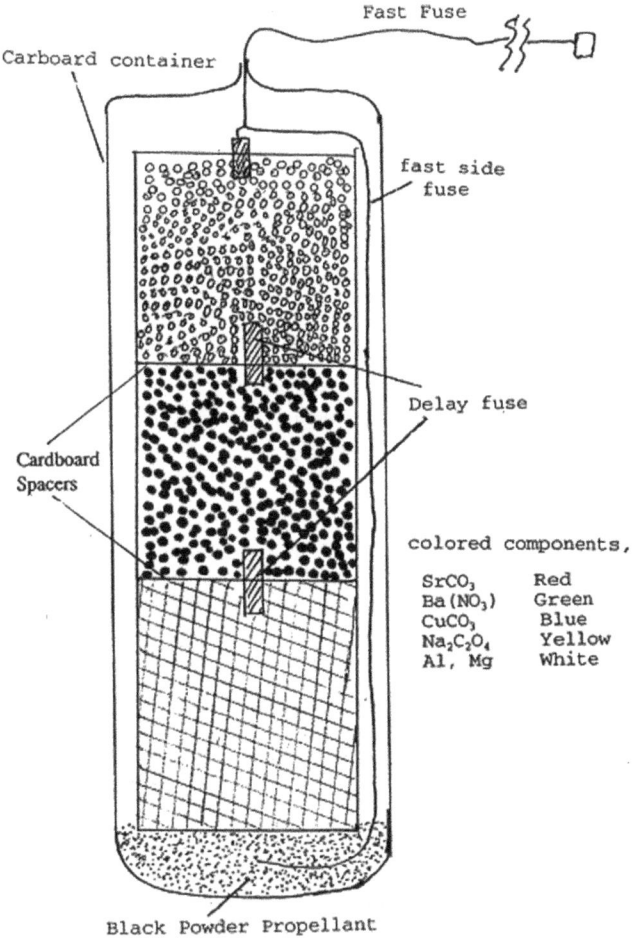

Fig. 14.1 Typical structure of an aerial shell for fireworks display. The compartments of the rocket usually produce different colored fireworks and patterns

Fig. 14.2 Typical arrangement for explosive metal forming

Fig. 14.3 Schematic diagram of the explosive cladding of a metal

Fig. 14.4 Typical sequence of a riveting system

cladding metal onto the base metal under plastic flowing conditions, producing a metallic fluid jet at the point of contact (see Fig. 14.3). This removes the surface oxides leading to intimate welding of the two metals. The critical speed of the explosive weld is 200–300 m/s, and granular explosives are formulated to meet these requirements.

14.6.4 Riveting

Explosive *riveting* has been used for over 60 years and was originally designed for rivets where backup space was insufficient for normal processing. Applications include automobile brake shoes, aircraft industry, tanks, and other military vehicles.

The typical rivet is hollow and filled with explosive which can be detonated either thermally or by an electric charge. A typical sequence of explosive riveting is shown in Fig. 14.4.

14.7 Accidental Explosions

From the early days of black powder usage to the industrial revolution and the extensive mining of coal, *accidental explosions* have been a constant reminder of technological development. The dust of grain causes 30–40 grain elevators to explode each year in the USA. Coal mine explosions caused about 100 deaths/year during the period 1931–1955, and in spite of precautions and improved safety regulations in coal mines and grain elevators, accidents still occur which often take lives and cause millions of dollars in losses.

Table 14.6 Classification of accidental explosion

Type of explosion	Typical system and results
1. Detonation of condensed phase systems	
(a) Light or no confinement	(a) Blastwave: Manufacture, transport, and handling of explosive chemical reactors, distillation columns
(b) Heavy confinement	(b) Less damaging because of the design
2. Combustion explosions in enclosures	
(a) Fuel vapor	(a) Ignition source: building shops, boilers, low energy blast when length-diameter (L/D) = 1
(b) Dust ($d < 75$ μm)	(b) Usually ignited by another explosion, coal mines, grain elevators, pharmaceutical industries
3. Pressure vessels	
(a) Simple failure	(a) Vessel fails due to heat or corrosion, explosion in boiler due to steam or combustion
(b) Combustion generated failure	(b) Contaminated compressed air lines
(c) Combustion and failure	(c) Fireball due to stored fuel
(d) Failure after runaway chemical reaction	(d) Exothermic reaction (Bhopal)
(e) Failure after runaway nuclear reaction	(e) Chernobyl
4. Boiling liquid expanding vapor explosion (BLEVE)	Ductile vessel (tank car) with high vapor pressure combustible liquid. Fireball
5. Unconfined vapor cloud explosion	Ignited spills of combustible fuels
6. Physical vapor explosion	Island volcanoes, liquid propane on water

Table 14.7 Estimation of fuel mass from spherical fireball size

Fireball diameter D (m)	Fuel mass M (kg)	Duration t (s)	Power (GW)
25	80	2	1.6
50	640	4	6.9
75	2,160	5.8	15
100	5,125	7.8	28
125	10,000	9.7	43
150	17,300	11.6	62
200	41,000	15.5	110

There are many varied types of explosions which are accidental. A common classification is given in Table 14.6.

A fireball is the result of a fire causing a container, holding a large quantity of flammable liquid, to burst. The sudden release of the liquid which is vaporized by the heat results in a fireball of diameter, D, lasting for a duration of time, t. Independent of the fuel or heat of combustion, the relation between the size of the spherical fireball (diameter D in meters), the mass of combustible substance (mass M in kg), and the duration (time t in sec) is given by

$$D = 5.8M^{1/3} \tag{14.7}$$

and

$$t = 0.45M^{1/3} \tag{14.8}$$

which are based on previous accidental explosions. Some calculated values are shown in Table 14.7 and plotted in Fig. 14.5 on a semilog scale. The progress of the fireball and the structure is shown in

Fig. 14.5 Semilog plot of (14.7) from the data in Table 14.7 showing the diameter of a fireball as a function of the mass of fuel ignited

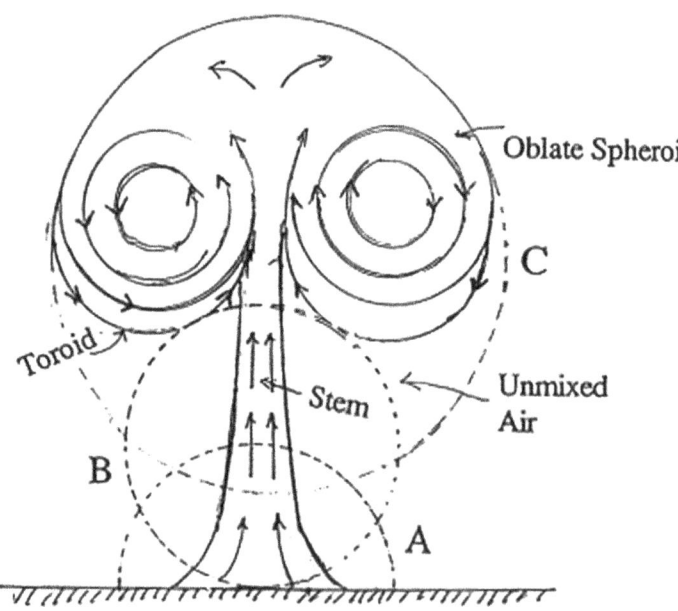

Fig. 14.6 Typical development of a fireball from *A* to *B* to *C* showing the stem through which the fuel is funneled

Fig. 14.6. As the pressure of the detonation products decreases to ambient pressure, the density of the gas decreases to less than that of the surrounding air, resulting in a buoyant force causing the fire to rise. The spherical shape is formed as the fuel continues to burn and becomes more buoyant. The spherical fireball then lifts from the ground, bringing in more air by convection and vortex motion.

Fig. 14.7 An apparatus used to determine relative VOD

A stem is formed from the fuel spilled on the ground, creating a mushroom-type cloud. The spherical fireball changes into an oblate spheroid and then into a toroid, eventually consuming all the fuel and ultimately dissipating its energy.

Combustible vapors must be treated with care and respect. Standard precautions in design and construction must be followed, accidents anticipated, and provisions provided.

Exercises

1. Distinguish between an explosive and a detonator.
2. What chemical functional groups impart explosive properties to a compound?
3. Distinguish between a propellant and an explosive.
4. What is meant by the sensitivity of an explosive?
5. How is sensitivity to impact determined?
6. Define brisance and VOD.
7. How may the strength of an explosive be determined?
8. Write a short note on the application of explosives to metalworking.
9. Gunpowder consisted of 75% KNO_3, 15% charcoal, and 10% sulfur. Based on these properties, write a chemical reaction in which K_2SO_4 is formed instead of K_2S_2 in reaction.
10. The VOD of explosive X is determined relative to the standard explosive S (see Fig. 14.7). The explosive is initiated at O. At A, the standard explosive ignites as X continues to burn. When the burn of X has proceeded to B, then S ignites to meet the other flame front at C which is marked in a lead plate P. From the distances AB, AC, and BC, show that

    show that
    $$(VOD)_X = \frac{AB(VOD)_S}{AC - BC}$$

11. Calculate the specific volume of gas produced at STP per g of AgN_3.
12. The explosive HMX (cyclotetramethylene tetranitramine), $C_4H_8N_8O_8$, has molecular mass of 296.17 g/mol. Calculate its oxygen balance.
13. Compare relative calculated brisance from (14.2) with the values in Table 14.4.
14. What blend of RDX and NH_4NO_3 would give a positive value for an overall oxygen balance?
15. Construct a log-log plot of (14.8) using the data in Table 14.7.
16. A recent fireball due to the explosion of a solvent recycling plant was estimated to have a diameter of 75 m and to have lasted about 6 s. Estimate the mass of solvent ignited.
17. A train derailment in which liquid propane exploded formed a fireball of 250 m in diameter. Estimate the mass of propane burned and the duration of the fireball.
18. Calculate the standard enthalpy of combustion of liquid by mass and by volume: (a) hydrogen and (b) fluorine (see Table 14.5).

Further Reading

1. Bailey A (1998) Explosives, propellents, and pyrotechnics, 2nd edn. Brasseys, Dulles
2. Cooper PW (1997) Explosives engineering. Wiley, New York
3. Cooper PW (1996) Basics of explosives engineering. VCH, New York
4. Cooper PW, Kurowski SR (1996) Introduction to the technology of explosives. VCH, New York
5. Suceska M (1995) Test methods for explosives. Springer, New York
6. Grady SM (1995) Explosives: devices of controlled destruction. Lucent Books, San Diego
7. Cheret R (1992) Detonation of condensed explosives. Springer, New York
8. Meyer R (1992) Explosives, 4th edn. VCH, New York
9. MacDonald GW (1991) Historical papers on modern explosives. Revisionist Press, Brooklyn
10. Barnard A, Bradley JN (1984) Flame and combustion, 2nd edn. Chapman and Hall, London
11. Brauer KO (1974) Handbook of pyrotechnics. Chemistry Publication, New York
12. Tooley P (1971) Fuels, explosives and dyestuffs. J. Murray, London
13. Urbanski T (1965) Chemistry and technology of explosives (trans: English), vol 3. Pergamon Press, Oxford.
14. Penner SS, Ducarme J (eds) (1960) The chemistry of propellants. Pergamon Press, London
15. Cook MA (1958) Science of high explosives. Reinhold, New York
16. Pennie AM (1958) RDX, its history and development. Chem Can 10:11
17. International Society of Explosive Engineers. www.isee.org/
18. Sandia National Lab.. http://www.sandia.gov/explosive/componts.htm
19. Canadian Explosive Research Laboratory. http://www.nrcan.gc.ca/mms/explosif/incerle.htm
20. U.S. Army Engineering Center. www.hnd.usace.army.mil
21. Explosives Ltd., cookbook. http://www.explosives.com/
22. The Ordnance shop, Types of explosives. www.ordnance.org/classifi.htm

Chapter 15
Water

15.1 Introduction

In recent years, strenuous efforts have been made to conserve and recycle resources. Of major concern has been *water* due to its rising consumption and resultant shortage. Any discussion of water must include its domestic and industrial use. The treatment of waste sewage is the concern of sanitation engineers and will not be considered here.

Some selected basic properties of water are given in Table 15.1. The unique characteristics of water, although not obvious, will become apparent as we learn more about it. The three-dimensional phase diagram for water, up to the pressure of 10,000 atm and for the temperature range of −50°C to + 50°C, is shown in Fig. 15.1. The density of ice (ice I) is less than that of water up to about 2,200 atm. Above this pressure, ice exists in various different crystalline modifications. The two-dimensional phase diagram for water is shown in Fig. 15.2, where the triple point, 0.0100°C, consists of solid ice, water, and water vapor at 4.579 torr in equilibrium. Also shown is the critical point above which liquid water cannot exist in the liquid state. The negative slope of the *P-T* line is due to the difference in molar volumes of liquid and solid, that is, $(V_t - V_s) < 0$, and because

$$dP/dT = dH_{fus}(V_t - V_s/T) \tag{15.1}$$

This situation exists for only four known substances, water, bismuth, iron, and gallium, where the solid floats on the liquid at the melting point. Water is unique, however, since its maximum density is not at 0°C but at 4°C. This is illustrated in Fig. 15.3 where the partial structure of dimers and trimers due to hydrogen bond formation allows for a more structured arrangement of the water clusters. This structure is disrupted as salt is added, as shown in Fig. 15.4, where the maximum density drops from 4°C to −1.33°C at 24.7 ppt (parts per 1,000). At higher salt concentrations, the density of water increases with decrease in temperature. Ice formed from saltwater is normally free of salt and has the normal density of ice, that is, 0.9170 g/mL.

Water is essential to humans. The average consumption per person is 2 L/day. The average household uses about 200 L/day. Water is also necessary for many industries, for example, 250 L is required for the manufacture of 1 kg of steel and 1 kg of aluminum requires about 1,300 L of water. Ten liters of water is needed for the production of each liter of gasoline, 250 tonnes for each tonne of paper, and 450 L to grow enough wheat to bake a loaf of bread, grow 1 kg of potatoes, or produce one weekend newspaper.

The amount of water on the earth's surface is about 1.4×10^9 km^3 of which only 3% is non-ocean water and of which two-thirds is in the form of ice on the polar ice caps. The remaining 1% is surface and underground water in a relative ratio of about 1:3. These ratios are changing slowly due to the

O.V. Roussak and H.D. Gesser, *Applied Chemistry: A Textbook for Engineers and Technologists*, DOI 10.1007/978-1-4614-4262-2_15, © Springer Science+Business Media New York 2013

Table 15.1 Selected properties of water

Molar mass (g/mol)	18.0153	
Melting point (°C)	0.0000	
Boiling point (°C)	100.000	
Triple point (°C)	0.0100	
Density liquid at 25°C (g/mL)	0.997044	
Density liquid at 0°C (g/mL)	0.999841	
Density solid at 0°C (g/mL)	0.9170	
$\delta H°_f$ (liquid) (kJ/mol)	−285.8	Standard enthalpy of formation
$\delta H°_f$ (gas) (kJ/mol)	−241.8	Standard enthalpy of formation
$\delta G°_f$ (liquid)	−237.2	Standard free energy of formation
$\delta G°_f$ (gas) (kJ/mol)	−228.6	Standard free energy of formation
$S°$ (liquid) (J/mol K)	69.9	Standard entropy
$S°$ (gas) (J/mol K)	188.7	Standard entropy
δH_{fus} (kJ/mol)	5.98	Enthalpy of fusion
δH_{vap} (kJ/mol)	40.5	Enthalpy of vaporization
Heat capacity (solid) (J/mol K)	37.6	
Heat capacity (liquid) (J/mol K)	75.2	
Thermal conductivity (solid) (J/s m K)	2.1	
Thermal conductivity (liquid) (J/s m K)	0.58	
Dipole-moment (Debye)	1.84	
H—O—H angle	104°	
Bond length, H—O (pm)	96	
γSurface tension (mJ/m^2) at 25°C	72	
h_iViscosity (poise, 10^{-1} kg m^{-1} s^{-1}) 25°C	0.01	
T_c, Critical temperature (°C)	347.15	
P_c, Critical pressure (atm)	218.3	
V_c, Critical volume (cm^3/mol)	59.1	
E (H—O) (kJ/mol)	464	Bond energy
D (HO—H) (kJ/mol)	498.7	Bond dissociation energy
D (O—H) (kJ/mol)	428.0	Bond dissociation energy
K_{ion} at 25 °C	1.002×10^{-14}	Ionization constant
pK_w at 25 °C	13.999	
Vapor pressure		

T(°C) 0 10 20 25	50 75	90 100
P(torr) 4.6 9.2 17.5 23.8	92.5 289.1	525.8 760

evidence showing that the north polar ice is receding. It will result in the rise of sea level and the possible flooding of costal land and cities.

Seawater consists of 3.5% by weight of dissolved solids, mostly NaCl with Mg^{2+}, Ca^{2+}, K^+, and SO_4^{2-} as the main ions with many more trace components in the ig/mL and pg/mL range. For example, uranium is present in seawater at around 3 pg/mL, and gold is about 1,000 times less concentrated (i.e., 3 pg/L). Distribution of water on the earth's surface is shown in Table 15.2. Water quality (as determined by standard chemical analysis) of the drinking water from some Canadian cities and some commercially available spring waters is listed in Table 15.3. Also included are the guidelines recommended by the World Health Organization (WHO). Table 15.3 lists the WHO guidelines for the limits of organic substances which are considered a health hazard in potable water. These substances are seldom analyzed, but readily show their presence by ultraviolet absorption spectrophotometry.

Attempts to alleviate the water shortage which exists in many parts of the world have included extensive efforts to desalinate seawater. This continuing effort will be discussed later.

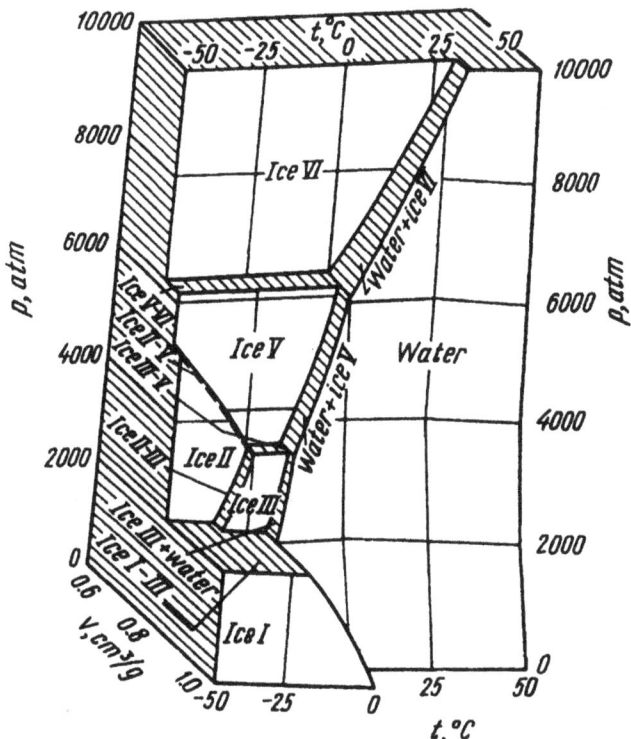

Fig. 15.1 The 3-D phase diagram of water showing the various modifications of ice (Ice IX is not shown in the diagram)

Fig. 15.2 The 2-D phase diagram of water showing the triple point, critical temperature, and critical pressure

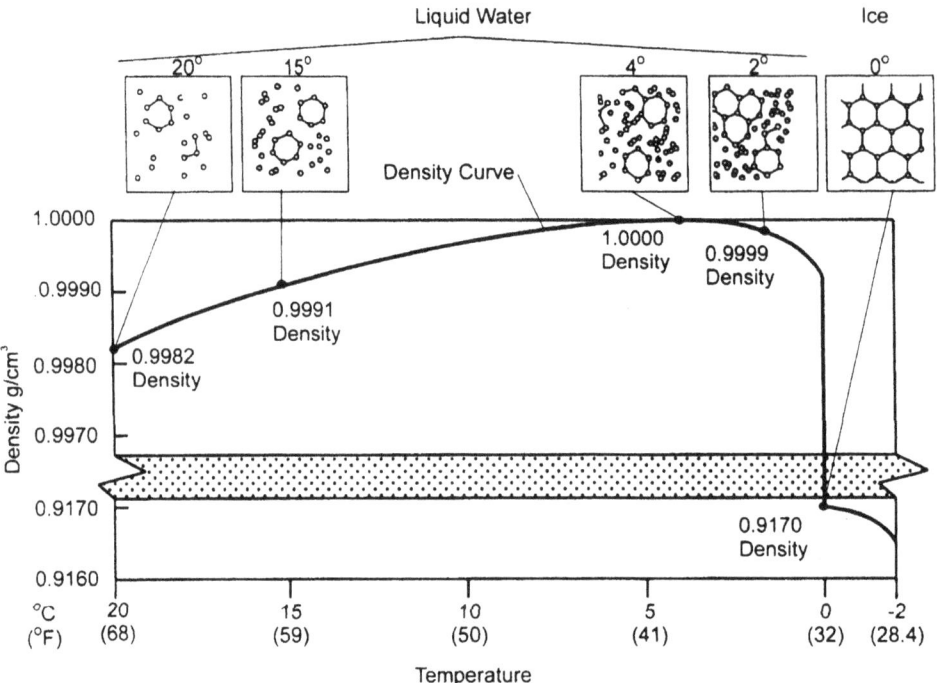

Fig. 15.3 A pictorial description of the changes in water structure near the position of maximum density, and as water changes to ice, it shows the formation of ice clusters in freshwater. The density of freshwater ranges from 0.9982 g/cm^3 at 20°C to a maximum of 1.000 g/cm^3 at 4 °C. However, the density of ice (*solid water*) is only 0.9170 g/cm^3, so it floats on water

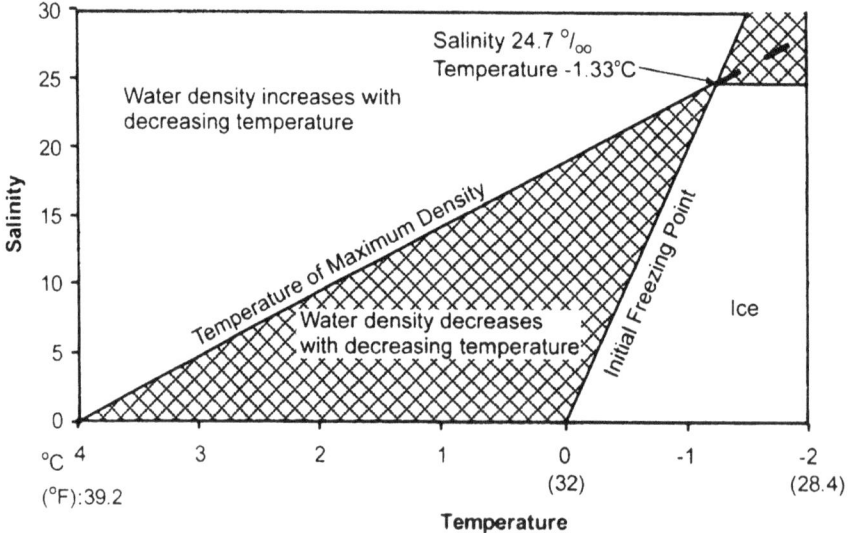

Fig. 15.4 Freezing point and temperature of maximum density as a function of salinity (grams of salt per 1,000 g of solution)

One elaborate scheme, proposed by Prince Mohamed Al Faesal of Saudi Arabia, involves the towing of icebergs, which are essentially salt free, from the Antarctic to desert countries. It is calculated that at a speed of 3 knots, the trip could take over 100 days and that a 1-km-long iceberg would melt before reaching the Gulf of Arabia. Hence, it was proposed to insulate the iceberg to

Table 15.2 Approximate water distribution on the earth's surface

Source	(%)
Oceans	96.1
Ice and snow	2.2
Underground	1.2
Surface and soil	0.4
Atmosphere	0.001
Living species	0.00003
Total = 1.36 × 10⁹ km³	*100*

Table 15.3 WHO guidelines for organic constituents in drinking water (1984)

Constituent	Concentration (µg/L)
Aldrin and dieldrin	0.03
Benzene	10
Benzo(a)pyrene	0.01
Carbon tetrachloride	3
Chlordane	0.3
Tetrachloroethane	10
Trichloroethene	30
Chloroform	30
1,2-Dichloroethane	10
1,1-Dichloroethane	0.3
2,4-D	100
Heptachlor and heptachlor epoxide	0.1
Hexachlorobenzene	0.01
Lindane	3
Methoxychlor	30
Pentachlorophenol	10
2,4,6-Trichlorophenol	10

reduce its melting rate. The ice could also be used for air conditioning while being converted into potable water. The estimated cost of the water was determined to be about $0.06 per 1,000 US gal.

Another recent scheme involved the transportation of water by sea from areas of plentiful supply in floating inflatable bags—called the *Medusa bag*—to areas where water is in short supply. This is considered feasible because freshwater is less dense than seawater and will float on the ocean surface.

15.2   Natural Water

The main source of domestic water is *natural water* which includes lakes, rivers, and wells. It is characterized by a variety of measurable quantities which include color, turbidity, odor, bacteria, suspended solids, total solids, pH, conductivity, and most importantly, hardness.

15.2.1   Turbidity

Suspended solids and colloids cause water to be turbid. Various turbidity scales have been established for the determination of the relative *turbidity* of a water sample. Active carbon filtration can often clarify water by adsorbing colored components and removing colloids.

15.2.2 Color

The presence of dissolved organic substances in water gives rise to coloration. The *color* of water can be determined by various methods that either involve a comparison with standards which are either colored glass or standard solutions of a blend of K_2PtCl_6 and $CoCl_2$, or by the more comprehensive approach, namely the spectrophotometric method, where the transmission of visible light is determined as a function of the wavelength of light.

The color value of water depends on pH and usually increases as the pH increases. Hence, the specification of color must also include the pH value of the water.

15.2.3 Odor and Taste

The presence of both organic and inorganic compounds in water contributes to the "chemical" senses—*odor* and *taste*. Pure distilled water is free of contaminants and has neither odor nor taste. The odor and taste of water can be removed by passing water through a column of activated carbon. Good-grade drinking water is free of odor and toxic substances and has a pleasant taste.

Aeration also improves odor and taste and is often the standard treatment during the conversion of natural waters to potable grade. This full treatment involves filtering the intake water to remove debris, fish, and other large items. The water is then treated with trivalent cations such as alum [$KAl(SO_4)_2 \cdot 12H_2O$] and ferric chloride ($FeCl_3$) that hydrolyze to amorphous, gelatinous precipitates, $Al(OH)_3$ and $Fe(OH)_3$, respectively, and as the floe settles by gravity, it clarifies the water by trapping and adsorbing the suspended impurities. Aeration is used to improve the color, odor, and taste of water. The soluble organic compounds and salts are adsorbed on activated carbon followed by disinfection by the addition of chlorine.

15.3 Water Sterilization

15.3.1 Chlorine

Water for drinking purposes—potable water—is treated to make it acceptable and free of harmful bacteria. This is most often accomplished by adding *chlorine* to the water which forms the strong oxidizing hypochlorous acid, HOCl:

$$Cl_2 + H_2O \rightleftharpoons HOCl + H^+ + Cl^- \qquad (15.2)$$

$$HOCl \rightleftharpoons H^+ + OCl^- \qquad (15.3)$$

Depending on pH and products formed, the effective residual concentration of chlorine [free available chlorine (FAC)] HOCl or "OCl" is 0.1–0.2 mg/L. Higher concentrations tend to give water a definite taste.

A more persistent source of the hypochlorite ion is the chloramines (made from the reaction of chlorine with ammonia at pH 4.5–8.5):

$$Cl_2 + NH_3 \rightleftharpoons NH_2Cl + HCl \qquad (15.4)$$

$$NH_3 + HOCl \rightleftharpoons NH_2Cl + H_2O \qquad (15.5)$$

Dichloramine, pH 4.5, and trichloramine, pH 4.4, are formed by successive additions of Cl_2 or HOCl to the monochloramine. Chloramines impart a green color to water. As the hypochlorite ion is consumed by oxidizing the contaminants, the chloramines supply more HOCl and so maintain a "safe" level of primary disinfectant. Excess chlorine is removed by reaction with sodium bisulfate:

$$NaHSO_3 + H_2O + Cl_2 \rightarrow NaCl + H_2SO_4 + HCl \tag{15.6}$$

The chlorine demand is the difference between the chlorine added and the residual concentration after a designated reaction time of approximately 10 min. Total residual chlorine is determined by the oxidation of N,N-diethyl-p-phenylenediamine (DPD) to produce a red-colored product. Addition of iodine then catalyzes further reaction with chloramines, and it is possible to obtain concentration values for all chloramines and free chlorine.

The extensive use of chlorine to purify water has recently been shown to result in the formation of chlorinated hydrocarbons. Low molecular weight compounds, such as the haloforms (HCX_3), also called trihalomethanes (THM), are volatile and have been shown to be carcinogenic. They have been detected in drinking water and in the air of enclosed swimming pools. Thus, several alternate disinfectants (such as ozone, chlorine dioxide, UV, and ferrates) have been considered as alternates to chlorine. Of these, the use of ozone has been most developed.

15.3.2 Ozone

Ozone is a triatomic molecule of oxygen, O_3, with a triangular structure where the O—O—O angle is 116° 49′. It has been used as a water disinfectant since 1903, and water treatment is the single major use of ozone. Next to fluorine (F_2), it is the most powerful oxidizing agent available. Its widespread use is due to its ability to add readily across a carbon–carbon double bond to form an ozonide

$$\tag{15.7}$$

ozonide

which subsequently decomposes to form aldehydes, RCHO, and ketones.

Ozone also decomposes to form an oxygen atom that is also a powerful oxidizing agent. Ozone is prepared either by the photodissociation of oxygen (O_2)

$$\tag{15.8}$$

$$O_2 + h\upsilon \xrightarrow{\lambda < 200\,nm} 2O \tag{15.9}$$

$$O + O_2 \longrightarrow O_3$$

or by a high-voltage ac electrical (silent) discharge through oxygen (or air). A typical ozone generator is shown in Fig. 15.5.

A mixture of ozone and air can be bubbled through water to oxidize impurities. Ozone is toxic and can be detected by its odor at about 0.01 ppm, whereas its TLV is 0.1 ppm.

Fig. 15.5 Silent discharge reactor tube for ozone production. The tube is rilled with electrolyte and placed in a container full of electrolyte solution which holds the high-voltage ac electrodes. The electrolyte is usually copper sulfate or sodium sulfate in water. The electrodes are usually copper

Unlike chlorine, ozone does not produce known carcinogens as a byproduct of its water treatment and, therefore, is gaining increased use for domestic as well as industrial water supplies. In North America, chlorination must follow ozonation in public water supply because ozone decomposes rapidly, and a chlorine residual may be carried throughout a distribution system. Chloramines are often used.

15.4 Infectious Agents

The final criterion of satisfactory sterilization of domestic water is the reduction in bacterial concentration to very low values. Bacteriological examination of drinking water uses the coliform bacteria (*Escherichia coli*—often referred to as *E. coli*) as an indication of the purity of the water since these bacteria are the normal inhabitant of the intestinal tract and constitute about 30% of the dry weight of adult human feces. Water suitable for human consumption should contain less than one viable coliform per 100 mL.

Three different types of standard tests are used:

1. *Multiple-type fermentation method.* The various aliquots of samples are incubated for 48 h at 35°C in a culture medium. The absence of gas formation indicates a negative test for coliform.

2. *Membrane filter technique.* Water is filtered through a 0.45-µm sterile filter which is then placed in a culture medium and incubated at 35°C for 24 h. The growth of visible colonies on the filter indicates the number of bacteria present.
3. *Standard plate count method.* Various diluted volumes of the sample are added to a solid agar culture medium and incubated at 35°C for 72 h. The colonies are counted and recorded as the number per milliliter.

Common bacterial concentration is designated as the "most probable number" (MPN) per 100 mL and represents a statistical interpretation of the results of replicate analysis.

Prior to the 1950s, most houses and buildings used galvanized iron pipes to distribute water. However, these pipes corroded and usually started leaking after about 25 years. Copper has since replaced iron, and because of the solder joints which contain lead, copper is being replaced by plastic for cold-water systems and CPVC and polybutylene are commonly used for both hot- and cold-water lines. There are, however, a large number of older houses which still have lead water pipes or lead connectors from the city water line. This metal is appreciably soluble in water containing carbon dioxide or organic acids. Chloride of lead ($PbCl_2$) is also soluble to a considerable extent. Cases of chronic lead poisoning have been observed due in some instances to the habitual use of water that remained in the lead pipes overnight. The fall of ancient Rome is attributed by some scholars to the use of lead casks and vessels to store wine.

It was shown in the 1930s that fluoride in natural waters in concentrations greater than 4 ppm caused a brownish discoloration or mottling of tooth enamel. Prior to World War II, it was demonstrated that such teeth were free from cavities. In 1945, fluoride was, for the first time, added to the public water supply in Grand Rapids, Michigan. Since then it has been confirmed throughout the world that fluoride added to drinking water reduces cavities in tooth enamel. Fluoride is now usually added to domestic water supplies as Na_2SiF_6 to bring the F^- level to about 1 g/mL. Though fluoride in high concentrations is harmful, the benefits of reduced dental cavities offset the potential hazards. However, many people disagree with this and often resort to drinking only distilled water.

15.5 Water Quality: Hardness

The quality of water varies considerably from one city to another and from one natural source to another. The results of analysis of several water supplies are given in Table 15.3. The water for three Canadian cities and some common spring bottled waters are compared with freshly fallen snow. Specific conductivity is a rough measure of the total ion concentration in the sample. Metallic ions are determined simultaneously by their emission spectra in a plasma or flame. Anions usually require special colorimetric methods. Spring waters listed in Table 15.3 show differences which do not truly demonstrate the organic content of the water. Only by a measure of the UV transmission through at least a 10 cm path length of water over the wavelength range of 200–360 nm is it possible to show which of the waters is the cleanest. Surprisingly, the fresh fallen snow is not the purest water.

One of the most important characteristics of domestic and industrial water, especially in boiler water systems, is a measure of its *hardness*. The hardness of water is due to the bicarbonate ion (HCO_3^-) which, in the presence of calcium and magnesium ions (Ca^{2+}, Mg^{2+}), will form insoluble carbonates when heated:

$$Ca^{2+} + 2HCO_3^- \rightarrow CaCO_3 \downarrow + CO_2 + H_2O \qquad (15.10)$$

$$Mg^{2+} + 2HCO_3 \rightarrow MgCO_3 \downarrow + CO_2 + H_2O \qquad (15.11)$$

Hardness due to carbonate is formed when carbon dioxide (CO_2) in air is dissolved in water which is in contact with calcium carbonate ($CaCO_3$ or limestone):

$$CaCO_3 \downarrow + CO_2 + H_2O \rightarrow Ca^{2+} + 2HCO_3^- \tag{15.12}$$

This reaction is reversed at high temperatures, precipitating the $CaCO_3$, which forms as scale on the surfaces of heaters or other hot surfaces.

Hardness of a water supply is measured in terms of the equivalent amount of $CaCO_3$ that would precipitate. Thus, 40 mg/L of Ca^{2+} (1 mM) would form 100 mg/L of $CaCO_3$ (1 mM). Similarly, 24.3 mg/L of Mg^{2+} (1 mM) would also produce 1 mM equivalent of $MgCO_3$, and its hardness (in mg/L) would also be 100.

Soap consists of the sodium salt of a fatty acid that ionizes in water:

$$RCOONa \rightarrow RCOO^- + Na^+ \tag{15.13}$$

The cleaning agent is the fatty acid anion $RCOO^-$ which is precipitated by divalent or trivalent metal ions such as Ca^{2+}, Mg^{2+}, and Al^{3+}:

$$2RCOO^- + Ca^{2+} \rightarrow (RCOO)_2Ca \downarrow \tag{15.14}$$

The presence of calcium or magnesium bicarbonate in water is called temporary hardness and requires the use of excess soap to precipitate the free Ca^{2+} and Mg^{2+} before the soap can work. Detergents are synthetic ionic compounds, for example, sodium salts of sulfonic acids,

$$(C_nH_{2n+1})-C_6H_5-SO_3^- +Na^+$$

where $n = 10$–20.

The detergent anion does not form an insoluble salt with divalent or trivalent cation, and hence, it can function in hard water almost as efficiently as in soft water, if in very hard water, the Ca^{2+} and Mg^{2+} are complexed by "builders" such as sodium pyrophosphate ($Na_4P_2O_7$) or sodium tripolyphosphate ($Na_5P_3O_{10}$). The extensive use of phosphates with detergents has resulted in contamination of rivers and lakes since the detergents are not as readily biodegradable as fatty acids, though recent modification has improved this. Phosphates which are released supply a required nutrient to aquatic plants, namely, algae, which can overrun a lake with algal bloom (eutrophication), depriving the water of oxygen needed by living organisms. Lakes can die if corrective measures are not instituted. Thus, nonphosphate detergents have been developed which now include "builders," such as sodium citrate, to complex the calcium and magnesium ions in hard water.

Hardness is determined by an analysis of the Ca^{2+} and Mg^{2+} ions in solution, together with the remaining multivalent ions. Titration can be easily performed using a chelating reagent ethylenediaminetetraacetic acid (EDTA).

EDTA combines with many multivalent cations such as Ca^{2+} to form an octahedrally coordinated chelate complex CaH_2EDTA.

15.6   Water Softening

Hard water can be softened by several methods which vary in efficiency and cost. Distillation will result in water free of dissolved salts and nonvolatile organic substances. This can be carried out using solar energy, and many devices have been described whereby seawater can be converted into potable drinking water, usually on a small scale.

Water frozen slowly will be free of dissolved organic and inorganic compounds. The freeze-thaw cycle is used to purify semiconductors by the process called *zone refining*. Fog and dew have also been used as sources of freshwater.

The removal of the Ca/Mg hardness can be effected by the lime process. In this process, lime (as CaO or Ca(OH)$_2$) is reacted with Ca(HCO$_3$)$_2$ or Mg(HCO$_3$)$_2$ to form CaCO$_3$ and Mg(OH)$_2$, respectively:

$$Ca(HCO_3)_2 + Ca(OH)_2 \rightarrow 2CaCO_3 \downarrow +2H_2O \tag{15.15}$$

$$Mg(HCO_3)_2 + Ca(OH)_2 \rightarrow Mg(OH)_2 \downarrow + Ca(HCO_3)_2 \tag{15.16}$$

Lime to be added is calculated from the measured Ca/Mg hardness and adjusted to a pH of 9–9.5 for calcium removal, but a pH of about 12 is needed to remove most of the Mg^{2+}.

The CaCO$_3$ can be used to regenerate the lime by the reaction

$$CaCO_3 + heat \rightarrow CaO + CO_2 \downarrow \tag{15.17}$$

Lime treatment does not remove all of the hardness. However, should complete removal be required, then the final treatment involves ion-exchange resins or distillation.

15.6.1   Ion Exchange

Naturally occurring minerals, such as *zeolites* or *permulites*, are sodium aluminum silicates (NaAlSiO$_4$) which can exchange sodium for calcium or magnesium:

$$2NaAlSiO_4 + Ca^{2+} \rightarrow 2Na^+ + Ca(AlSiO_4) \tag{15.18}$$

Zeolite can be regenerated by the addition of salt (NaCl) which reverses reaction (15.18).

More efficient ion exchangers such as synthetic polymers of substituted polystyrene are referred to as *ion-exchange* resins. These are either cation or anion exchangers and when mixed together in a single bed, will deionize water to a distilled water grade if the cation exchanger is initially in the protonated or acidic form (H$^+$) and the anion exchanger is in the basic or hydroxide (OH$^-$) form. If the cation exchanger is in the sodium form (R.Na), then the resin will exchange Ca^{2+} and Mg^{2+} for Na$^+$:

$$2R.Na + Ca^{2+} \rightarrow (R)_2Ca + 2Na^+ \tag{15.19}$$

Regeneration of the single-bed ion exchangers is the reverse reaction, that is, adding a solution of NaCl for reactions (15.17) and (15.18). In the case of the complete deionization process of water, acid (HCl) must be used to regenerate the cation-exchange resin, and base (NaOH) is used to regenerate the anion-exchange resin.

Fig. 15.6 Electrolysis cell
used in the electrocoagulation
process of converting raw
water into potable water

15.6.2   Reverse Osmosis

Another popular method which gives a high-quality water is *reverse osmosis* (RO), often called *ultrafiltration* or *hyperfiltration*. Though often considered too expensive for industrial use, RO has found extensive applications in domestic water supplies. The production of highly efficient osmotic membranes has made RO competitive with distillation for the production of salt-free water. RO does not, however, remove volatile organic compounds (VOC) from the water supply. Treatment with granulated activated carbon (GAC) can be very effective for this purpose (see Chap. 16 and Appendix C).

15.6.3   Electrocoagulation

Recently, it has been reported that a simple patented process can convert raw water into potable water. Raw water is passed through a specially designed electrolysis cell shown in Fig. 15.6. Anodes and iron rods are centered in perforated cylindrical stainless steel cathodes with a spacing gap of about 1.5 mm between anode and cathode. The DC voltage across the electrodes is 3 V, drawing a current of about 0.2 A, producing an electrical field of up to 2,000 V/m. Raw water flows upward carrying the evolved gases, namely, oxygen and hydrogen (though ozone has also been reported), to an exhaust vent. Single units have 10–20 cells in parallel.

Fig. 15.7 Schematic representation of the desalination of water by electrodialysis. "A" refers to the anion-exchange membrane, and "C" refers to the cation-exchange membrane. The applied voltage attracts the ions from the low-flow water stream into the high-flow wastewater

This process has been used to prepare potable water from natural, industrial-contaminated, and sewage-effluent water. Bacteria such as *E. coli* and Coliform are removed, organic material is oxidized, and heavy metals are plated out or precipitated. The chemical oxygen demand (COD) and biochemical oxygen demand (BOD) are reduced, as well as the hardness of the water. The biochemical oxygen demand (BOD) is defined as the amount of oxygen needed by bacteria in aerobic stabilization of organic matter in water. Chemical oxygen demand (COD) is the oxygen needed to completely oxidize the organic matter in water.

How does the system work? It has been proposed that the highly reactive hydroxyl (OH) radical is responsible for the strong oxidizing power. The OH radical could be formed by Fenton's reaction:

$$Fe^{2+} + H_2O_2 \rightarrow Fe^{3+} + OH^- + {}^*OH \tag{15.20}$$

Since O_3 may be formed, it is likely that hydrogen peroxide may also be produced. Fe^{2+} is readily formed from the iron anode.

Another possible process which may be occurring depends on the presence of chloride (Cl^-) ions in the treated water. Chloride could be oxidized to chlorine (Cl_2), which can oxidize the organic matter and kill the pathogenic bacteria. In either case, the electrolysis relies on reactants which are formed during the electrolysis. The simplicity of this process and its reported effectiveness make it an attractive method of recycling water.

15.6.4 Electrodialysis

An electrode potential can direct the motion of ions that depends on the polarity of the voltage and the sign of the charge of the ions. This is illustrated in Fig. 15.7 where the flow of cations is directed through cation-exchange membranes (C), moving toward the negative terminal, and anions will move in the opposite direction through the anion-exchange membrane (A), depleting the solution of ions. The alternate channels, however, become enriched in salt. The flowing current is an indication of the desalination process. When freshwater is alternated with saline water and separated by the anion- and cation-exchange membranes, the movement of ions from the saline to the freshwater channels will result in a current flow and represent a means of generating electrical energy. This is called *reverse electrodialysis.*

Some water-softening (conditioning) units which are sold for domestic use are advertized as "no saltwater conditioners." These units are reported to remove $CaCO_3$ from hard water by using a catalyst, which by epitaxial nucleation, and the reduction of pressure by virtue of a change in water velocity, converts the $Ca(HCO_3)_2$ into $CaCO_3$ and CO_2. These units are very attractive and are advertised to work with detergents but are not intended for use with soap. This can be interpreted to mean that the calcium ions are not removed from the water system and that a precipitate will form from the calcium salt of the fatty acid from the soap (Eq. 15.14).

15.7   Boiler Scale

There are several types of scales that deposit on the hot surfaces in a boiler. The most common is calcium carbonate, which is formed from the hardness of water. Scale is also formed from the deposition of insoluble salts such as calcium sulfate, calcium phosphate, or insoluble silicates. The relation used to calculate the influence of scale on the temperature drop across a boiler tube is

$$\Delta T = QL/K \tag{15.21}$$

where Q is the heat transferred in W cm^{-2}, L is the thickness of the scale in cm and K is the thermal conductivity in $Wcm^{-1} K^{-1}$, and ΔT is the temperature drop across the pipe in K or °C.

The thermal conductivity of scale such as $CaCO_3$ is approximately 0.03 W $cm^{-1} K^{-1}$ and that of $CaSO_4$ is around 0.003 W $cm^{-1} K^{-1}$. Production of steam at 600 psi (40 atm) in a 4-cm OD tube of SA 210 carbon steel 3.4 mm thick $(K = 0.41$ W/cm K) that requires a heat flux of 40 W/m^2 has a temperature gradient across the steel tube (32 K) and a water boundary film of about 2 mm (40 K). Water temperature is approximately 250°C, and outside-tube temperature is around 325°C, both well below the safe limit of 525°C.

With a $CaSO_4$ scale of 0.15 mm thick, the temperature gradient across the scale is approximately 200 K in order to maintain the rate of formation of steam. This raises the external temperature of the steel from 325°C in the absence of any scale to about 525°C. This exceeds the safe temperature for steel in an air atmosphere, and at the steam pressure generated, it will result in tube failure.

Scale formation is thus a major problem of failure in steam boilers where hard water is used. Thermal conductivity of silicate scale is so low (0.0008 W/cm K) that a buildup of as little as 0.05 mm can cause boiler failure.

Scale formation can be prevented by removing salts which cause the scale or by adding substances which prevent the formation of scale in the boiler. A common additive is Na_3PO_4, though other phosphates such as NaH_2PO_4, Na_2HPO_4, and $Na_2P_2O_7$ are also used depending on the acidity of the water. Phosphate is added to the water at 25–50 mg/L to precipitate the calcium as the insoluble $Ca_3(PO_4)_2$. In small plants, additives such as tannin or starch are added to soften the scale, making it easier to remove after it has formed. EDTA can also prevent calcium and magnesium scale by complexing the cations. Similarly, EDTA can remove scale once formed.

A simple but as yet unexplained process of reducing scale formation is the use of a magnetic field of up to 7,000 G. The method was initially proposed in 1865, and several patents have been issued for what is also referred to as a magnetohydrodynamic effect. The most successful applications are those in which the magnetic field is at 90° (orthogonal) to the water flow. The crystal size and morphology of the $CaCO_3$ formed are influenced by the magnetic field. Various claims have been made by manufacturers of magnetic treatment devices, and these include not only the reduction in scale formation but also its removal.

Water treated with ozone tends to reduce scale formation because of the chelating compounds formed by the oxidation of the organic substances in the water. This aspect, though significant, is often of secondary consideration in choosing ozone for water treatment.

The cost of scale formation to US industry is estimated to be $10 Billion annually, and even a small reduction in boiler scale will result in large savings.

15.8   Wastewater

Industrial and municipal *wastewater* treatment is designed to permit the safe disposal of the discharged water. The important components of the contaminants are:

1. Suspended solids
2. Biodegradable organics
3. Pathogens
4. Nutrients
5. Industrial metals and organics

The cleanup is classified into primary and secondary treatment.

Primary treatment includes sedimentation and filtration which reduces the BOD by about 30%. The secondary treatment contains about 40% of the suspended solids and all of the dissolved metals and organic substances. Microorganisms can be used to remove the organics by aerobic digestion where pure oxygen is often used to accelerate the biological rate. The remaining dissolved organics and metals are removed by physical processes such as adsorption, by microporous membrane filtration, and by oxidation and precipitation. In some cases, further tertiary treatment is used to remove nitrogen and phosphorus compounds and other plant nutrients.

An interesting and novel approach to the treatment of wastewater has recently been developed by Delta Engineering (Ottawa) called the *Snowfluent* method. The process consists of three steps:

1. During freezing conditions, with outdoor temperatures below $-6°C$, the contaminated raw wastewater (from a storage lagoon) and compressed air are sprayed into the air forming fine droplets $(d < 200\ \mu m)$ which freeze and produce snow. The volatile components, CO_2, NH_3, H_2S, and some VOCs are released into the atmosphere, while the freezing of bacterial cells causes membrane ruptures, thus sterilizing the snow. When conducted during daylight, the UV contributes to an additional disinfection effect. At distances of 30 m from the spray, the odors are significantly reduced and considered not to be a nuisance.
2. The snowpack is aged during which time the BOD decreases further, and the pH increases while nutrients, such as phosphorus precipitate, do not redissolve when the snow melts. Organic particles also separate out forming a residue on melting.
3. The remaining process leaves runoff and residue as the snow melts. Meltwater can be released into neighboring freshwater systems or used for irrigation. In the case of animal waste treatment, the runoff can be stored in a holding pond for recycling as a flushing system for barns.

The few remaining surviving pathogens in the snowpack are destroyed by exposure to sunlight (UV) or are unable to reproduce due to ice damage. A diagram of the process is shown in Fig. 15.8. The cost of the Snowfluent process is from 10 to 25 cents/m^3 compared to $0.50–$1.50/m^3 for conventional wastewater treatment.

Fig. 15.8 Diagram of the basic components of the Snowfluent process which converts wastewater into sterile snow

A United Nations report has predicted that by 2025, two-thirds of the world's population will be facing water shortage. The need to conserve and recycle water cannot be overemphasized if mankind is to prosper in our limited global environment.

Exercises

1. What are the physical characteristics that determine water quality?
2. How are the quantities in Exercise 1 evaluated?
3. How is water made safe for drinking?
4. What six substances (in order of decreasing importance) would you want to have analyzed in your well water?
5. Write the hydrolysis reactions for alum and $FeCl_3$ used to clarify water.
6. What is hardness and how can it be determined?
7. How can the hardness of water be reduced?
8. Draw a diagram showing the EDTA complex for Al^{3+}.
9. Calculate the temperature gradient across a silicate scale of 0.1 mm thick if a heat flux of $40 \ W/cm^2$ is required.
10. Speculate on the possible mechanism whereby a magnetic field influences the formation of scale.
11. Why is reverse osmosis so popular?
12. Calculate the amount of gold present in the oceans.
13. Explain how a nuclear-powered ship could successfully transport icebergs to desert lands.
14. What is the BOD and COD of a water sample?
15. Describe how gas hydrates (see Chap. 6) can be used in the desalination of seawater.

Further Reading

1. Manahan SE (2009) Fundamentals of environmental chemistry, 3rd edn. CRC Press, Taylor and Francis Group, Boca Raton, p 1233
2. Manahan SE (2010) Water chemistry. Green science and technology of nature's most renewable resource. CRC Press. Taylor and Francis Group, Boca Raton, p 398
3. Aery NC (2010) Manual of environmental analysis. CRC Press, Taylor and Francis Group, Ane Books Pvt. Ltd., Boca Raton/London/New York, p 413
4. Katz SA, Bryan JC (2011) Experiments in nuclear science. CRC Press, Taylor and Francis Group, Boca Raton, p 168
5. Salem MA, Ali MH (2011) Practices of irrigation & on-farm water management, vol 2. Springer, New York
6. Dore M (ed) (1996) Chemistry of oxidants and treatment of water. VCH, New York
7. Droste RL (1997) Theory and practice of water and wastewater treatment. Wiley, New York
8. Lewis SA (1996) The sierra club guide to safe drinking water. Sierra Club Books, San Francisco
9. Nunn RG (1996) Water treatment essentials for boiler plant operation. McGraw-Hill, New York
10. Greenberg AE (ed) (1995) Standard methods for the examination of water and wastewater, 19th edn. APHA, AWWA, WPCF, Washington, DC
11. Wrobel LC (ed) (1995) Water pollution III, modelling, measuring and prediction. Computation Mechanics, Billerica
12. Guidelines for Canadian Drinking Water Quality, Health and Welfare Canada (1993)
13. Pilson ME (1992) Introduction to the chemistry of the sea. Prentice-Hall, Englewood Cliffs
14. Postel S (1992) The last oasis: facing water scarcity. Norton, New York
15. Stewart JC (1990) Drinking water hazards. Envirographics, Hiram
16. Hem JD (1990) Study and interpretation of the chemical characteristics of natural water. State Mutual, New York
17. Faust SD, Aly OM (1983) Chemistry of water treatment. Butterworths, Boston
18. Sawyer CN, McCarty PL (1978) Chemistry for sanitary engineers, 3rd edn. McGraw-Hill, New York
19. Coulston F, Mrak E (eds) (1977) Water quality. Academic, New York
20. (1975) Water quality parameters. ASTM Publication #573 ASTM, Philadelphia
21. Camy TR, Meserve RL (1974) Water and its impurities. Dowden, Hutchinson and Ross Inc., Stroudsburg
22. Ciaccio LL (ed) (1971) Water and water pollution handbook, vol 4. Dekker, New York
23. Sourirajan S (1970) Reverse osmosis. Academic, New York
24. Spiegler KS (ed) (1966) Principles of desalination. Academic, New York
25. Speigler KS (1962) Salt-water purification. Wiley, New York
26. American Water Works Association. www.awwa.org/
27. American Water Resources Association. www.awra.org/
28. Water Quality Association. www.wqa.org/
29. National Ground Water Association. www.ngwa.org/
30. Water Environment Federation. www.wef.org/
31. U.S. EPA. www.epa.gov/ow
32. World water forum. www.worldwaterforum.org/
33. Water info. www.ecoworld.com/water/ecoworld—water—home.cfm
34. US water resources. http://water.usgs.gov/
35. University Water Info. Network. http://www.uwin.siu.edu
36. Scale formation. http://www.fuelefriciencyllc.com/

Chapter 16
Carbon-Based Polymers, Activated Carbons

16.1 Introduction

Carbon-based polymers and absorbents constitute a topic of huge scientific interest and great strategic importance in an interdisciplinary approach spanning applied physics, biology, pharmacy and medicine, mechanics, electronics, chemistry, and chemical engineering. Today carbon-based materials and adsorbents are widely used for energy storage and catalysis; for fabrication of electrodes and sensors operating in harsh environment; as hemoabsorbents of toxins and viruses in blood recovery; and as a biocompatible coating of medical implants, water purification, and wastewater treatment. Carbon is a surprisingly versatile element, able to hybridize in three different states, sp^1, sp^2, and sp^3. The changes in local bonding of carbon atoms account for the existence of extremely diverse allotropic phases, exhibiting a very broad range of physical and chemical properties. This element can crystallize as diamond (sp^3 hybridization) or graphite (sp^2 hybridization) and give rise to many noncrystalline phases (generally containing a mixture of sp^1, sp^2, and sp^3 hybridizations), such as fullerenes; carbon nanotubes; and disordered, nanostructured, and amorphous carbons. Strong tetrahedral δ bonds are responsible for the extreme physical properties of diamond, a wide gap semiconductor having the largest bulk modulus of any solid, the highest atom density, and the largest limiting electron and hole velocities of any semiconductor. Graphite, whose sheets (graphenes), featured by strong intralayer trigonal δ bonding, are held together by weak interlayer van der Waals' forces, is considered as an anisotropic metal.

16.2 Polymeric Carbon: Diamond

Natural diamonds are believed to have been formed millions of years ago when concentrations of pure carbon were subjected to great pressures and heat by the Earth's mantle (see Fig. 16.1). They are the hardest known material in nature. The majority of diamonds (nongem) are now man made. Most of the synthetic diamonds are no larger than a grain of common sand. The major use of synthetic diamonds is as industrial shaping and cutting agents to cut, grind, and bore (drill). By 1970, General Electric was manufacturing diamonds of gem quality and size through compressing pure carbon under extreme pressure and heat. It was found that addition of small amounts of boron to diamonds causes them to become semiconductors. Today, such doped diamonds are used to make transistors. While diamonds can be cut, shaping is done by trained cutters striking the rough diamond on one of its cleavage plates. These cleavage plates are called faces and represent sites of preferential cleavage and reflection of light. This balance between strength and flexibility, crystalline and amorphous regions is demonstrated to one extreme by diamonds that are very crystalline, resulting in a strong, inflexible, and brittle material [1].

O.V. Roussak and H.D. Gesser, *Applied Chemistry: A Textbook for Engineers and Technologists*,
DOI 10.1007/978-1-4614-4262-2_16, © Springer Science+Business Media New York 2013

Fig. 16.1 Representation
of diamond, where each
carbon is at the center
of a tetrahedron composed
of four other carbon atoms

16.3 Polymeric Carbon: Graphite

While diamond is the hardest naturally occurring material, the most common form of crystalline carbon is the much softer and flexible graphite. Graphite occurs as sheets of hexagonally fused benzene rings (see Fig. 16.2) or "hexachicken wire." The bonds holding the fused hexagons together are traditional covalent bonds. The bonds holding the sheets together are weaker than the bonding within the sheets consisting of a weak overlapping of pi-electron orbitals. Thus, graphite exhibits many properties that are independent on the angle at which they are measured. They show some strength when measured along the sheet, but very little strength if the layers are allowed to slide past one another. This sliding allows the graphite its flexibility, much like the bending of bundles of proteins sliding past one another, allowing our hair flexibility. The fused hexagons are situated such that the atoms in each layer lie opposite to the centers of the six-membered rings in the next layer. This arrangement further weakens the overlapping of the pi electrons between layers such that the magnitude of layer-to-layer attraction is on the order of ordinary secondary van der Waals forces. The "slipperiness" of the layers accounts for graphite's ability to be a good lubricant.

The variance of property with angle of applied force, light, magnetism, and so on is called anisotropic behavior. Calcite is anisotropic in its crystal structure, resulting in a dependency of its interaction with light with the angle of incidence of the light.

As with diamond, graphite's discovery and initial usage are lost in antiquity. It was long confused with other minerals such as molybdenite MoS_2. At one time, it was known as plumbago (like lead), crayon noir, silver lead, black lead, and carbon minerals. Werner in 1789 first named it graphite, meaning (in Greek) "to write."

The Acheson process for graphite production begins by heating a mixture of charcoal, or coke, and sand. Silica is believed to be reduced to silicon that combines with carbon forming silicon carbide, which subsequently dissociates into carbon and silicon. The silicon vaporizes, and the carbon condenses forming graphite. Graphite is also produced using other techniques.

Today, graphite is mixed with clay to form the "lead" in pencils. Graphite conducts electricity and is not easily burned, so many industrial electrical contact points (electrodes) are made of graphite. Graphite is a good conductor of heat and is chemically inert, even at high temperature. Thus, many crucibles for melting metals are graphite-lined. Graphite has good stability to even strong acids; thus, it is employed to coat acid tanks. It is also effective at slowing down neutrons, and thus, composite bricks and rods (often called carbon rods) are used in some nuclear generators to regulate the progress of the nuclear reaction (see Chap. 7). Its slipperiness allows its use as a lubricant for clocks, door locks, and handheld tools and skies. Graphite is also the major starting material for the synthesis of

Fig. 16.2 Representation
of graphite emphasizing
the layered and sheet
nature of graphite

C=0.67 nm

synthetic diamonds. Graphite is sometimes used as a component of industrial coatings. Dry cells and some types of alkali-storage batteries also employed graphite. Graphite fibers are used for the reinforcement of certain composites [1].

16.4   Graphene

The first material in this new class is graphene, a single atomic layer of carbon (see Figs. 16.3 and 16.4). This new material has a number of unique properties, which makes it interesting for both fundamental studies and future applications. Graphene is a single layer of carbon packed in a hexagonal (honeycomb) lattice, with a carbon-carbon distance of 0.142 nm. It is the first truly two-dimensional crystalline material, and it is representative of a whole class of 2D materials including, for example, single layers of Boron-Nitride (BN) and Molybdenum-disulphide (MoS_2), which have both been produced after 2004 [2]. The electronic properties of this 2D-material lead to, for instance, an unusual quantum Hall effect. It is a transparent conductor that is one atom thin. It also gives rise to analogies with particle physics, including an exotic type of tunneling which was predicted by the Swedish physicist Oscar Klein [2]. Graphene is practically transparent. In the optical region, it absorbs only 2.3 % of the light. In contrast to low-temperature 2D systems based on semiconductors, graphene maintains its 2D properties at room temperature. Graphene also has several other interesting properties, which it shares with carbon nanotubes. It is substantially stronger than steel and very stretchable.

The thermal and electrical conductivity is very high, and it can be used as a flexible conductor. Its thermal conductivity is much higher than that of silver. Graphene has a number of properties which makes it interesting for several different applications. It is an ultimately thin, mechanically very strong, transparent, and flexible conductor. Its conductivity can be modified over a large range either by

Fig. 16.3 Structure of graphene

Fig. 16.4 C60 fullerene molecules, carbon nanotubes, and graphite can all be thought of as being formed from graphene sheets, that is, single layers of carbon atoms arranged in a honeycomb lattice

chemical doping or by an electric field. The mobility of graphene is very high, which makes the material very interesting for electronic high-frequency applications. Recently it has become possible to fabricate large sheets of graphene. Using near-industrial methods, sheets with a width of 70 cm have been produced. Since graphene is a transparent conductor, it can be used in applications, such as touch screens, light panels, and solar cells, where it can replace the rather fragile and expensive indium tin oxide (ITO). Flexible electronics and gas sensors are other potential applications. The quantum Hall effect in graphene could also possibly contribute to an even more accurate resistance standard in metrology. New types of composite materials based on graphene with great strength and low weight could also become interesting for use in satellites and aircraft [2].

The Nobel Prize in Physics 2010 honors two scientists who have made the decisive contributions to this development. They are *Andre K. Geim* and *Konstantin S. Novoselov*, both at the University of Manchester, UK. They have succeeded in producing, isolating, identifying, and characterizing graphene [2].

A new form of molecular carbon is the so-called fullerenes (see Fig. 16.4). The most common, called C60, contains 60 carbon atoms and looks like a football (soccer ball) made up from 20 hexagons and 12 pentagons which allow the surface to form a sphere. The discoverer of fullerenes was awarded the Nobel Prize in Chemistry in 1996 [2].

A related quasi-one-dimensional form of carbon, carbon nanotubes, has been known for several decades, and the single-walled nanotubes, since 1993 [1, 2]. These can be formed from graphene sheets which are rolled up to form tubes, and their ends are half spherical in the same way as the fullerenes. The electronic and mechanical properties of metallic single-walled nanotubes have many similarities with graphene.

Thus, the difficulty was not to fabricate the graphene structures, but to isolate sufficiently large individual sheets in order to identify and characterize the graphene and to verify its unique two-dimensional (2D) properties. This is what Geim, Novoselov, and their collaborators succeeded in doing.

16.4.1 Density of Graphene

The unit hexagonal cell of graphene contains two carbon atoms and has an area of 0.052 nm^2. We can thus calculate its density as being 0.77 mg/m^2.

A hypothetical hammock measuring 1 m^2 made from graphene would thus weigh 0.77 mg.

16.4.2 Optical Transparency of Graphene

Graphene is almost transparent; it absorbs only 2.3% of the light intensity, independent of the wavelength in the optical domain. Thus, suspended graphene does not have any color.

16.4.3 Strength of Graphene

Graphene has a breaking strength of 42 N/m^2. Steel has a breaking strength in the range of 250–$1{,}200$ MPa $= 0.25-1.2 \times 10^9$ N/m^2. For a hypothetical steel film of the same thickness as graphene (which can be taken to be 3.35 Å $= 3.35 \times 10^{-10}$ m, i.e., the layer thickness in graphite), this would give a 2D breaking strength of 0.084–0.40 N/m^2. Thus, graphene is more than 100 times stronger than the strongest steel.

In our 1-m^2 hammock tied between two trees, you could place a weight of approximately 4 kg before it would break. It should thus be possible to make an almost invisible hammock out of graphene that could hold a cat without breaking. The hammock would weigh less than one mg, corresponding to the weight of one of the cat's whiskers.

16.4.4   Electrical Conductivity of Graphene

The mobility is theoretically limited to $\mu = 200{,}000$ cm^2 V^{-1} s^{-1} by acoustic phonons at a carrier density of $n = 10^{12}$ cm^{-2}. The 2D sheet resistivity, also called the resistance per square, is then 31 Ω.
   Our fictional hammock measuring 1 m^2 would thus have a resistance of 31 Ω.
   Using the layer thickness, we get a bulk conductivity of 0.96×10^6 Ω^{-1} cm^{-1} for graphene. This is somewhat higher than the conductivity of copper which is 0.60×10^6 Ω^{-1} cm^{-1}.

16.4.5   Thermal Conductivity of Graphene

The thermal conductivity of graphene is dominated by phonons and has been measured to be approximately 5,000 W m^{-1} K^{-1}. Copper at room temperature has a thermal conductivity of 401 W m^{-1} K^{-1}. Thus, graphene conducts heat ten times better than copper.

16.5   Activated Carbons

With its immense capacity for adsorption from gas and liquid phases, activated carbon is a unique material. It occupies a special place in terms of producing a clean environment involving water purification and water treatment; decolorizing; chemical, nuclear, and pharmaceutical processing; food processing; air and gas purification; solvent vapor recovery; and motor vehicle emission canisters. In these roles, it exhibits a remarkable efficiency, as the world production capacity is estimated to be around 400,000 tonnes per year (excluding China and other Eastern countries, for which figures are not accurately known), with perhaps two million tons being in continuous use. This is equivalent of 200 mg per person of the world population to be compared with the world use of fossil fuels of 2 tonnes per person of the world population [3].
   The largest producer of activated carbons is the USA (accounting for more than 40 wt.% of the total production capacity), followed by Europe, Japan, and Pacific Rim countries. The largest producer in the USA is Calgon Carbon Corporation, with plants in several locations. Its annual capacity represents around 42% of the total production capacity of the country; the main activation process is thermal, and the precursors used are bituminous coal, coconut shell, and charcoal. The second largest producer is Norit Americas, Inc., with around 23% of total capacity. Thermal activation is carried out with lignite and bituminous coal, and chemical activation (with phosphoric acid) is carried out with wood and peat. The third producer is Westvaco Corporation, with around 12% sawdust being the main precursor for both thermal and chemical (with phosphoric acid) activation. Regeneration capacity of spent activated carbon has increased considerably in the last few years, this being estimated to be over 50, 000 tonnes per year now.
   Consumption of activated carbon is estimated to be around 200, 000 tonnes per year (2005) in the USA, but the market is negatively affected by imports from the Asia-Pacific region. More than 80% of that amount was used in liquid-phase applications. Sweetener decolorization used to be the main end-use of activated carbon, but it has been displaced by water treatment as a consequence of the increasing severity of government legislation on pollution control. Potable water is now the largest end-use market for activated carbon, around 26% of total consumption, 47% of which is powdered and 53% granular.
   In the USA, the price of granular activated carbon ranged from $ 1.65 US to $ 9.90 US per kg, when most powdered activated carbon was priced in the range of $ 0.80 US to $ 2.00 US per kg. The largest difference in price is for carbons used in water treatment, with granular carbons being up to three times more expensive.

However, in such applications as gas and air purification, the costs of granular and powdered activated carbon are similar [3].

Usually untreated coal-based carbons contain up to 20 wt.% ash content, such as Si, Al, Fe, Mn, Ni, Cu, Sn, and Pb (according to the ASTM 2866–70 ash content test). Coconut shell carbons have less ash content about 1–3 wt.%. Also disarmament and partial substitution of military equipment and reorganization of some chemical production can release significant amounts of the used gas—proof carbons and chemical absorbents base on activated carbons. As a rule, these materials contain up to 20% nickel or iron chlorides or copper sulfate, and most of these components are water soluble. But some applications need very pure activated carbons, and to comply with this need, many manufacturers remove most of the ash components by washing the carbon with water or acids such as hydrochloric or nitric acid [4]. In some cases, a more exhaustive washing requires the use of hydrochloric and hydrofluoric acids to eliminate the aluminosilicate components of the ash. The washing must be particularly thorough when the carbon is to be used in pharmaceutical and medical preparations, in food industries, or in catalysis (when the carbon is used either as a hemoabsorbent or as a catalyst).

The main mechanism by which activated carbon removes impurities is one of physical adsorption, this being a reversible process. Consequently, one can expect that desorption of the impurities will render the carbon surface available again for adsorption. Regeneration of spent activated carbon is not only important from the point of view of restoring the adsorption capacity of carbon but also because in many cases, the recovery of the adsorbed species is important. If the adsorption is of chemical type (chemisorption), the formation of a bond between the carbon and the adsorbate makes the process nonreversible, and even if desorption is possible, the desorbed species will be different from those originally adsorbed. Additionally, adsorption (especially in liquid phase) is often accompanied by precipitation of species which cannot be removed by simple desorption.

Activated carbon is porosity (space) enclosed by carbon atoms. Porosity has the size of molecules and is probably slit shaped. Porosity within a porous solid is space which is accessible to molecules from the gas/vapor and liquid phases. Micropores have entrance dimensions less 2.0 nm. Mesopores have dimensions between 2 and 50 nm. Micropores have entrance dimensions higher than 50 nm. It is porosity within activated carbon which imparts their dominant characteristics of adsorption.

The *Langmuir equation* or *Langmuir isotherm* or *Langmuir adsorption equation* or *Hill-Langmuir equation* relates the coverage or adsorption of molecules on a solid surface to gas pressure or concentration of a medium above the solid surface at a fixed temperature (see the Langmuir equation and isotherm in Appendix C, p. 333).

Dubinin and his coworkers—Kaganer, Radushkevich—discovered adsorption using potential theory originally formulated by Polanyi. They developed adsorption methods for pore volume and specific surface (the methods of Dubinin and of Kaganer). Dubinin and his coworkers advance arguments favoring the view of the adsorption space may be expressed as a Gaussian function of the corresponding adsorption potential. Dubinin's method makes it possible to calculate the micropore volume from the low pressure part of isotherm, the region where the adsorption is still much below the plateau value. It also offers the possibility of using different adsorbates as molecular probes. Dubinin has plotted the results for adsorption isotherms of nitrogen, saturated hydrocarbons, benzene, and cyclohexane and has found equation for calculation of the micropore volume W_o (see the Dubinin equation in Appendix C, p. 333). Dubinin's treatment has been modified by Kaganer to yield a method for the calculation of specific surface from the isotherm [5, 6].

Regeneration of activated carbons is based on principle that carbon is a stable material that can withstand changes in temperature and it is resistant to acidic and basic media. The more common regeneration process is by passing a flow of superheated steam through the carbon bed at a temperature lower than the one used during activation. As the temperature increases during regeneration, there is a desorption of the more volatile adsorbed compounds; above 400°C, there is also decomposition of adsorbed organic material, thus leaving some less-organized carbon, which is eliminated by reaction with steam at around 600°C. Of course, although the less-organized carbon is more reactive, a portion of the activated carbon is also gasified, and regeneration usually means a loss of 8–15 wt.% of the original activated carbon. For this reason, when regeneration is carried out under these conditions, it is also called reactivations. It is important to note that the off-gases must be properly handled, passing them through a post-burner and a scrubber, where they are destroyed. If regeneration is properly carried out, the bulk density and the surface area of the resulting carbon will be similar to that of the original carbon. If the surface area is larger after regeneration, this means that a proper reactivation has taken place, with the subsequent reduction in bulk density.

Additional regeneration procedures include desorption by hot inert gases, desorption under vacuum, and, more important, use of conventional liquid solvents. The main problem for the latter is the slow desorption and the difficult solvent regeneration, and the last few years have seen the development of regeneration with supercritical solvents, mainly carbon dioxide and water. Complete recovery of components and environmental safety of the corresponding industrial operations require also development of low-waste or waste-free extraction procedures and the corresponding process apparatus.

One of directions of low-waste extraction is multiple treatment of a material with azeotropic nitric acid (13.4 wt.%) and water or hydrochloric acid (20.9 wt.%) and water in the Soxhlet laboratory apparatus invented in 1879 by Franz von Soxhlet and shown in the Fig. 16.5. The acids are boiled, and its vapor of azeotropic acid travels upward through the extraction tube into the condenser tube. The cool water flowing around the outside of the condenser tube condenses the vapor, which then drips into the thimble, containing the material. Because the mineral alkaline impurities are soluble in azeotropic acids, they move into the condensed acids as they accumulated in the thimble. The solution, now containing the acids and dissolved impurities, build up in the thimble. Once the liquid reaches the level of the bypass arm, it is siphoned back into the flask. This continuous condensation, buildup, and siphoning is known as the reflux event. The advantage of the Soxhlet is that once the acids and dissolved mineral impurities are brought into solution and siphoned back into the flask, they stay in the flask—so that the sample in the extraction thimble is continuously reexposed to fresh, heated extraction agent—thus greatly increasing the extraction rate.

Demineralization ratio **d** is calculated from the content of the mineral components (wt. %) before and after extraction treatment according to the formula: $\mathbf{d = (a - b) \times 100/a}$. The purification coefficient **K** is calculated from the formula $\mathbf{K = a/b}$, where **a** and **b** denote the contents of mineral components in the sample before and after extraction treatment, respectively. The azeotropic mixture of hydrochloric acid and water more effectively removes mineral impurities from commercial carbons than azeotropic nitric acid does due to the difference in the concentration of the corresponding acids (the concentration of nitric acid is reduced due to oxidation of the carbon sorbent) and to different nature of the acid anions. It has been experimentally found that demineralization ratio for most English carbons was higher than 41.5% and only in the case of NORIT RBXC carbon was it 6.5%, indicating profound preliminary demineralization of this sorbent. A single treatment with acid and water significantly decreases the total content of mineral components (Mn; Fe; Ni; Cu; Pb) and increases the exchange capacity of the resins too. For example, in case of VP-1AP anion-exchange resin, very large decontamination coefficients for nickel (K = 180,000) and iron (K = 1, 078) have been observed [7].

It is well known that the Soxhlet extractor underwent numerous improvements, and the most cardinal change was installation of a valve connecting extraction in the bottom section of the extractor, which allows operation both in mode of batch washing and in the of ideal displacement and vapor treatment (Figs. 16.6 and 16.7).

But there was lack of industrial application of the Soxhlet apparatus with large volume and capacity because glass extractors with a charge volume of up to 5 Liters were produced only in the United States. In 1984 Prof. Anatoliy I. Loskutov and Oleg V. Roussak designed and proposed industrial Soxhlet apparatus that allowed sampling of the solvent, charging and discharging of the

Water In →
Condenser Tube
Water
Water Out →
Bypass Sidearm
Extraction Tube
Reflux Sidearm
Cellulose Thimble
Glass Wool
Sample and
Sodium sulphate
Solvent Vapour
Flask
Organic Solvent
Heat Source

Fig. 16.5 Laboratory Soxhlet extractor

carbon material without complete dismantling of the setup, and stirring and treatment of the whole material with any solvent at its boiling temperature in the following four modes: vapor treatment, batch washing, ideal displacement, and heating (Fig. 16.8). An improved industrial Soxhlet apparatus has been tested for regeneration of granular activated carbons AG-3 and AG-5 as well as powdered activated carbons SKT-6A that are widely used in a solvent recuperation and wastewater treatment. The reagents of regenerated activated carbons were 20% aqueous solutions of ammonium hydroxide, potassium hydroxide, sodium hydroxide, distilled water, and azeotropic hydrochloric acid in various sequences of treatment: acid-water, alkali-water, and alkali-water-acid-water. It helped to solve problems of regeneration of powdered activated carbons (SKT-6A carbon) as well as regenerated calcium-containing activated carbons, which significantly increase safety, economical efficiency of regeneration, and reduce volume of liquid wastes [8, 9].

Extraction of activated carbons by a multiple treatment with a mixture of azeotropic hydrochloric acid and distilled water and large-scale production of high-purity activated carbon absorbents have been successfully applied to wastewater treatment and water purification and pharmaceutical and medical preparations.

Fig. 16.6 Improved vertical Soxhlet extractor

Fig. 16.7 Improved horizontal Soxhlet extractor

Fig. 16.8 Improved industrial Soxhlet extractor

It also can be used for the low-waste, environmentally friendly, clean and safe production of especially pure materials; ion-exchange resins of nuclear cleanliness; low-waste extraction of components from the solid materials (for instance, the extraction of platinum and palladium from the spent catalysts); and deactivation of materials and equipment of radiochemical enterprises [10, 11]. The industrial Soxhlet extractor and evaporator were made from steel covered with acid-resistant Ftoroplast coating sheets. The charge volume of the industrial Soxhlet apparatus could be significantly increased up to 16 cubic meters using stainless steel apparatus with acid-resistant ftoroplast coating sheets.

Exercises

1. What is activated carbon?
2. What is the size and shape of porosity?
3. What are activated carbons used for?
4. What is activation?
5. What are activated carbons made for?
6. Can any natural materials be used to make activated carbon?

7. Are all activated carbons very similar to each other?
8. What are adsorption isotherms?
9. Is there a common structure in activated carbons?
10. Is shape and size of porosity all important for activated carbons?

Further Reading

1. Carraher CE Jr (2010) Carraher's polymer chemistry, 8th edn. CRC Press/Taylor and Francis Group, Boca Raton/ London, 788pp
2. http://en.wikipedia.org/wiki/Graphene
3. Marsh H, Rodriguez-Reinoso F (2006) Activated carbon, 1st edn. Elsevier, Oxford, 536pp
4. Bruno TJ, Svoronos PDN (2011) Handbook of basic tables for chemical analysis, 3rd edn. CRC Press/Taylor and Francis Group, Boca Raton/London, 870pp
5. Bond RL (1967) Porous carbon solids. Academic, London/New York, pp 8–19
6. Cregg CJ, Sing KSW (1967) Adsorption surface area and porosity. Academic, London/New York, pp 223–231
7. Loskutov I, Slepchenkov IS, Murashkin YV, Kopyrin AA, Mordovina LA, Kuznetsov MS (1999) Demineralization of carbon sorbents and ion exchange resins. Russ J Appl Chem 72(11):1896–1901. Web site www.adsorba.narod.ru/ download/Socslet_angl.pdf
8. Roussak O, Loskutov A, Kuznetzov M (1984) Extraction of mineral impurities from carbonic materials. Latest achievements in the study of production and use of carbonic materials: selected theses, reports and talks delivered at the All-Union scientific conference, vol 103, Chelyabinsk
9. Loskutov AI, Slepchenkov IS, Kopyrin AA, Medvedskii NL (2003) Demineralization of industrial activated carbons in an extractor for removing components from solid materials. Russ J Appl Chem 76(8):1225–1229
10. Loskutov AI (2006) Possible applications of improved large-scale and industrial Soxhlet apparatus for technology and production of catalysts and sorbents. Catalytic processes and catalysts. St. Petersburg Technological Institute (Technical University) journal of scientific and research publications. St. Petersburg Technological Institute (Technical University), 2006, pp 118–126
11. http://www.adsorba.narod.ru/p59.htm

Chapter 17
Cement, Ceramics, and Composites

17.1 Introduction

Cement and its applications as concrete (a composite of cement and aggregate) are known throughout the world. The most common cement used today is Portland, named after the gray rock of Portland, England, which it resembled. World production of Portland cement increased from 133 million tonnes in 1950 to about 1,000 million tonnes in 1985 and over ten times the 1950 value in 1995. The energy usage during this period dropped from 9.6 MJ/kg to about 5.7 MJ/kg in 1990. Table 17.1 shows the world and US production of Portland cement from 1970 to 2009. Research continues in all aspects of cement from quick setting to increase in strength—the predictability of which is still a major problem.

The history of cement starts in the earliest times when the Assyrians and Babylonians used clays to bind stones into massive walls. The Egyptians used a lime and gypsum mortar as a binding agent for the Pyramids. The Romans perfected such mortar and concrete for use in their structures, some of which still stand. They mixed slaked lime with volcanic ash from Mount Vesuvius to form a cement which hardened under water. The Mongols and Aztecs had developed a similar technology. However, the word "cement" is derived from the Roman "caemenium," meaning building stone.

This skill was lost during the Middle Ages and was not rediscovered until the scientific approach was taken by John Smeaton in 1757 when he built the Eddystone lighthouse on the southwest coast of England. He found that a good hydraulic cement was formed when the limestone used had clay impurities. We now know that aluminosilicate clays, when calcined with lime, form the desired cement. Between 1757 and 1830, the essential roles of the lime and silica were established by Vicat and Lisage in France and by Parker and Frost in England.

In 1824, Joseph Aspdin, a bricklayer and mason from Leeds, obtained a patent for a superior hydraulic cement which he called *Portland cement*. His process required the mixing and pulverizing specific quantities of limestone and clay and heating the mixture to a required temperature forming clinkers. Unfortunately, all details are missing, including the kiln Aspdin used. Two major improvements were introduced about 1890: (a) the addition of gypsum ($CaSO_4 \cdot 2H_2O$) to the clinker grinding step to act as a set retardant and (b) higher burning temperatures to permit higher lime and silicate content which results in more rapid strength development in concrete.

Portland cement presently constitutes over 60% of all cement produced and is a carefully apportioned combination of the oxides of calcium, silica, aluminum, and iron.

O.V. Roussak and H.D. Gesser, *Applied Chemistry: A Textbook for Engineers and Technologists*, DOI 10.1007/978-1-4614-4262-2_17, © Springer Science+Business Media New York 2013

Table 17.1 The World and US production of Portland cement—million metric tons[a]

Year	US production	US imports	US exports	World production
1970	62.4	1.99	0.144	571.8
1980	68.2	3.04	0.169	883.1
1990	70.0	10.3	0.503	1,043
1995	76.9	10.9	0.750	1,450
2000	87.8	24.6	0.738	1,660
2005	99.3	30.4	0.766	2,350
2009	63.9	6.21	0.884	3,060

[a]US geological survey 2010

17.2   Cement Nomenclature

Special notation is used by cement chemists to denote the various ingredients in cement. These are listed in Table 17.2 with the approximate concentrations used. Thus, tricalcium silicate is represented as $C_3S(3CaO \cdot SiO_2)$. Portland cement is a mixture of minerals: C_3S, 42–60%; C_2S, 15–35%; C_3A, 5–14%; C_3AF, 10–16%; and C and M. The minerals are formed in the kiln during the burning of limestone and the silicate clay which are usually readily available and inexpensive materials. The relative energy expended in producing various materials is given in Table 17.3. The advantages of cement in an age of rising energy costs are obvious.

17.3   Manufacture of Portland Cement

Portland cement is made from readily available and cheap raw materials (limestone, sand, and clay). The components in an appropriate composition are mixed as a wet slurry and passed into the top end (500°C) of a rotary iron kiln which is lined with firebrick, about 150 m long, and at a 15° angle to the horizontal. A diagram of the flow process commonly used in the manufacture of cement is shown in Fig. 17.1. The following reactions occur in the kiln as the temperature increases:

1. Free water is evaporated.
2. Combined water from the clay is released.
3. Magnesium carbonate is decomposed, and CO_2 is released.
4. Limestone is decomposed to form lime and CO_2.
5. Lime and magnesia combine with the clays and silica to form "clinker."
6. Cooled clinker is then ground, and some 20% gypsum is added to prevent the cement from setting too rapidly.

A common accelerator which speeds up the hydration process is $CaCl_2$, but its corrosiveness makes it unacceptable in steel-reinforced concrete.

An alternate process which is gaining popularity is to introduce the ingredients into the kiln in the dry state and so reduce the energy required to drive off the water present in the wet slurry process.

Table 17.2 Components and nomenclature of Portland cement

Symbol	Compound	Cement (%)	Clinker (%)
C	CaO	60–67	67
S	SiO_2	18–24	22
A	Al_2O_3	4–8	5
F	Fe_2O_3	2–5	2.6
N	Na_2O	0.1–1	0.2
K	K_2O	0.1–1.5	0.5
M	MgO	1–2	
S	SO_3	2–3	
C	CO_2	3	
H	H_2O	3	

Table 17.3 The relative energy content of various materials

Material	Relative Energy (vol.)
Portland cement	1.0
Flat glass	3.0
Polyvinyl chloride	3.8
Polyethylene (low density)	4.2
Polyethylene (high density)	4.4
Polystyrene	6.0
Steel	19.2
Stainless steel	28.8
Aluminum	31.8
Zinc	34.8

17.4 Setting of Cement

The *setting* of Portland cement consists of the hydration of the various silicates and aluminates as well as the compound $3CaO \cdot Al_2O_3 \cdot 3CaSO_4 \cdot 32H_2O$ (a process which occurs over a long period). Attempts to reduce setting time and time to maximum strength have included resistive heating, accelerated microwave, and RF drying as well as the use of additives. Evidence of gel formation with fibrillar growth during the dehydration has also been obtained.

Hardened cement is a porous solid with a density of about 2.5 g/mL and a surface area of 100–300 m^2/g. Such pores can be fully sealed by polymer impregnation that has numerous applications.

The compressive strength of cement increases from 10^2 N/m^2 when initially set to 10^7 N/m^2 after a few days, reaching 70% of its final strength value (10^8 N/m^2) after 28 days. Further, small increases in strength are observed even after 1–2 years. Thus, the normal working stress, with the appropriate safety margin, is restricted to about 10 MPa. The tensile strength, σ, of a solid (measured in bending) can be given by the Griffith relationship

$$\sigma = (ER/\pi c)^{1/2} \tag{17.1}$$

where E is Young's modulus of elasticity, R is the surface fracture energy, and c is the crack length. A reduction in particle size of the cement and the removal of all the air bubbles create a macro-defect-free (MDF) cement which results in a strength of up to 150 MPa, a value comparable to aluminum and its alloys. The high setting strength of MDF cement is illustrated by the coiled cement spring in Fig. 17.2 which is shown in the relaxed and extended states.

Fig. 17.1 Flow diagram for the manufacture of Portland cement by the wet process. The limestone is crushed, mixed with wet clay, and ground to a fine slurry in a mill. This raw material is stored and corrected for composition by blending before being fired in a rotary kiln where the process of water evaporation, mineral dehydration, limestone dissociation, and chemical reaction proceeds. Clinker formation forms finally at 1,450 °C and is cooled and ground with additives before storage

Fig. 17.2 Coiled high-strength micro-defect-free (MDF) cement spring in relaxed and extended positions

17.5 Concrete

Concrete is composed of cement and filler, or aggregate, which can be gravel or stones. A proper concrete is composed of aggregate of different sizes which permits better packing and fewer voids. For lightweight concrete, the aggregate can be vermiculite, perlite, or other low-density filler. The concrete must be vibrated as it is poured. The tensile strength of concrete can be increased by embedding iron rods, called *rebar*, and iron wire mesh in the slurry before it sets. The coefficient of expansion of concrete and steel are similar, and thus, temperature changes do not disrupt the structure—though corrosion of the iron must be avoided if long-term strength is to be maintained.

The strength of polymer-impregnated concrete is determined by the glass transition temperature, T_g, of the polymer. Above the T_g, the reinforcement is lost and the concrete is less resistant to salt penetration.

A patented invention (1995) has described how a conducting concrete can be made by adding conducting carbon fibers and particles to the cement. An electrical current through the concrete can heat the material. This can have a broad application for deicing in areas where freezing temperatures occur and cause traffic accidents. Airport runways, bridges, and highway intersections are other important locations where ice-free surfaces are desirable. Another application is as a secondary anode in existing cathodic protection systems.

The cement content of 28-day dried concrete is normally determined by dissolving the cement in hydrochloric acid (HCl), leaving the insoluble aggregate. A simpler and rapid determination can be carried out using differential thermal analysis (DTA) of a thoroughly ground and mixed sample (mesh <80). A typical DTA trace is shown in Fig. 17.3 where the area of the cement peak is directly proportional to the amount of cement in the concrete. It remains to determine the ultimate strength of concrete while the cement is still slurry in the cement truck. Though several methods have been proposed, none has, as yet, had any success.

17.6 Ceramics

Ceramics are inorganic solids, usually oxides, which contain ionic and covalent bonds. The material, formed by sintering at high temperatures, ranges from amorphous glass-like material to highly crystalline solids, from insulators to conductors or semiconductors. They include earthenware, which is fired at 1,100–1,300 K and a porosity of about 8%; fine china or bone china, fired at 1,400–1,500 K with a porosity of less than 1%; stoneware, fired at over 1,500 K with a porosity of about 1% before glazing; and porcelain which is fired at over 1,600 K and has a much finer microstructure than either stoneware or bone china.

One of the most noteworthy ceramics is the high-temperature superconductors that were first described is 1988 and show superconductivity at temperatures as high as 90 K.

The superconductor $YBa_2Cu_3O_7$ ($T_5 < 90$ K), known as the "1-2-3" compound, is prepared by mixing powdered Y_2O_3, $BaCO_3$, and CuO in the appropriate stoichiometry and heating to 950°C. The cooled solid is then pressed and sintered again at 950°C to bond the grains and to increase the density of the pellet. Final heating is in the presence of oxygen at 500–600°C followed by slow cooling to room temperature.

The glass or vitreous state of matter is a solid with the molecular random structure similar to that of a liquid. Such a state is not limited to inorganic substances since many organic compounds and mixtures can form transparent glasses. Glass can be translucent or opalescent due to the presence of small crystalline material in the glass. However, the term "glass" normally refers to a solution of inorganic oxides with silica (SiO_2) as the basic material.

Fig. 17.3 Differential
thermal analysis (DTA)
of concrete sample ground
to <80 mesh. Area of
cement peak is proportional
to amount of cement
in sample

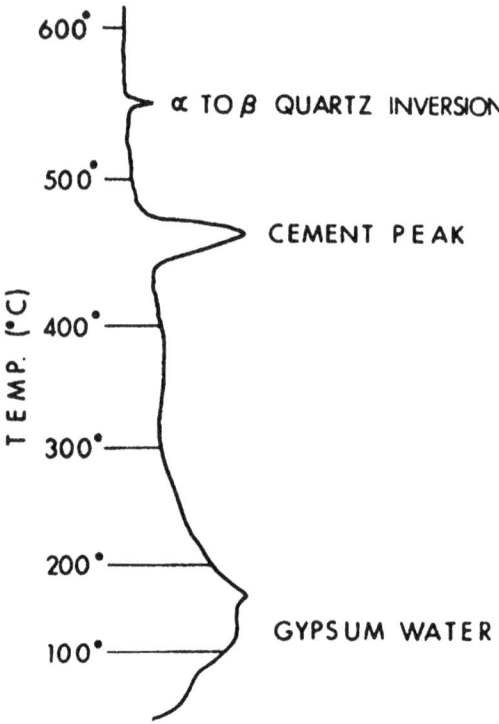

Ceramics can be cast prior to heat treatment or, like glass, can be formed by the sol-gel process, a recent development whereby a glass can be formed without first forming the melt. The oxides are converted into a colloidal gel which can also be formed from the organic alkoxide and M(OR), where n is the valency of the element M. The controlled hydrolysis reaction in an organic solvent such as methanol forms the hydroxide gel:

$$M(OR)_n + nH_2O \rightarrow M(OH)_n + nROH \qquad (17.2)$$

When hydroxide gel is dried slowly in air, a xerogel is formed which is porous, often transparent, and usually smaller and more dense than the original gel. If the solvent is removed under supercritical conditions (temperature and pressure above the critical point), the shrinkage is minimal and the solid is called an *aerogel*. This is shown in Fig. 17.4. Aerogel has a very low thermal conductivity (10 mW/ m, K), which is slightly lower than that of a xerogel and much lower than polyurethane foam (28 mW/ m, K). A recent silica xerogel produced in Norway had a bulk density of 0.26 g/cm^3 and a thermal conductivity of 18 mW/m, K. These materials are being studied for possible use as insulating windows because of their high degree of transparency.

The possibility of blending, doping, or mixing various materials in the gel state makes the resulting homogeneous solids very easy to prepare to exact specifications. When a xerogel or aerogel is heated, the pores collapse, forming the oxide solid at a temperature much lower than that required if formed from the melt.

An interesting glass ceramic, MACOR, manufactured by Corning, is a two-phase machinable crystalline solid composed of mica and glass. Heat treatment converts the amorphous glass into a crystalline structure (please see http://www.ortechceramics.com/). A comparison of the properties relative to other ceramic materials is given in Table 17.4. Its machining characteristics are compared to some metals and nonmetals in Table 17.5. Such excellent properties are unique to ceramics and ceramic-like materials.

Fig. 17.4 Photograph of two aerogels prepared from the hydrolysis of tetraethylortho silicate in methanol and dried under supercritical conditions

Table 17.4 A comparison of macor with other common ceramic materials

Property	Units	MACOR™ machinable glass ceramic	Boron nitride 96% BN	Alumina nominally 94% Al_2O_3	Valox thermoplastic polyester
Density	g/cm^3	2.52	2.08	3.62	1.31
Porosity	%	o	1.1	0	0.34
Knoop hardness	NA	250	<32	2,000	NA
Maximum use temp, (no load)	°F	1,832	5,027	3,092	204°
	°C	1,000	2,775	1,700	140°
Coefficient of thermal expansion	in/in (°F)	52×10^{-7}	23×10^{-7}	39×10^{-7}	530×10^{-7}
	in/in (°C)	94×10^{-7}	41×10^{-7}	71×10^{-7}	934×10^{-7}
Compressive strength	psi	50,000	45,000	305,000	13,000
Flexural strength	psi	15,000	11,700	51,000	128,000
Dielectric strength (ac)	Volts-mil	1,000	950	719	590
Volume resistivity	Ω-cm	$>10^{14}$	$>10^{14}$	$>10^{14}$	$>10^{14}$

17.6.1 Composites

Composite materials were described in the Bible (Exodus), where straw was required and used in the preparation of reinforced bricks. The Inca and Maya people mixed plant fiber with their pottery to reduce cracking while being dried. Concrete is a composite structure, and most children are familiar with papier maché of paper and a glue made of flour and water. The Egyptians mixed their old papyrus manuscripts with pitch to wrap their mummies.

Table 17.5 Machinability index for various materials[a]

Material	Machinability index (MI)
Graphite	1
Teflon TFE	2.5
MACOR (glass ceramic)	25
Free-machining brass	36
Aluminum 2024-T4	50
Copper alloy no. 10	97
Cold-rolled 1018 steel	111
AISI 4340 steel (Rc46)	206
304 stainless steel	229

[a]The index unit (MI) is arbitrary and increases with difficulty of machining. Its approximate value is 1 HP/in^3/min = 200 MI

Fig. 17.5 Photograph of a short section of a polystyrene-fiberglass window frame prepared by the continuous exuding process of styrene-fiberglass and fiberglass mats

A durable and popular composite, *transite,* was asbestos with up to 15% cement. In the form of boards and sheets, it was a substitute for wood, and in severe climates, was used for roofing, fence materials, and other structures because it weathers as extremely well. *Transite* with asbestos is no longer available. Its replacement, *silica*, has limited applications because of health concerns. Other composites are continuously being developed to serve the needs of industry. Other *composites* include linoleum (linseed oil and jute), Bakelite (phenol-formaldehyde resin and cellulose fiber), plywood, and vehicle tires.

The more recent composites are the fiber-reinforced plastics and resins. The fibers include glass, Kevlar, and carbon fiber. Glass fibers used to reinforce plastics were introduced during World War II. The fibers are from 5 to 10 μm in diameter and have a tensile strength of about 3,000 MN/m^2. The glass fibers are treated with coupling agents to prevent the fibers from bundling and to bond the fibers to the plastic or resins. Because the fibers are about 100 times stronger than the plastic, the strength of the composite is proportional to the fiber content. An example of an extruded section of a polystyrene window frame which is reinforced with glass fiber is shown in Fig. 17.5. Molded fiberglass-reinforced polyester is the material of choice for boats though reinforced cement boats have also been built.

Fig. 17.6 A comparison of the diagonal-ply construction and the Michelin X radial tire

Carbon fibers were used about 100 years ago by Edison for his electric lamps. In 1964, the carbonized polyacrylonitrile (Orion) fibers were first produced with a tensile strength of about 2,000 MN/m^2 and a high modulus of over 400,000 MN/m^2. Thus, carbon fiber-reinforced resins are very stiff and have found wide application in artificial limbs, golf clubs, tennis rackets, skis, and many aircraft parts. However, the composites are not especially strong in tension.

Vehicle tires are another example of a composite, which are composed of rubber cord plies and beads, which hold the tire to the wheel. The cord plies are usually nylon, rayon, polyester, fiberglass, and steel wire. The construction of the tire was initially bias belted, with the cords running from bead to bead, crossing the tread at an angle with the number of plies determining the strength of the tire. Addition of two stabilizing belt plies below the tread increased traction and resistance to punctures. The radial-ply tires, manufactured first by Michelin in 1948, had cords, which ran from bead to bead with no bias angle (i.e., 90° to the longtitudal tread). This is shown in Fig. 17.6. Tread design has reduced planing on wet pavement and increased the life of a tire to approximately 80,000 miles.

New composites are constantly being developed to meet specific requirements with specified strengths, many to satisfy the aerospace industry.

Exercises

1. What are the primary components of cement?
2. What reactions occur in the kiln?
3. What reactions occur during the hydration process?
4. The lime (C), silica (S), and alumina (A) phase diagram is shown in Fig. 17.1a as mol%, and where the area of Portland cement and high alumina cement are indicated. Identify on the phase diagram the location for the following minerals found in (a) Portland cement clinker, C_3S, C_2S, and C_3A; (b) found in high alumina clinkers, CA, and CA_2.
5. In Fig. 17.1a, locate the positions of the substances listed in Exercise 4a and 4b assuming that the scale of the diagram is in units of wt.%.
6. Identify the composition of points x, y, and z on the phase diagram (Fig. 17.1a) for both a mol% and wt.% scales.
7. The silicate garden in which metallic salts grow colored "trees" when crystals are dropped into water glass (aqueous sodium silicate) is analogous to the fibrillars developed during the hydration process in cement. Explain.

8. In what way is polymer-impregnated concrete an important improvement?
9. It has recently been shown that the treatment of concrete with linseed oil (in dilute solutions of an organic solvent) extends the life of the concrete. Explain.
10. What advantages would there be in using fatty acid methyl esters of linseed oil instead of oil (triglyceride) to treat concrete?
11. Distinguish between cements, ceramics, and composites.
12. What is the difference between xerogels and aerogels and how are these materials used?
13. How is a ceramic or glass formed from the xerogel or aerogel?
14. What are the ideal properties of a fiber for reinforcement used in a composite?
15. Explain how the modern radial tire is an example of a composite structure.
16. Why is the addition of $CaCl_2$ to steel-reinforced concrete not to be recommended?

Further Reading

1. Munz D, Fett T (1999) Ceramics: mechanical properties, failure behavior, materials selection. Springer, New York
2. Matthews FL (1999) Composite materials. CRC Press, Boca Raton
3. Chawla KK (1998) Composite materials: science and engineering, 2nd edn. Springer, New York
4. Taylor HF (1997) Cement chemistry, 2nd edn. T. Telford, London
5. Bach H (ed) (1996) Low thermal expansion glass ceramics. Springer, New York
6. Dean-Mo L (ed) (1996) Porous ceramic materials: fabrication, characterization, application. LPS District Center, Lebanon
7. Mallhotra VM (1996) Pozzolanic and cementitious materials. Gordon and Breach, Newark
8. Watchman JB (1996) Mechanical properties of ceramics. Wiley-Interscience, New York
9. Yanagida H (1996) The chemistry of ceramics. Wiley, New York
10. Harper CA (ed) (1996) Handbook of plastics, elastomers and composites, 3rd edn. McGraw-Hill, New York
11. Hull D, Clyne TW (1996) An introduction to composite materials. Cambridge University Press, New York
12. Brandt AM (1995) Cement-based composites. Chapman and Hall, New York
13. Pope EJ et al (eds) (1995) Solgel science and technology. In: Proceedings of the international symposium on solgel. American Ceramic, Westerville
14. Struble RJ (1994) Cement research progress. American Ceramic, Westerville
15. Kelly A et al (eds) (1994) Concise encyclopedia of composite materials, Revisedth edn. Elsevier, New York
16. Ashbee K (1993) Fundamental principles of fiber-reinforced composites, 2nd edn. Technomic, Lancaster
17. Balaguru P, Shah SP (1992) Fiber-reinforced cement composites. McGraw-Hill, New York
18. Brook RJ (ed) (1991) Concise encyclopedia of advanced ceramic materials. Elsevier, New York
19. Chao VK (1990) Fundamental of composite materials. Knowen Acad., La Cruces, New Mexico
20. Taylor HFW (1990) Chemistry cement. Academic, London
21. Perry K (1986) Rotary cement kiln, 2nd edn. Chemistry Publishing, New York
22. Bye GC (1983) Portland cement. Pergamon Press, New York
23. Lubin G (ed) (1982) Handbook of composites. Van Nostrand Reinhold, New York
24. Davidge RW (1979) Mechanical behavior of ceramics. Cambridge University Press, Cambridge
25. www.wikipedia.org
26. Portland Cement Association, http://www.portcement.org/
27. Cement and Concrete Institute, cnci, http://www.cnci.org/
28. Cement & Concrete Association, NZ, http://www.cca.org.nz
29. Comprehensive Procurement Guideline Program, http://www.epa.gov/cpg/products/cement.htm
30. Corning Ltd, http://www.corning.com/lightingmaterials/products/macor
31. Ceramic sources, http://www.ceramics.com
32. American Ceramic Society, www.acers.org/
33. Silica Aerogels, a history, http://www.eetd.lbl.gov/ECS/aerogels/satoc.htm
34. Aerogels, Lawrence Livermore Nat'l Lab., http://www.llnl.gov
35. New Virtual Library, http://www.ikts.fhg.de/VL.ceramics.html
36. All about composites, http://www.owenscorning.com/owens/composites/about/

37. Honeycomb composites, http://www.euro-composites.com/
38. Composites Fabricators Assoc., www.cfa-hq.org/
39. Aerospace Composite Products, www.acp-composites.com/
40. Composite applications, http://www.netcomposites.com/
41. Worldwide composites search engine, www.wwcomposites.com/
42. Composites Registry, http://www.compositesreg.com/

Chapter 18
Semiconductors and Nanotechnology

Semiconductors and nanotechnology have become interconnected, and their development has resulted in new and interesting devices and methodologies.

18.1 Semiconductors

18.1.1 Introduction

The basic electronic theory of the semiconductor was developed in two papers in the Proc. Royal Society in 1931. It was based on the existence of potential wells in which the electron can reside. With the application of additional voltage, the electron can leave the well and migrate on the surface of the solid. It did not reach a practical stage until the World War of 1939 when the physicist, R. Ohl, at Bell Laboratories tried to use the cat's whiskers as an electronic amplifier. It was found to work in an unreliable manner and was about to be abandoned when a cracked crystal showed excellent but unpredictable results. This was eventually traced to the junction caused by the crack in the crystal. The research team that was formed developed the solid-state diode—composed of two joined crystals. The diode was made of germanium and became known as the semiconductor.

18.2 The Triode

After the war, the team at Bell Laboratories (Shockley, Bardeen, and Brattain) started to develop the amplifying triode. The difficulty lays in the need to have the emitter and collector close to each other without failing. This was eventually achieved by slicing a gold foil on a crystal with a razor blade and adding the needed contact points. The development of the germanium triode was given the name "transistor," and a team of the three researchers was awarded the Nobel Prize in Physics in 1956. The replacement of the germanium by silicon simplified the production of the transistor by allowing the use of zone melting to purify the silicon and thereby improve upon the product.

The early model for an electron in a metal is shown in Fig. 18.1 and more recently in Fig. 18.2 where ε_w is the work function; the difference between the energy of an electron in the Fermi level of the metal, E_F; and the energy of an electron, e_o, outside of the metal:

$$\varepsilon_w = e_o - E_f \tag{18.1}$$

O.V. Roussak and H.D. Gesser, *Applied Chemistry: A Textbook for Engineers and Technologists*, DOI 10.1007/978-1-4614-4262-2_18, © Springer Science+Business Media New York 2013

Fig. 18.1   Schematic diagram of energy levels in a one-dimensional structure

$$\phi = E_0 - E_F.$$

Fig. 18.2   Square-well potential model of a metal. The work function φ is equal to the difference in energy between the Fermi level and the energy of an electron outside the material

The semiconductor has been defined in several ways. Two of the most common definitions are based on the limits of two types of measurements: (a) electrical resistance with values between 10^{-2} and 10^9 Ω cm or 10^3 and 10^{-8} siemens/cm and (b) the energy gap with electronic excite energy between 0 and 3 electron volts (eV). However, some exceptions to these limits do occur. The best known material for the semiconductor is silicon (Si), but many others include naturally occurring minerals such as zinc blende (ZnS), cuprite (Cu_2O), and galena (PbS).

A recent award to a company producing semiconductors was made by Global Semiconductor Alliance to SiGe Semiconductor, Inc. in recognition of their efforts in advancing the work on silicon-based RF front-end solutions for wireless connectivity applications in consumer electronics and their lead-free initiative for the industry.

Recent developments in semiconductors are in the production of photovoltaic devices that convert solar intensity into electrical energy. There are two classes of semiconductors: the **n-** and the **p-**types. When the doped material releases free electrons, an **n**-type semiconductor is formed with more electrons than holes. When the reverse occurs, that is, more holes than electrons, then the conduction is a **p-**type. When these two materials are brought together, they form the **p–n** junction that favors the flow of electrons in one direction and is referred to as a semiconductor diode.

18.3 Applications

Semiconductors are the basic components of modern electronic systems such as radio, computers, telephone, lighting, heating, and cooling as well as communication. Figure 18.3 illustrates the band gap for an electron in three different states of matter. The electrons are considered to be in potential wells from which they can leave when given sufficient energy. When the energy of the wells is not uniform in depth, then the conductor would vary with temperature which when increased permits more electrons with higher energy to leave the well and to move freely on the surface.

The energy of the electron varies with the wave vector k. However, a more simplified band structure can be used for semiconductors without referring to the value of k. This approximation is acceptable for limits comparison of semiconductors of different structures and distinguishes between the direct and indirect band gap semiconductor. This is illustrated in Fig. 18.4.

In intrinsic semiconductors, the conductivity, σ (Ω^{-1} m^{-1}), increases as the temperature increases and approaches a maximum in conduction at the highest temperatures, Fig. 18.5. A plot of the log σ

Fig. 18.3 Simplified band structure diagrams of a metal, semiconductor, and insulator. Typical values of the band gap are 0 eV in metals, 0.5–5.0 eV in semiconductors, and 5 eV or greater in insulators

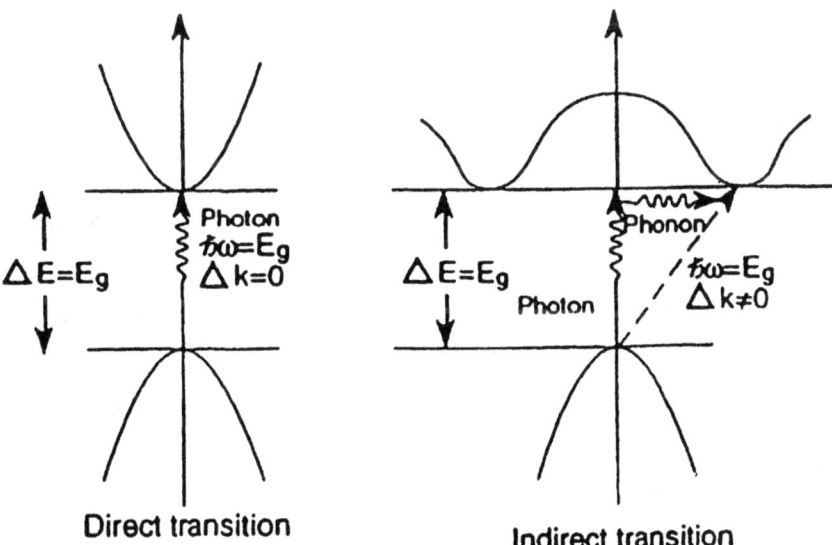

Fig. 18.4 Direct and indirect band gaps in semiconductors

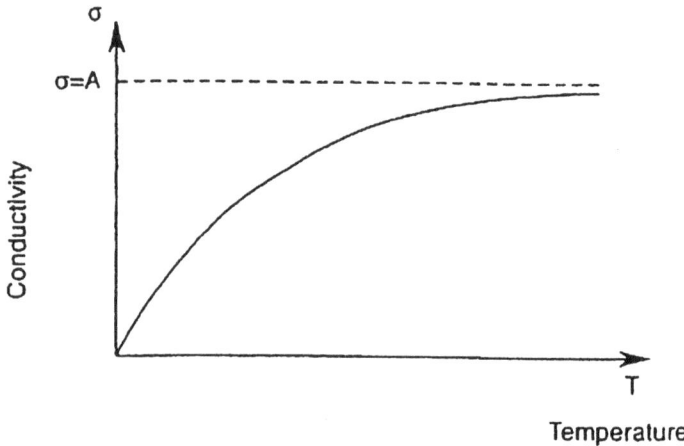

Fig. 18.5 Variation of the conductivity of an intrinsic semiconductor as a function of temperature

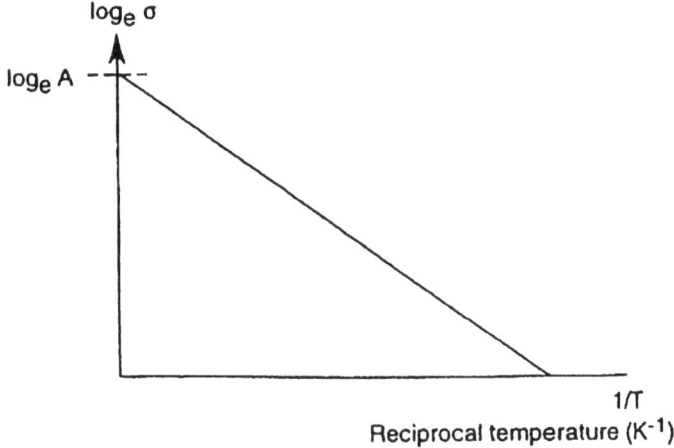

Fig. 18.6 Variation of log σ with 1/T for an intrinsic semiconductor

against 1/T is shown in Fig. 18.6. In the extrinsic semiconductor trace, impurities of the Group 5 elements form n-type semiconductors or when doped with Group 3 elements such as B, Al, Ga, or In, will form p-type semiconductors. The effect of temperature on the number of charge carriers as a function of temperature of extrinsic semiconductors is shown in Fig. 18.7. The effect of temperature on the conductivity is shown in Fig. 18.8, where comparison is also made with the intrinsic semiconductor. At higher temperatures, the extrinsic conductor usually converts to an intrinsic semiconductor.

18.4   Optical Effects

Light is emitted from an object when heated to high temperatures. The light bulb is a simple example of such a source. Luminescence is an alternate method of generating light by allowing an electron in a higher energy state to return to a lower ground state emitting the excess energy in the form of photons.

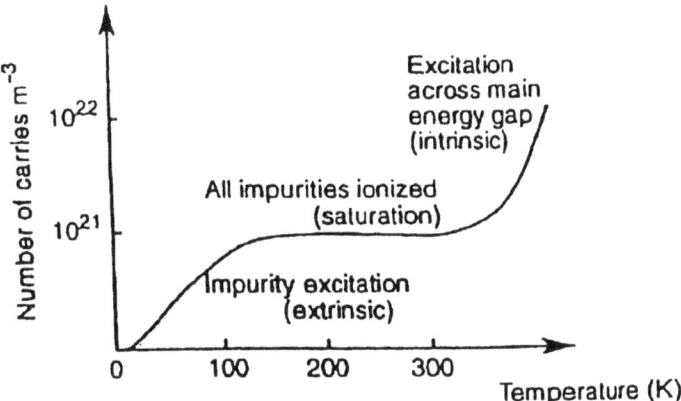

Fig. 18.7 Variation of conductivity of an extrinsic semiconductor with temperature

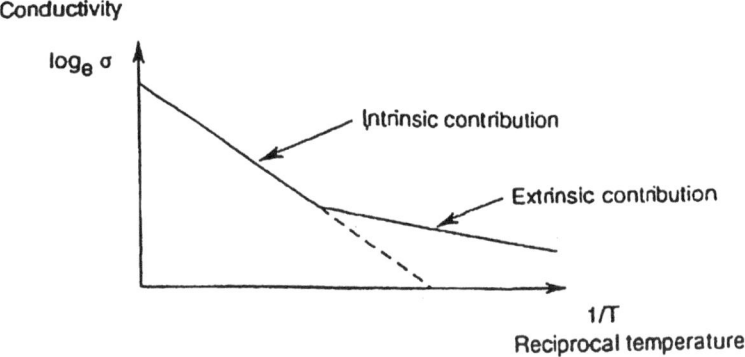

Fig. 18.8 Variation of log σ with 1/T for an extrinsic semiconductor

Such diodes are called LEDs, light-emitting diodes, and are usually formed when a current provides electrons that combine with holes and return to a lower energy level and emit photons. Figure 18.9 shows the emitted spectra of several semiconductor systems that have been studied.

18.5 Nanotechnology

Nanotechnology, unlike semiconductors, is a recent addition to the catalog of scientific fields. However, the two topics have become closely linked and appear to be complimentary to each other. The subject matter of nanotechnology was first discussed by Richard Feynman in a speech given at a scientific meeting in Dec. 1959. He predicted that the scientific trend would be moving to the study of smaller groups of atoms and molecules with tools that will be following the transition by the observation of individual atoms and molecules.

The term "nanotechnology" was first used in a paper presented by Norio Taniguchi of Tokyo University in 1974. This title was further developed in the 1980s by K. Eric Drexler in his 1991 MIT Ph.D. thesis (nanosystems, molecules, manufacturing, and computation) in speeches and in books such as "Engines of Creation: The Living Era of Nanotechnology (1986)."

Fig. 18.9 Absorption coefficient versus photon energy for Ge, Si, GaAs, and selected other III–V binary semiconductors at T = 300 K

A nanometer (nm) is 10^{-9} m or 1 nm = 10 Å. In terms of human parameters, the diameter of the double helix is 2 nm, bacteria are about 200 nm long. Another comparison that has been made is that the ratio of a nanometer to a meter is similar to that of a marble to the diameter of the earth.

The subject matter that can be included in the topic "nanotechnology" is limitless and much has been written on the various subject areas. However, it is not possible to treat each of the topics covered by nanotechnology. It will be useful to itemize the fields that have been subject to nanotechnology treatment

One recent study by Dr. Nayoo in China has shown that the effect of nanoparticles on plant cells such as rice was not invariably beneficial. In fact, the nanomaterial induced the death of plant cells. Caution is advised when nanotubes are injected into plants.

The 2010 Nobel Prize in Physics was awarded to two British physicists at the University of Manchester for groundbreaking experiments regarding the two-dimensional material "graphene" shown in Fig. 16.4 as the two-dimensional sheet of one atomic layer thick. As an electrical conductor, it is as good as copper, and as a conductor of heat, it is superior to most other materials. It is almost completely transparent but impervious to penetration by all gases including the smallest gas–helium atoms.

The future of graphene has been predicted to include its application to electronics, solar cells, and panels. It has been predicted to find applications in strengthening thin materials such as solar panels. When added to plastics, it is argued that the new materials could become good electrical and heat conductors as well and serve in several capacities in satellites and airplanes where low weight is important.

It has recently been shown that it is possible to make super-hydrophobic surfaces that repel water and possess a contact angle of about 150° with water. Normal non-wetting surfaces of glass or plastic can have contact angles with water that can vary from 60° to 90°. The high values of 100° are similar to that obtained for the lotus plant leaves and are attributed to air pockets that form between the water and the rough surface. One immediate application is for wind turbine blades with one caveat—the effect decreases at low temperature—a regular problem facing Canadians.

Examples of nanomaterials can be classified in different ways. Some of the common classes are based on application and consist of the following: carbon nanotubes (fullerenes), nanoparticles, nanorods, and nanoelectronic devices. Other more specific are, for example, medical applications, silicon solar cells, semiconductors, nanoelectromechanical systems (NEMS) or microelectromechanical systems (MEMS), and nanolithography. Another major potential application is in the medical field as nanorobotics.

This new and exciting field of nanotechnology and its many applications have recently been critically evaluated and debunked by showing that the many applications that have been proposed are impractical, excessively expensive, and energy consuming. Ms Georgia Miller at Friends of the Earth (FOE) Australia has been leading the effort of exposing the deception that nano-based developments will reduce the release of greenhouse gases. Her report "Nanotechnology, Climate and Energy: Overheated Promises and Hot Air?" raises some important factors that have yet to be resolved.

Hence, the future of nanotechnology, though full of enthusiastic hope and expectation, may be packed with problems and could become a great disappointment.

Exercises

1. Describe the physical properties of insulators, semiconductors, and conductors and show how some interchange can be made to occur.
2. Chemical potential wells exist. In what way do these differ with electronic energy wells?
3. In simple chemical reactions at low temperatures, it is possible to observe "tunneling" through the energy barrier. Is there any evidence for this with the electron?
4. To what extent does temperature plays in the current–potential interaction of the electronic systems?
5. In Fig. 18.9, what major characteristics are not mentioned in determining the properties of the semiconductors?

Further Reading

1. Wonbong Choi, Jo-won Lee (2012) Graphene. Synthesis and applications, CRC Press/Taylor and Francis Group, Boca Raton/London. Wikipedia: Semiconductors (& Devices)
2. Grayson M (ed) (1984) Encyclopedia of semiconductor technology. Wiley, New York
3. Yu PY, Cardon M (1999) Fundamentals of semiconductors. Springer, Berlin
4. Smith RA (1978) Semiconductors, 2nd edn. Cambridge University Press, Cambridge
5. Jiles' D (2001) Introduction to electronic properties of materials, 2nd edn. Nelson Thornes, Cheltenham
6. Bar-Lev A (1979) Semiconductors & electronic devices. Prentice Hall, London
7. Phillips JC, Lucovsky G (2010) Bonds & bands in semiconductors, 2nd edn. Momentum Press, New York
8. Semiconductors & Nanotechnology: www.nanosemi.co.uk.about.html www.qub.ac.k/research-centre/NISRC/, The Role of Nanotechnology in the Semiconductor Industry.<keithley Instruments inc>. Wikipedia Nanotechnology
9. Korvink JG, Greiner A (2002) Semiconductors for micro- and nanotechnology: an introduction for engineers. Wiley, Weinheim
10. Balandin AA, Wang KL (2006) Handbook of semiconductor nanostructures and nanodevices. American Scientific Publishers, Stevenson Ranch
11. Levy D (2007) Electron Dance: Nanoscale Semicondurtors, Stanford Report, the University of Stanford, http://news.stanford.edu/news/2007/march14/kondo-031407.html
12. Frnwick O et al (2009) Thermochemical nanopatterning of organic semiconductors. Nat Nanotechnol 4:664–668

Chapter 19
Epilogue

World resources which must be preserved are being depleted, and our upgraded renewal and repair of processes are now necessary and desirable. The world's population is continually increasing and recently reached 7 billion. It is estimated to reach to 9.1 billion by 2050.

First, human beings will face new challenges of food problems. It means that we need to optimize food production as well as to optimize the distribution of food on the Earth. Second, in future, the world's resources will be depleted and the environment will be deteriorated. Though oil resources are in decline, the remaining sources will be found in less desirable locations, e.g., under deep waters such as the Gulf of Mexico where a recent spill occurred.

Nuclear power seems to be potentially the best for the time being with its huge capacity to supply, even though the cost of constructing, maintaining, and disassembling a nuclear power plant is high. The recent nuclear accident in Japan has caused many countries to reconsider their choices of alternate energy that includes nuclear energy. According to publication in the Associated Press in August 26, 2011, Japan's Nuclear Agency said that the amount of radioactive cesium that has leaked from a tsunami-hit nuclear plant is about equal to 168 of atomic bombs dropped on Hiroshima at the end of the World War II. The damaged plant has released 15,000 tera becquerels of cesium-137, which lingers for decades and could cause cancer, compared with the 89 tera becquerels released by the US uranium bomb.

Solar and wind energy are becoming major alternatives to hydro-electrical energy. There is a very ambitious proposal named the Sahara Solar Breeder Plan that is based on assumption that if there comes a time when solar energy utilization, nanotechnology, and power distribution technologies using superconductivity make further progress. It will become possible to build solar photovoltaic power generation facilities in desert and meet the global demand for electricity. Wind power stations also become a highly effective technologies including new construction materials in turbines' magnets made from 500 pounds or more of rare earth metal-like neodymium.

The electrical and hybrid vehicles are becoming a serious alternative to the internal combustion engine. But hybrid green cars would not exist without special materials like rare earth elements – lanthanum for their batteries, neodymium magnets for their electric motors, advanced piezoelectric materials for their engines, and various polymers and plastics for automotive appliances. For example, the battery in a single Toyota Prius contains more than 20 pounds of the rare earth element lanthanum.

Potable water (as distinct from seawater) has become a valuable commodity which is generally used in homes and industry that include irrigation. It has recently been shown that over 70% of fresh water is usually used to irrigate our food sources. Much of this water is supplied by either surface drip or sprays as well as by subsurface drip. All such sources require appropriate knowledge of the plants' requirements as well as rainwater deliveries. The water on demand plant watering system using the microporous hydrophilic membranes was developed by Dr. Lue Errede at 3M in Minnesota.

O.V. Roussak and H.D. Gesser, *Applied Chemistry: A Textbook for Engineers and Technologists*,
DOI 10.1007/978-1-4614-4262-2_19, © Springer Science+Business Media New York 2013

When offered both water and nutrients, the plants distinguish between the two sources. A water/nutrient ratio that was three changed to five when the concentration of the nutrients was doubled. This system has been extended to grow crops by one of the authors (HDG) by using the subsurface microporous hydrophilic tubular membrane for water delivery. Extensive studies on different plants have verified that self-watering control by plants is commercially viable.

Further Reading

1. Conditions for Survival (2010) Toward a "solar energy – based society" full of vibrant life. In: Special round table conference on global environmental problems. The Asahi Glass Foundation. Tokyo, 272pp
2. Smil V (2010) Energy myth and realities. Bringing science to the energy policy debate. The AEI Press, Washington, DC, 215pp
3. Mari Yamaguchi (2011) Government estimates cesium leak from Japan's nuke plant equals 168 Hiroshima bombs. The Associated Press, 26 Aug 2011 http://www.winnipegfreepress.com/breakingnews/government-estimates-cesium-leak-from-japans-crippled-nuke-plant-equals-168-hiroshima-bombs-128463423.html
4. Kenji Uchino (2010) Advanced piezoelectric materials. Woodhead Publishing Limited, Oxford/Cambridge/Philadelphia/New Delhi. Uchino. Advanced piezoelectric
5. www.eco-ag.us.photo.php

Appendix A: Fundamental Constants and Units

Speed light (in vacuum)	$C = 2.99792458 \times 10^8$ m s^{-1}
Planck's constant	$h = 6.6260755 \times 10^{-34}$ Js
Gas constant	$R = 0.0820584$ L atm mol^{-1} K^{-1}
	$R = 8.314510$ J mol^{-1} K^{-1}
Gravitational constant	$g = 9.8066$ m/s^2
Faraday constant	$F = 9.6485309 \times 10^4$ C mol^{-1}
Avogadro number	$N_A = 6.0221367 \times 10^{23}$ mol^{-1}
Electron charge	$e^- = -1.60217733 \times 10^{-19}$ C
Electron mass (rest)	$m_e = 9.1093897 \times 10^{-31}$ kg $= 0.00054858$ amu
Proton mass (rest)	$m_p = 1.67262 \times 10^{-27}$ kg $= 1.007276$ amu
Neutron mass (rest)	$m_n = 1.67493 \times 10^{-27}$ kg $= 1.008665$ amu
Solar constant (sea level)	1,370 W/m^2
Solar luminosity	3.85×10^{26} W

Nomenclature for powers of 10

Prefix	Symbol		Prefix	Symbol	
atto	(a)	10^{-18}	kilo	(k)	10^3
femto	(f)	10^{-15}	mega	(M)	10^6
pico	(p)	10^{-12}	giga	(G)	10^9
nano	(n)	10^{-9}	tera	(T)	10^{12}
micro	(μ)	10^{-6}	peta	(P)	10^{15}
milli	(m)	10^{-3}	exa	(E)	10^{18}

Length

$\overset{\circ}{A} = 10^{-10}$ m

inch $= 2.54$ cm

mile $= 1.609$ km

knot $= 1$ nautical mile/hr

$\quad = 1.1516$ mile/hr

$\quad = 1.853$ km/hr

O.V. Roussak and H.D. Gesser, *Applied Chemistry: A Textbook for Engineers and Technologists*, DOI 10.1007/978-1-4614-4262-2, © Springer Science+Business Media New York 2013

Area
  acre $= 43{,}560$ ft^2
       $= 4{,}047$ m^2
  hectare $= 10^4$ m^2
  square mile $= 640$ acres
             $= 259$ hectares

Volume
  1 ft^3 $= 28.317$ L
  1 m^3 $= 1{,}000$ L
  1 gal (US) $= 3.785$ L
           $= 4$ quarts
           $= 128$ fluid oz.
  1 gal (Imp) $= 4.546$ L
            $= 4$ quarts
            $= 160$ fluid oz.
  1 barrel (US) $= 31.5$ US gal
  1 barrel (Imp) $= 36$ gal (Imp)
  1 barrel (oil US) $= 42$ US gal
                  $= 158.9873$ L

Mass
  1 lb. $= 453.6$ g
  1 ton $= 2{,}000$ lb.
  1 tonne $= 1{,}000$ kg
         $= 2204.6$ lb.
  1 kg $= 2.2046$ lb.
  1 amu $= 1.66054 \times 10^{-24}$ g
       $= 931.4874$ MeV

Pressure
  1 atm $= 1.01325$ bars
       $= 760$ torr
       $= 101{,}325$ Pa

Energy
  Joule (J) – 1 volt (V) $\times$ 1 amp
  1 eV $= 1.6021 \times 10^{19}$ J
      $= 4.450 \times 10^{-26}$ kWh
  1 MeV $= 9.65 \times 10^{10}$ J/mol
  Watt (W) $= 1$ J/sec
  Curie (Ci) $= 3.7 \times 10^{10}$ disintegration/sec
  Becquerel (Bq) $= 1$ disintegration/sec
  1 calorie (cal) $= 4.183$ J
  1 horsepower (HP) $= 745.7$ W
  1 kWh $= 3{,}412$ Btu
  1 Btu $= 1055.06$ J
       ≈ 1 kJ $= 2.931 \times 10^{-4}$ kWh
  1 Quad $\approx 10^{15}$ Btu
        $\approx 10^{15}$ kJ
        $= 10^{12}$ ft^3 (CH_4)
        $= 2.93 \times 10^{11}$ kWh

$\approx 40 \times 10^6$ tons coal

$= 170 \times 10^6$ bbl crude oil

$= 8.0 \times 10^9$ US gal gasoline

1 bbl oil $= 42$ gal (US)

$\qquad = 5.8 \times 10^6$ Btu

$\qquad = 1.65 \times 10^3$ kWh

$\qquad = 159$ L

$\qquad = 136$ kg

$\qquad = 5.8$ M Btu

1 ft^3 Natural gas (CH$_4$) $= 1{,}035$ Btu

$\qquad\qquad\qquad\quad = 0.310$ kWh

1 lb. coal $= 3.84$ kWh

1 gal (US) gasoline $= 36.7$ kWh

1 ton oil equiv. (toe) $= 4.19 \times 10^{10}$ J

$\qquad\qquad\qquad = 10^7$ kcal

1 ton coal equiv. (toe) $= 2.93 \times 10^{10}$ J

$\qquad\qquad\qquad = 7 \times 10^6$ kcal

1 therm $= 10^5$ Btu 1 kg TNT $= 10^9$ J

On July 23, 1983 a new Boeing 767 (Air Canada Flight No. 143) refueled in Montreal. The fuel gauge was not working so it was decided to refuel by the manual use of a dipstick which correctly showed the fuel in the aircraft to be 7,682 L. The required fuel for the trip was 22,300 kg. The mechanics, using 1.77 as the density conversion factor, calculated the required fuel necessary,

$$7682 \times 1.77 = 13{,}597 \text{ kg on board}$$

$$22{,}300 - 13{,}597 = 8703 \text{ kg to be added}$$

$$8703 \div 1.77 = 4916 \text{ L of fuel to be added}$$

They believed 1.77 represented the conversion of liters to kilograms, in fact it was the conversion of liters to pounds, i.e., 1.77 Ib/L is the density of the jet fuel. The density of the jet fuel in proper units is 0.803 kg/L and the amount of fuel which ought to have been added was 20,163 L. Using 1.77 without any units led to a near disaster.

Further Readings

1. Carey WM (1985) Out of fuel at 26,000 feet. Readers' Digest, 126 (May), p 213

2. Hofler W, Mona M (1989) Freefall a true story. St. Martin's Paperback, New York

3. The Gimli Glider http://www.acs.org/VC2/2my/my2—143.html

4. International System of Units, Constants, Units, Uncertainty http://www.physics.nist.gov/

Appendix B: Viscosity

Introduction

Viscosity is that property of a fluid that opposes the relative motion of adjacent portions of the fluid and can consequently be regarded as a type of internal friction. Viscosity can be defined as the force required to move a layer of fluid of unit area with a velocity 1 cm/s greater than the velocity of another layer 1 cm away (see Fig. B.1). Since force is proportional to the velocity difference between the layers and inversely proportional to the distance between the layers, then

$$F \propto \frac{VA}{x} \tag{B.1}$$

where F is force, V is velocity difference, x is distance between layers, and A is area of layer. Therefore,

$$F - \frac{\eta V A}{x} \tag{B.2}$$

where η, the proportional constant, is the viscosity.

The units of viscosity are dyne second per square centimeter or gram per second per centimeter (1 dyne second per square centimeter is simply called 1 poise, after Poiseuille). The reciprocal of viscosity is called the fluidity and is often represented by ϕ; it is a measure of the ease with which a liquid flows.

In a gas the viscosity increases as the temperature increases, whereas in a liquid the converse is true. The interpretation of viscosity in a gas utilizes the high kinetic energy of the molecules and involves the transfer of momentum for one layer of the gas to another, leading to a relationship of the form $\eta \propto v$, where v is the average speed of the gas molecules. In liquids, a completely different interpretation is required, since the molecules are closely packed except for the presence of holes. Over half a century ago it was found that the fluidity of a substance at its melting point is proportional to $V - b$, where V is the volume of liquid and b is the van der Waals constant. This is the effective space occupied by the molecules. $V - b$ is, therefore, the free volume of the liquid. Since most solids expand approximately equal at this temperature the free volumes of liquids are approximately the same; therefore, we could expect the viscosities of most liquids at their melting points to be approximately equal. This is correct within an order of magnitude. The theory of the significant structures of liquids has been able to relate the viscosity mechanism to fluidized vacancies and to

O.V. Roussak and H.D. Gesser, *Applied Chemistry: A Textbook for Engineers and Technologists*, DOI 10.1007/978-1-4614-4262-2, © Springer Science+Business Media New York 2013

Fig. B.1 Schematic representation of the viscosity of a liquid. (**a**) Liquid near the walls of a tube moves slower than liquid near the center or furthest from the walls. Velocity profile depends on the viscosity of the liquid. (**b**) Viscosity is the force required to maintain a unit velocity gradient for 1 cm^2 plates 1 cm apart

show that the effect of temperature on viscosity is associated with the energy required to create a hole in the liquid. The influence of temperature on the viscosity of a liquid is often represented by the following equation:

$$\eta = Ae^{E_{vis}/RT} \tag{B.3}$$

where E_{vis} is the energy of activation of viscosity, R is the gas constant, T is the absolute temperature, and A is a constant depending on the substance.

It is interesting that the average heat of vaporization of a liquid is approximately three times the activation energy of viscosity. This means that three times as much energy is required to remove a surface molecule as to move a bulk molecule past a neighbor. The ratio $n = E_{vap}/E_{vis}$ was shown byEyring to be equal to the ratio of the size of a molecule to the size of a hole needed for viscous flow. It has been found that, since a hole of molecular dimensions is not required if, for example, two molecules rotate about their point of contact, the value of n is about 3 for a spherically symmetric nonpolar molecule and increases to 5 as the deviation from spherical symmetry increases.

Measurement of Viscosity

The viscosity of a liquid can be measured by several methods. The most convenient for laboratory work are the Ostwaid viscometer and the falling ball methods.

Ostwald Viscometer

In 1846 the viscosity of a liquid was related by Poiseuille to the rate of flow of a liquid through a tube under a pressure differential. The Poiseuille equation is as follows:

$$Q = \frac{\pi P r^4}{8\eta l} \tag{B.4}$$

where Q is the volume of liquid flowing per unit time through a tube of radius r and length l, across which there is pressure drop P. If the pressure drop is due to gravity, then $P \sim hdg$ where h is the height of the liquid, d is the density, and g is the acceleration due to gravity. By substituting hdg and P in (B.4), we obtain

Fig. B.2 *Ostwald viscometer.* The quantity of liquid (volume *V*) required is such that when the liquid is drawn into bulb A, the level in reservoir C is not below level L. The time of flow *t,* from level *x* to y, is recorded and compared with other liquids of identical volume *V* and of known density *d*

$$\eta = \frac{\pi r^4 h d g}{8 Q l} \tag{B.5}$$

The measurement of absolute viscosity by this method requires an exceptional amount of care and patience. For most purposes, it is sufficient if relative viscosities are known. Hence, if the time for a fixed volume of liquid to flow through a capillary is measured, then the comparison of its time of flow with that of another liquid enables us to calculate a relative viscosity. Since

$$Q = V/t \tag{B.6}$$

where *V* is the volume of liquid, and *t* is the time of flow, then

$$\eta = \frac{\pi r^4 h d g t}{8\, V\, l} \tag{B.7}$$

For a fixed apparatus as shown in Fig. B.2, *V, h, g, r, I* are constant. Therefore, for one liquid,

$$\eta_1 \propto d_1 t_l \tag{B.8}$$

Hence, for two liquids,

$$\frac{\eta_1}{\eta_2} = \frac{d_1 t_1}{d_2 t_2} \tag{B.9}$$

The viscosities of several substances at different temperatures, which can conveniently be used as references, are given in Table B.1.

Table B.1 Viscosities of several substances at various temperatures, in units of centipoise

| Name | Formula | Temperature (°C) | | | | | | |
		0	10	20	30	40	50	60
Water	H_2O	1.79	1.31	1.00[a]	0.801	0.656	0.549	0.469
Ethanol	C_2H_5OH	1.77	1.47	1.20	1.00	0.834	0.702	0.592
»-Propanol	$n\text{-}C_3H_7OH$	3.88		2.25	1.72	1.41	1.13	
Isopropanol	$iso\text{-}C_3H_7OH$	4.56		2.37		1.33		
Benzene	C_6H_6		0.758	0.652	0.564	0.503	0.442	0.392
Toluene	$C_6H_5CH_3$	0.772		0.590	0.526	0.471		
Chlorobenzene	C_6H_5Cl	1.03		0.799		0.631		
Ethyleneglycol	$C_2H_4(OH)_2$			19.9		9.13		4.95

[a]Viscosity of H_2O at 20.20°C is 1.0000 centipoise

Falling Bail Method

The viscosity of a liquid may be determined from Stokes' law. In 1850 Stokes showed that a sphere of radius r under a constant force F will move with constant

$$\frac{\eta}{\eta_2} = \frac{d_1 t_1}{d_2 t_2}$$

velocity v in a viscous liquid of viscosity η according to the following relation:

$$F = 6\pi r \eta v \tag{B.10}$$

If the sphere is acted on by gravity alone, then

$$F = 4/3\pi r^3 (d - d')g \tag{B.11}$$

where d is the density of sphere, d' is density of fluid, and g is acceleration due to gravity.
Therefore

$$\eta = \frac{2r^2(d - d')g}{9v} \tag{B.12}$$

Since $v = I/t$ then

$$\eta = \frac{2r^2 g}{9l}(d - d')t \tag{B.13}$$

Since r, d, g, I are constant, then the relative viscosity can be easily determined from the following equation:

$$\frac{\eta_1}{\eta_2} = \frac{(d - d_1)t_1}{(d - d_2)t_2} \tag{B.14}$$

For very viscous liquids the falling ball viscometer is generally more suitable than the Ostwald viscometer.

The viscosity and the activation energy of viscosity of a pure liquid substance can tell us very little about the substance. However, the viscosity of a solution is much more informative. Thus, if acetone and chloroform are mixed and the viscosities of the various solutions are measured, it can be determined whether or not "compound" formation can exist. These two substances form a hydrogen bond about 11.3 kJ/mol, and the appearance of a maximum about 1:1 molecular ratio supports this view. Many such compounds, first detected by viscosity anomalies, are now being characterized by more elegant methods. It is interesting to note that substances that show this viscosity effect generally show a volume change when mixed. This suggests a change in the free volume of the solution compared with the free volume available for each of the individual substances.

It must be pointed out that a viscosity effect in the mixture of two or more substances does not necessarily prove the formation of a compound, since other effects, such as steric hindrance[1] and the association of one of the components may give rise to a viscosity anomaly.

Applications of Viscosity

Molecular Weight of Polymers

From the measurements of viscosity of solutions of polymeric substances, it is possible to determine the average molecular weight of the solute. The empirical relation is as follows:

$$[\eta] = KM^a \tag{B.15}$$

where K and a are constants that depend on the solute, solvent, and temperature, M the average molecular weight, and [q] the intrinsic viscosity and fractional change in the viscosity of a solution per unit concentration of solute at infinite dilution. This is represented as

$$[\eta] = \lim_{C \to 0} \left[\frac{1}{C} \left(\frac{\eta - \eta_0}{\eta_0} \right) \right] \tag{B.17}$$

where η_0 and η are the viscosity of the solvent and the solution, respectively, and C is the concentration of the polymer, usually in wt.%. Although other methods for determination of molecular weight of polymers are more accurate and more reliable, the intrinsic viscosity can give relative molecular weights with reasonable accuracy and facility. It is often the first determination made for the molecular weight of a polymer.

Drag Reducers

As the flow of a fluid in a tube increases as a result of increasing applied pressures, the Reynolds number (Re) increases to the onset of turbulence where

$$\text{Re} = \text{u}d/v \tag{B.17}$$

---

[1] Steric hindrance is the spatial interference experienced in molecules because of the specific location of bulky chemical groups.

Fig. B.3 Reduction of
turbulence by the addition of
a drag reducer (DR) to a fluid
OAB, fluid only; OX, with
DR

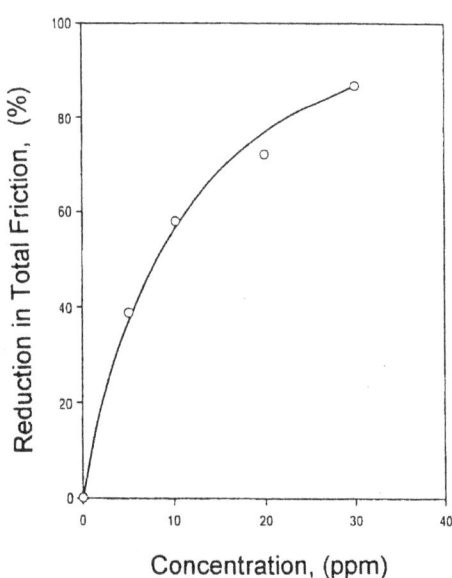

Fig. B.4 Effect of
polyethylene oxide on the
flow rates of water through
1,000 ft of a 1.5 in. hose at
various pressures

u is the velocity of flow, d is the diameter of the tube, and v is the kinematic viscosity[2] and equal to η/ρ, where ρ is the density of the fluid.

This is shown in Fig. B.3 where at point A, turbulence occurs, i.e., Re > 2,000. When a low concentration of a long chain soluble polymer is dissolved in the fluid it prevents the onset of turbulence and the flow rate-pressure line is extended from O to X along the straight line plot. This effect is called *drag reduction* and has been applied to the pumping of oil in pipelines, blood in arteries, and free flow of fluids. Figure B.4 shows the reduction in friction (viscosity) as the concentration of polyethylene oxide (PEO, mol. wt. = 5 MD) in water is increased. Figure B.5 shows the increased flow of water through a 1.5 in. hose as a function of pressure when PEO is added to the water. This is illustrated in Fig. B.6 where a 1.5 in. fire hose can deliver water at a rate equal to that of a 2.5 in. hose if polyethylene oxide has been added to the water.

---

[2] $v = 0.0100$ cm^2/s (Stokes) at 20°C for water; $v = 6.8$ cm^2/s (Stokes) at 20°C for glycerol.

Fig. B.5 Effect of low concentrations of polyethylene oxide on friction (viscosity) reduction

Fig. B.6 Effect of high molecular weight polyethylene oxide) in increasing water flow through fire hoses. Hoses had been adjusted to throw water the same distance and then 30 ppm of polyethylene oxide) were injected into the hose on the left

Electrorheological Fluids

Some liquids with a suspension of solids or colloids can, under the influence of an applied potential, show an increase in viscosity by a factor as high as 10^5, i.e., the liquid is converted to what is essentially a solid. This is called the *Winshw effect* and the liquid is an electrorheologica! (ER) fluid.

Winslow (1949) reported that silica gel in a low-viscosity oil showed this effect under an electric field of 3 kV/mm. The fluid can be sheared with a force proportional to the square of the electrical field. For example, a 25% by volume of hydrophobic colloidal silica spheres of 0.75 pm diameter in 4-methylcyclohexanol showed ER responses at 40–4,000 Hz, although dc fields are also viable. Dispersants are often added to the suspension in order to prevent the settling of the solids.

The ER fluids can have applications as clutches, speed controllers, and valves. Other applications can be expected as work on the subject continues.

Exercise

1. From the data shown in Table B.1 convert the value of the viscosity of ethylene glycol at 20°C to SI units, i.e., mNsec/m^{-2}.
2. The viscosity of hexadecane is 3.6 mNsec/m^{-2} at 22°C. The flow of water (20 mL) in an Ostwald viscometer took 47 s at 22°C. Calculate the time it would take 20 mL of hexadecane (density = 0.7751 g/mL) to flow through the same viscometer at the same temperature.
3. Discuss the differences in viscosity expected for H_2O, H_2O_2, and D_2O.
4. Of the two propanols which one would you expect to have the higher viscosity? Give reasons for your answer.
5. (a) Calculate E_{vis} for trinitroglycerol (TNG) which has the value of $\eta = 360$ mNsec/m^{-2} at 20°C and $\eta = 13.6$ mNsec/m^{-2} at 40°C, (b) calculate η for TNG at 60°C.
6. The fluidity of a liquid explosive doubles between 10° and 20°C. Determine (a) E_{vis} and (b) η_{60}/η_{10}.

Further Readings

1. Yaws CL (1994) Handbook of viscosity. vol 1–3, Gulf, Huston
2. Viswanath DS, Natarajan G (1989) Data book on the viscosity of liquids. Hemisphere, Bristol
3. Barns HA et al (1989) An introduction to rheology. Elsevier, New York
4. Viscosity and Stokes' Law www.math.mcmaster.ca
5. Norcross Ltd http://www.viscosity.com/
6. Viscosity and Surface Tension http://www.physics.bu.edu/pyl05/notes/Viscosity.htmI
7. http://www.pe.utexas.edu
8. Drag reducers http://www.liquidpower.com/about/awhatis.htm

Appendix C: Surface Chemistry

C1.1.1 Surface Tension

A surface is the boundary between two phases. The chemistry of this interface is of great importance to a variety of subjects such as adhesion, corrosion, surface coatings, and many others. The differential attraction of surface molecules in a liquid results in a surface energy that is also called *surface tension,* and accounts for the tendency of a free or suspended liquid to assume a spherical shape in droplets or to expose as small a surface area as possible. Thus the surface of a liquid in contact with air or another liquid phase in which it is immiscible may be considered similar to a rubber elastic membrane or balloon that assumes a spherical shape. To distort this to any other shape would require the expenditure of energy, since the surface area would increase; i.e., the rubber membrane would stretch. A surface energy can therefore be associated with a liquid–gas interface and can be defined as the energy or work required to increase the surface area of a liquid by 1 cm^2 by bringing bulk molecules to the surface; i.e., work per square centimeter. Since the surface is under a tension, a force called the surface tension y can be defined as the force applied to increase the surface area of a liquid when acting on 1 cm of surface; i.e., force per centimeter. The units of surface energy are

$$\frac{ml^2t^2}{l^2}$$

and those of surface tension as force are

$$\frac{mlt^{-2}}{l}$$

The surface energy and surface tension both have units of mt^{-2} and are equivalent (see Fig. C.1).

When a drop of liquid is placed on a solid, a definite angle of contact exists at the point where the liquid meets the solid. This is shown in Fig. C.2 where the angle θ is called the *contact angle.*

The contact angle has one of two values, advancing or receding, depending on whether the liquid–solid interface area is increasing or decreasing. This is commonly called a *hysteresis* effect. Until recently, it was believed that the ratio of these two angles depended only on the roughness of the surface. It is now reasonably well established that the hysteresis is due to the penetration of the liquid molecules into surface discontinuities and, therefore, depends on the size of the molecules relative to the intermolecular pores of the solid surface. The contact angle thus depends on the type of solid surface, liquid, and to a lesser degree, temperature. Table C.1 lists the values of the contact angles for various substances as well as some surface tension values.

O.V. Roussak and H.D. Gesser, *Applied Chemistry: A Textbook for Engineers and Technologists,* DOI 10.1007/978-1-4614-4262-2, © Springer Science+Business Media New York 2013

Fig. C.1 Thin film of liquid *{ayb}* is stretched on an inert wire frame *{xyz}*. A wire barrier *(ab)* of length/is pulled toward *xz* with force *F*. Since the liquid increases in area on both sides of the frame, then at balance, *F = 2ly*, where *y* is the surface tension of the liquid

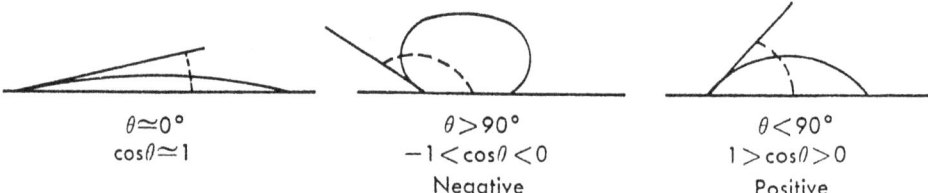

Fig. C.2 A drop liquid is placed on a solid surface. The angle formed at the interface is called the contact angle. If the volume of the drop is increased, the advancing angle is measured. If the volume of the drop is decreased, the receding angle is measured

Table C.1 Contact angles and surface tensions for various substances

Contact angle θ^a					
Water on siliconized glass:				Water on:	
$T(°C)$	γ(dyne cm^{-1})	θ_A	θ_R	Glass	0°
4	75	104°	75°	Paraffin	108°
22	72	106°	76°	Polyethylene	94°
75	63	104.5°	76.5°	Teflon	110°
				Graphite	86°
				Kel F	90°
Standard surface tension in dynes per centimeter, versus air at 20°C					
Water	72.8				
Acetone	23.7				
Benzene	28.9				
Toluene	28.4				
Choloroform	27.1				
Carbon tetrachloride	26.8				
n-Hexane	18.4				
n-Octane	21.8				
n-Octanol	27.5				

[a]θ_A is advancing angle and θ_R is receding angle. When one value of θ is given, it is usually the arithmetic average of θ_A and θ_R

The contact angle depends on the relative bonding between liquid and solid molecules (B_{ls}), compared with the bonding between liquid molecules (B_u). Thus, if B_{us} is greater than B_u, 6 is usually less than 90° and the solid is considered to be wetted by the liquid. However, if B_{ls} is less than B_u, then θ is usually greater than 90° and the solid is not wetted by the liquid. It must be pointed out that the angle of 90° chosen for the demarcation between wetting and nonwetting is quite arbitrary.

Fig. C.3 Measurement of contact angle. A solid slide of the material is immersed in the liquid and the meniscus x at the air interface is noted. The slide is then rotated about the point of contact with the liquid surface until the meniscus flattens *(y)*. The advancing and receding angles are associated with decreasing and increasing contact angles, respectively

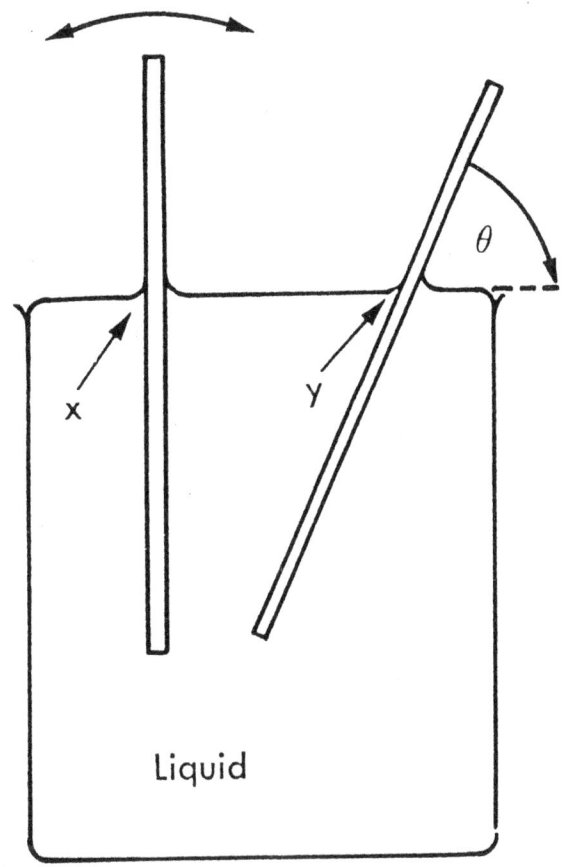

The contact angle can be measured by direct observation with the use of magnification of the liquid drop on a flat surface of the solid material, or by inclination of a slide of the solid in the liquid until the meniscus flattens out. The latter method is shown in Fig. C.3.

The wetting of a solid by a liquid is exceedingly important for several applications, including soldering, welding, adhesion and gluing, painting, and dyeing. The presence of fluxes to remove oxide layers in soldering, and the need of special wetting agents in epoxy glues, adhesives, and paints are related to the contact angle and surface tension.

Measurement of Surface Tension

The surface tension of a liquid can be measured in a variety of ways, including capillary tube rise, du Noüy ring method, bubble pressure, and drop weight. These methods vary greatly in their applications.

Capillary Tube Rise Methods

The simplest and most common method of determining the surface tension of a liquid is the *capillary tube rise method*. If a glass capillary tube is immersed in a liquid such as water, the liquid in the capillary tube will rise above the outside level of the liquid. This is due to the greater liquid–solid force than the liquid-liquid intermolecular forces, and the liquid tends to wet as much of the solid

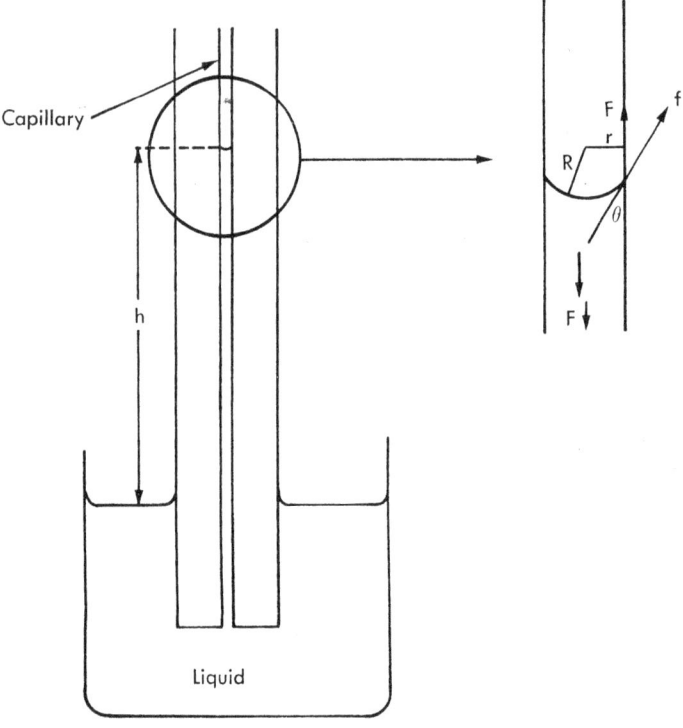

Fig. C.4 Measurement of surface tension by capillary rise. The liquid rises in the capillary until equilibrium is reached where the gravitational force is balanced by the upward surface tension force. The radius of the meniscus R is assumed to be approximately equal to the radius of the capillary, r

as possible until the gravitational pull of the column of liquid is equal to the surface tension force (see Fig. C.4).

If the radius of the capillary tube is r and the density of the liquid is d, then the force down, $F\downarrow$, which is due to the column of liquid, is defined as follows:

$$F \downarrow = \pi r^2 h d g \tag{C.1}$$

The force f along the contact angle θ is equal to the surface tension x the length of the liquid–solid contact, f; i.e., $f = \gamma l$. Since f is equal to $2\pi r$, then f is defined as follows:

$$f = 2\pi r y \tag{C.2}$$

However, the vertical force $F\uparrow$ is defined as follows:

$$F \uparrow = f cos\theta \tag{C.3}$$

Therefore,

$$F \uparrow = 2\pi r y \cos\theta \tag{C.4}$$

When the column of liquid has reached the same equilibrium position from either a lower or a higher height. (The effect of contact angle hysteresis may make these two heights slightly different.)

$$F \downarrow -F \uparrow \tag{C.5}$$

$$\pi r^2 hdg = 2\pi r y \cos \theta \tag{C.6}$$

or

$$\gamma = \frac{rhdg}{2 \cos \theta} \tag{C.7}$$

If $9 = 0$, then $\cos 9 = 1$ and

$$\gamma = \frac{rhdg}{2} \tag{C.8}$$

If $\theta = 90°$, $\cos \theta$ is negative, and h becomes a negative value; i.e., instead of a rise in the capillary tube, there is a depression. This is observed for the mercury-in-glass system.

Ring or du Noüy Method

Just as it is possible to float objects heavier than water on the surface of water, it is possible to pull the surface upward (increase the surface area) by the application of a suitable force, and thereby calculate the surface tension. The *du Noiiy* method makes use of a clean platinum ring of radius r that is placed on the liquid surface. The liquid, which wets platinum, tends to adhere to the ring, which is slowly raised by the application of a force, which is previously calibrated, until the net force pulling the ring upward exceeds the surface tension and the ring breaks from the surface. At that point the surface tension force, F, is $2ly$, where l is the circumference of the ring, and since $F = mg$,

$$2l\gamma = mg \tag{C.9}$$

where m is the weight calibration for the system and g is the acceleration due to gravity. Since $l = 2\pi r$, then

$$\gamma = \frac{mg}{4\pi r} \tag{C.10}$$

The factor 2 appears in (C.9) because two liquid surfaces (one on the inside and one on the outside of the ring) are formed as the ring is raised.

In practice, the absolute value is in error because of the diameter of the wire, the density of the liquid, as well as other terms, and it is often calibrated by a variety of substances to minimize such errors. The most important application of this method is the determination of interfacial tension between two liquids, where other methods do not apply very readily. The ring method is shown in Fig. C.5.

Bubble Pressure Method

If a tube is immersed in a liquid, and the gas pressure in the tube is increased slowly, the liquid level in the tube will drop until the end of the tube is reached, then a further increase in pressure will create bubbles. If P is the maximum pressure measured and P_h the hydrostatic pressure hdg, then

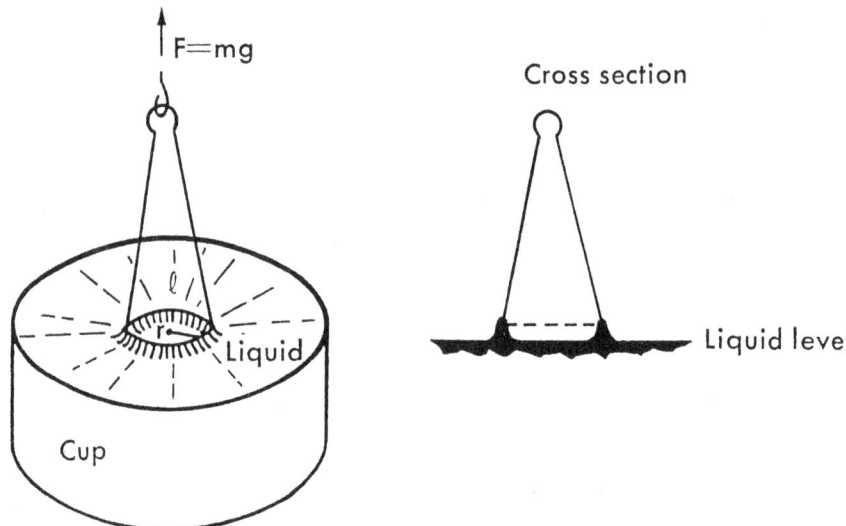

Fig. C.5 Measurement of surface tension by du Noüy method. The platinum ring of radius r is allowed to touch and be wetted by the liquid surface. It is then raised by a force F (effected by a torsion type of balance) until the ring breaks away from the liquid

$$P - P_h = P_s \qquad (C.11)$$

$P_S \Delta V$ is the work done to increase the volume of the bubbles. However, if the volume increases, the area of the bubble increases, and the energy required to increase the surface area of the bubble is $\gamma \Delta A$ (ΔA is the area increase) (see Fig. C.6). Therefore, at equilibrium,

$$\gamma \Delta A = P_s \Delta V \quad \text{or} \quad \gamma dA = P_s dV \qquad (C.12)$$

If r is the radius of curvature of the bubble, then

$$A = 4\pi r^2, V = 4\pi r^3/3, \ dA = 8\pi r dr, dV = 4\pi r^2 dr$$

By substituting these values in (C.12), we obtain

$$\gamma 8\pi r dr = P_s 4\pi r^2 dr \quad P_s = 2\gamma/r \qquad (C.13)$$

The value of r is equal to the radius of the tube when the bubble just breaks away from the tube; i.e., when the radius of the bubble is at a minimum, and the pressure is at a maximum. Since P_h — hdg, then if h, d, r, and p are measured, it is possible to calculate γ. It should be noted that γ, so evaluated, is independent of contact angle.

Drop Weight Method

When a liquid is allowed to flow slowly from a vertical capillary tube, it forms drops that grow and finally become detached from the end of the capillary tube. The size of the drops will depend on the outside radius of the capillary tube and the surface tension of the liquid. At the moment when a drop falls, the gravitational pull mg, or Vdg, is equal to the surface tension $2\pi r\gamma$, where m is the mass of the

Fig. C.6 Measurement of surface tension by the bubble pressure method. The maximum pressure is the pressure above which the bubble breaks away from the tube. This value is corrected for the hydrostatic pressure P_h and gives P —
$P_h = F_s = 2\gamma/r$

$$P-P_h = P_s = \frac{2\gamma}{r}$$

drop, V is the volume of liquid, d is the density of the liquid, r is the radius of the drop, and y is the surface tension. Thus, if we weigh and count the drops of a liquid, we can determine the surface tension from the following equation if the density is known.

$$2\pi r\gamma - mg \qquad (C.14)$$

In general, a correction is required, since not all the liquid forming the drop leaves the tip of the capillary tube and the surface tension does not act exactly vertical. Thus

$$\gamma = \frac{mg}{2\pi r\phi} \qquad (C.15)$$

where $\emptyset$ depends on the ratio $r/V^{1/3}$ and has the value of about 1.7 when $r/V^{1/3}$ is between 0.75 and 0.95.

For a given tip, it is possible, with reasonable accuracy, to determine the relative surface tension from the following relation:

$$\frac{\gamma 1}{\gamma 2} = \frac{m1}{m2} \qquad (C.16)$$

The *drop weight method* is independent of contact angle and is suitable for the determination of interfacial tension if the tip is immersed into a second liquid immiscible with the heavier dropping fluid. Thus, if the dropping liquid can be collected free from the second liquid and weighed, the liquid-liquid interfacial tension can be determined.

Surfactants reduce the interfacial tension of water either between the two liquids or between liquid and solid. The important properties include:

1. The solubility of the surfactant in at least one of the phases.
2. The surfactant is composed of both hydrophilic and hydrophobic groups on the molecule.

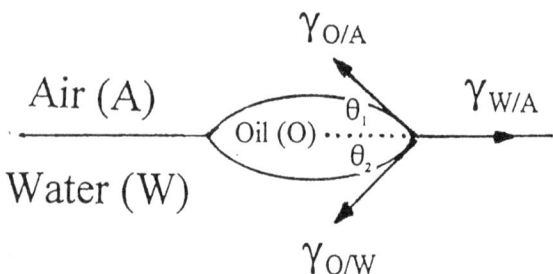

Fig. C.7 Cross section of a drop of an organic liquid such as oil (O) on the surface of water (W) at equilibrium

3. The surfactant molecules tend to orientate on the surface of the liquid.
4. The surfactant tends to concentrate at the interface.
5. The surfactant usually lowers the surface tension of the liquid.

The study of surface tension is really a branch of surface chemistry, and its development has been exceedingly rapid in the last decade. Thus, adhesion can be considered to be partially an exercise in wetting and spreading of liquids on solid surfaces. The flotation of ores is accomplished by gravity differences as well as by the adhesion of the solid particle to an air bubble, and it involves solid–liquid-gas interfaces. It is possible to reduce the vaporization of water from bodies of water with large surfaces such as reservoirs and lakes, by adding a monolayer of a substance such as hexadecanol or other surface-active agents. The action of soaps produces a decrease in surface tension on water. Many other applications in our modern environment can be readily identified.

The Spreading Coefficient

Consider a drop of an organic liquid (O) on a water surface (W) as shown in Fig. C.7. At equilibrium the forces exerted at the three interfaces (O-W, O-A, and W-A), where A represents the air phase will be given by

$$\gamma w/A = \gamma o/A \cos\theta_2 \qquad (C.17)$$

when spreading occurs θ_1 and θ_2 must approach zero (cos $\theta = 1$) and

$$\gamma w/A \geq \gamma o/A + \gamma o/w \qquad (C.18)$$

A spreading coefficient is defined as

$$S = \gamma w/A - (\gamma o/A + \gamma o/W) \qquad (C.19)$$

and when $S \geq 0$ spreading occurs. When $S < 0$ a lens will form on the water surface.

The Solid-Gas Interface

There are two types of gas-solid interaction: (a) physical adsorption, which is due to van der Waals bonds and is reversible, (b) the chemisorption, where the gas forms chemical bonds with the solid surface and results in irreversible adsorption.

If the gas A forms a monomolecular adsorption layer on the solid surface it can be treated as a dynamic process

$$A(g) \underset{k_2}{\overset{k_1}{\rightleftharpoons}} A \quad \text{(adsorbed layer)} \tag{C.20}$$

If P = pressure of gas (A) above the surface and ∂ = fraction of surface covered by A, then at equilibrium, the rate at which the molecules leave the surface R_l is equal to the rate at which they condense onto the surface R_2, i.e.,

$$R_1 = R_2$$
$$R_1 \propto \theta \quad \text{or} \quad R = k_1\theta$$
$$R_2 \propto P(1 - \theta) \quad \text{or} \quad R_2 = k_2P(1 - \theta)$$
$$k_1\theta = k_2P(1 - \theta) \tag{C.21}$$

Hence,

$$\theta = \frac{k_2P}{k_1} + k_2P = \frac{P}{a + P} \qquad \text{where} \quad a = \frac{k_1}{k_2} \tag{C.22}$$

If y = amount of adsorbed gas, ym = a monolayer of adsorbed gas (the maximum amount adsorbed), then $y/ym = \theta$. From (C.23) when P is small, $y \propto P$ and when P is large, y = constant. A plot of y versus P is shown in

$$y = \frac{ymP}{a + P} \tag{C.23}$$

and

$$P/y = a/ym + P/ym \tag{C.24}$$

Thus a plot of P/y versus P is a straight line with slope = $1/ym$ and intercept = a/ym. This is called the *Langmuir adsorption isotherm*. For multilayer adsorption, the more complicated treatment developed by Brunauer, Emmet, and Teller (BET) allows for the determination of surface areas (Fig. C.8).

If the solid is composed of narrow capillaries then, at high pressures, the gas will condense in the capillaries. This results in adsorption-desorption hysteresis.

Dubinin and his co-workers advance arguments favouring the view of the adsorption space may be expressed as a Gaussian function of the corresponding adsorption potential. For the standard vapour we have

$$W = W_o\exp(-k(\varepsilon_0/\beta)^2) \tag{C.25}$$

where W_o is the total volume of all the micropores; ε_0 – the Polanyi potential; k is a constant characterizing the pore size distribution; β is a constant related to sorption free energy.

Dubinin has plotted the results for adsorption isotherms of nitrogen, saturated hydrocarbons, benzene, and cycloxehane and has found equation for calculation of the micropore volume W_o:

$$\log x = \log(W_o \rho) - D (\log \{p_{sat}/p\})^2 \tag{C.26}$$

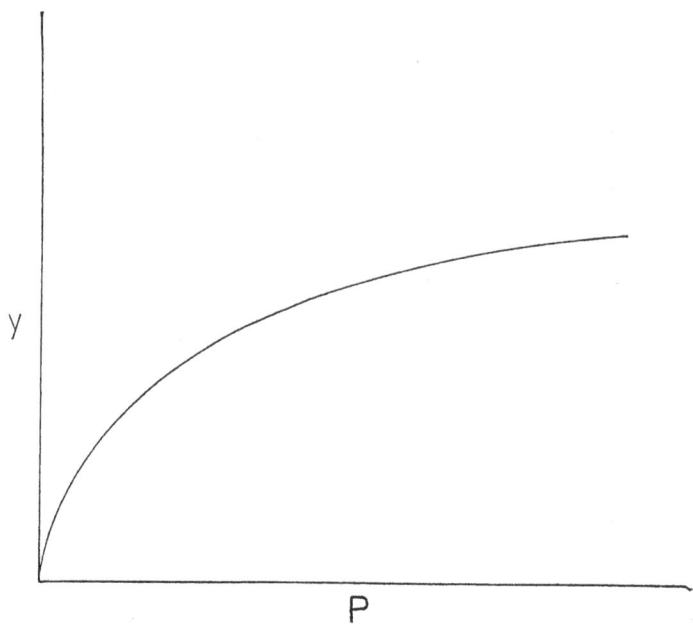

Fig. C.8 A Langmuir plot of y, the amount of gas adsorbed (per unit mass of solid) on a solid as a function of equilibrium pressure, p of the gas at constant temperature

where W_o is the total volume of all the micropores; $E = \beta \times E_o$ is the adsorption energy of another chemical; p_{sat} and p are vapor pressure at saturation and vapor pressure at equilibrium; R is universal gas constant; T – temperature, K; ε – the Polanyi potential; ρ is constant; D is a slope.

The slope D is equal:

$$D = 2.303(k/\ \beta^2)(RT)^2 \tag{C.27}$$

Thus a plot of log x against $[\log (p_{sat}/p)^2]$ should give a straight line, of slope D and with intercept log $(W_o\ \rho)$.

The Dubinin's method makes it possible to calculate the micropore volume from the low pressure part of isotherm, the region where the adsorption is still much below the plateau value. It also offers the possibility of using different adsorbates as molecular probes.

The Solid–Liquid Interface

It is possible to treat the adsorption of a liquid onto a solid surface in a manner analogous to that given for the solid-gas system. However, in the liquid state it is usually a solute which is removed by adsorption onto the solid surface. The Langmuir equation is applicable in most cases. In other cases where the surface is heterogeneous the Freundlich adsorption isotherm

$$\frac{x}{m} = kCn \tag{C.28}$$

will usually fit the data better than the Langmuir equation, where x is the mass of material adsorbed on the solid, m is the mass of solid, C is the concentration (at equilibrium) of the solute being adsorbed, n is a constant with a value usually between 0.1 and 0.5, and k is a constant which depends on the system.

The empirical constants n and k can be determined from a plot of $\log x/m$ versus $\log C$, where the resulting straight line has a slope equal to n and an intercept equal to $\log k$.

Adsorption of impurities from solution is used to purify water, decolorize sugar in solution, and many other systems.

Exercises

1. It has been suggested by Bikerman that a solid can have no surface energy and that all phenomena attributed to the surface energy of a solid is due to impurities which are adsorbed on such large surfaces. Comment on Bikerman's viewpoint.
2. The traction of an automobile tire in snow can be increased by changing the wetting properties of the surface of the tire. Explain!
3. The contact angle of mercury on glass is about $120°$ ($\cos\theta = -0.5$). What is the significance of the -0.5 for the $\cos 9$ in terms of the capillary method of determining surface tension.
4. The motion of a ship, boat, or torpedo through water is significantly influenced by the surface coating. Would friction be affected by changing a hydrophobic surface to hydrophilic and give reasons for your opinion.
5. The solubility of benzene in water changes the surface tension of water, $y_{W/A}$ from 72.8 to 62.2 dynes/cm for a saturated solution. The solubility of water in benzene has only a small effect on the $\gamma_{O/A}$ (28.8 dynes/cm when $\gamma_{O/W} = 35.0$ dynes/cm). If the value of $S = 9.0$ dynes/cm for benzene on water as an initial value—what will eventually occur as the water becomes saturated with the benzene?
6. The dissipation of oil slicks on water was initially effected by adding detergents. What does this do and why has this method ceased to be employed?
7. Methane can be encapsulated in molecular sieves 3A (zeolite) under high pressure and high temperature. The experimental data is given in Table C.A and can be shown to follow the Langmuir adsorption isotherm [(C.24)]. Plot the data in Table C.A and determine the maximum storage capacity by a plot according to (C.25).
8. Determine the parameters of the Freundlich adsorption isotherm [(C.28)] for the adsorption of acetic acid by active charcoal. The charcoal is used to absorb the acid from aqueous solutions of different concentrations. When equilibrium is reached the amount of acetic acid was determined by titration with standardized NaOH (0.1189 M). The data is given in Table C.B.

Table C.A Methane uptake in a 3A molecular sieve as a function of pressure

Pressure (Pa) $\times 10^{-8}$	$\%CH_4$ (w/w)[a,b]
0.12	1.5
0.69	3.7
0.69	3.8
1.38	4.7
2.07	5.7
2.76	5.8
3.79	6.6
4.14	6.4
4.14	6.9

[a]Based on weight of zeolite after activation under vacuum at 350 °C
[b]2 hr encapsulation period at 350 °C

Table C.B Results for the adsorption of acetic acid by active charcoal

A^a	B^b	C^c
50.0	10.0	42.23
25.0	10.0	20.3
10.0	25.0	17.8
5.00	50.0	14.3
2.50	50.0	4.70
1.00	50.0	0.22

[a]The volume of acetic acid (1.06 M) diluted to 100 mL for adsorption by 1.00 g of active charcoal
[b]The volume of solution at equilibrium taken for titration with standardized NaOH (0.1189 M) to determine the residual acetic not adsorbed by the active charcoal
[c]The volume of base used to neutralize the acetic not adsorbed by the carbon

Plot the mass of acetic acid adsorbed per gram of carbon (x/m) versus the equilibrium concentration of acetic acid.

Determine the Freundlich parameters from a plot of log x/m versus log C.

Further Readings

1. Birdi KS (1999) Surface and colloid chemistry handbook. CRC Press, Boca Raton

2. Brady PV (ed) (1995) Physics and chemistry of mineral surfaces. CRC Press, Boca Raton

3. Noguera C (1995) Physics and chemistry at oxide surfaces. Cambridge University Press, New York

4. Shaw DJ (1992) Introduction to colloids and surface chemistry. Butterworth-Heinemann, Woburn

5. Christmanmn K (1991) Introduction to Surface Physical Chemistry. Springer, New York

6. Hudson JB (1991) Surface science. Butterworth-Heinemann, Woburn

7. Adamson A (1990) Physical chemistry of surfaces, 5th edn. Wiley, New York

8. Castro GR, Cartlona M (1987) Lectures on surface science, Springer, New York

9. Aveyard R, Haydon DA (1973) An introduction to the principles of surface chemistry. Cambridge University Press, Cambridge

10. Parfett GG (1996) Surface and colloid chemistry. In: Trotman-Dickerson AF (ed) Chemical kinetics, Pergamon Press, Oxford. www.jools.com/

11. Introduction to Surface Chemistry http://www.Chem.qmw.ac.uk/surfaces/scc/

12. Dynamic surface tension http://www.firsttenangstroms.com/

13. Aspects of surface tension http://hyperphysics.phy-astr.gsu.edu/hbase/surten.html

14. DemiLab www.ilpi.com/genchem/demo/tension/

Appendix D: Patents[1]

Introduction

It is easy to obtain a patent of your invention. It is more difficult to license, sell and profit from its implementation and use. Anyone can file a patent application in any country provided it is completed in the required language. Lawyers or patent agents may be useful but they are also costly and they could be engaged at a later stage when the application is ready to be filed or appealed.

With the availability of the provisional patent (in U.S.A., Australia, New Zealand, U.K.) it is possible to have a full year of protection before a formal patent has to be filed. During the 1 year period it would be possible for the inventor to discuss the invention (preferably under a non-disclosure agreement which ensures confidentiality and does not constitute a public disclosure and therefore, does not invalidate the filing of patents in foreign countries) with potential buyers without fear of losing rights or control of the invention unless, or course, the patent is not filed. Any public disclosure or sale of the product or operation of the invention for profit could invalidate the filing of patents in some foreign countries. A typical confidential non-disclosure agreement is shown in Fig. D.1.

If a patent is not filed in a specific country then it is possible for the invention to be made and sold there without fear of infringement. The product or process cannot be marketed in a country where it is covered by a patent except by the patent holder or a licensee.

Filing Fees

The cost of obtaining a patent (see Table D.1) varies from one country to another and the judicious choice of countries in which to file depends on the returns to be expected relative to the filing fees and subsequent maintenance fees.

Foreign patents can be filed under the Patent Cooperation Treaty (PCT). The cost depends on the number of pages (30 pages @ \$455 + \$10 for each additional page) plus a designation fee (about \$1,000) and some additional fees including transmittal fee, \$240, and search fee, \$1,002, making a total of over \$2,700 US excluding the Patent Agent's fee (about \$500). More fees (\$2,500) are required within 19 months for the examination of the PCT filing. Costs continue to mount as the countries are selected for specific attention, especially if translations are required.

Patent rules and regulations change and it is important to verify costs and time limitations as well as other items given in this description of patents.

O.V. Roussak and H.D. Gesser, *Applied Chemistry: A Textbook for Engineers and Technologists*,
DOI 10.1007/978-1-4614-4262-2, © Springer Science+Business Media New York 2013

FIGURE D-1

CONFIDENTIALITY AGREEMENT

You,_____ (XYZ) (of _____) the Recipient, produces, manufactures, sells or is interested in _____
_____and

John D Student (JDS) of _____ the Discloser, has successfully performed preliminary experiments or has tested or invented a new method or a new application of the Product which information will be disclosed on the following terms and conditions:-

1. The Discloser (JDS) maintains his rights to patent the application (or has applied for a patent) or has Know-How* in relation to the Product which will be disclosed.

2.1. The Recipient shall limit dissemination of Confidential Information within its Organization to those of its employees who need to receive it for the purposes specified and shall ensure that such employees are made aware of the Recipient's obligations hereunder and are bound to uphold them.

2.2 A detailed description of any tests which are performed to assess the application or process and the results obtained shall be air mailed, couriered or faxed to JDS as they are obtained.

3. Recipient's obligations hereunder shall not apply or shall cease to apply to any information which:-

   (a) Recipient can demonstrate by written records was known to it prior to disclosure hereunder otherwise than as a result of a previous confidential disclosure by the Discloser

   (b) is in the public domain or come into the public domain through no fault of Recipient;
   *(except in the case of Know-How which may already be in the public domain)

   (c) is disclosed to Recipient with restriction on disclosure by a third party under no obligation of confidentiality to the Discloser with respect thereto.

4. No right or license is granted hereby under an intellectual property to which a party is entitled or to use any Confidential Information except as specified herein.

5. The Effective Date of this Agreement is the date by which it has been signed by both parties as specified below.

6. All obligations of the parties under this Agreement expire after a period of seven (7) years following the Effective Date.

7. This Agreement is subject to the Law of the (country, state or province)

Agreed for John D. Student (JDS)   Agreed for   (XYZ)

_____   _____

Date_____

*The agreement with respect to Know-How may be more complicated because the subject matter may be buried in the literature and is therefore in the public domain.

Fig. D.1 Confidentiality agreement

The invention usually starts with an idea that the inventor wishes to protect and which he or she often would like to discuss on a confidential basis with colleagues, friends and even potential buyers without fear of being robbed of the idea.

A search of the literature and patents would be necessary to determine if the idea is new and not already patented or published in the open literature. This can be done by a review of patent abstracts or by a computer search of appropriate data banks. Patent Agents can do this at a cost of from $100 to $500.

Having established the feasibility of the idea the inventor must now decide if a patent will be filed and where. For $125 (small entity) a provisional patent (PP) can be filed in the U.S.A. though other countries may be preferred if the formal patent will be filed there. This provisional patent can be in the form of a publication to be submitted, Za report or a preliminary draft of the patent. There is no requirement to include claims, though it is important to present the supporting data and results as well as the object of the invention and its novelty. The improvement of the invention over previous versions should be stressed. The life of a PP is 1 year.

After filing the provisional patent the inventor should be trying to license, sell or otherwise exploit the invention on a confidential non-disclosure agreement with the view that an interested company would within the 1 year period of grace, file the necessary patents in various countries on behalf of the inventor and, or course, pay all the application fees and maintenance fees. If this does not transpire, then it is up to the inventor to pay the application fees which would be classed as a small entity status and is therefore usually half of the regular filing fee.

A license is a means by which an owner (the licensor) grants to the user (the licensee) a license (which can be exclusive, non-exclusive or limited by time or district) under the patent to use the product or technology in exchange for annual (or semiannual or quarterly) royalties plus or including a minimum fee. Minimum royalty fees guarantee the inventor a return on the license and some diligence on the part of the licensee to market the invention. The royalty is usually based on a percentage of gross sales (e.g., 5%) or profits. Profits, however, can be manipulated and reduced by paying high consulting fees to sister companies thereby lowering or eliminating royalties. Other topics covered by the license include the duration, notice of cancellation, improvements, infringement suits and settlement of disputes by arbitration, etc.

Components of a Patent

A patent consists of the following main sections:

1. Title of Invention with the Names of the Inventors
2. Background of the Invention

   (a) Field of the Invention
   (b) Description of Prior (or Related) Art

3. Summary of the Invention
4. Brief Description of the Drawings
5. Description of the Preferred Embodiment
6. Claims.
7. Abstract

Though this topic will be discussed in sequence they usually appear in another order.

Title of Invention with the Names of the Inventors

A snappy title is worth some thought and can tell the reader exactly what the invention is all about. The home address(es) of the inventor(s) must be supplied.

Background of the Invention

The contents appear under two headings.

(a) *Field of the Invention*

The field may be broad as well as narrow and may differ from what the patent office may select. Other patents of the subject can be of great help and are essential for this, and the general jargon to be used.

Table D.1 Small entity fee schedule[a] for U.S.A. and Canada. (2011)

	U.S.A.	Canada
Basic filing fee	$380/190	$400/200
Independent claims in excess of 3, each	$250/125	
Claims in excess of 20, each	$60/30	
Utility examination fee	$250/125	$800/400
Patent post-allowance fees	$1,740/870	$200
Provisional application filing fee[a]	$250/125	
Electronic filing[a]	$105	

[a]Complete fee schedules are available from the respective Patent Offices and from the Web
http://www.uspto.gov/web/offices/ac/qs/ope/fee092611.htm#comp (U.S.A.)
http://www.cipo.ic.gc.ca/eic/site/cipointernet-internetopic.nsf/eng/wr01103.html (Canada)

(b) *Description of Prior (or Related) Art.*

This outlines the need for the invention and the problems to be solved. This means that it is important to describe the technology as it is at present and how the invention can change and improve the world. This is analogous to the review of the literature with references and examples wherever possible.

Summary of the Invention

This section now describes how the invention solves the problems mentioned in and how the invention improves on previously available units. It explains how the state of the art will be advanced and how it will be of benefit to mankind. The Summary is not a rewording of the Abstract but an anticipation of the Claims of the patent. This section is often referred to as Summary of Disclosure.

Brief Description of the Drawings

This section lists the captions or legends to the figures used to illustrate the various aspects of the invention. The components of the drawings are usually clearly numbered and referred to by number in the preferred embodiment, the next section.

If the drawings are inadequate then the examiner might ask for additional figures or clearer versions for some of the drawings. The patent office will supply detailed requirements and specifications for the drawings for a fee, if asked.

Description of the Preferred Embodiment

This is the heart of the patent where the inventor must now give all the details and basis of the invention. In principle, the details should be sufficient for anyone "skilled in the art" to duplicate the invention. This is not always adhered to and occasionally an important and essential step is omitted or a further improvement is kept secret to be included in another patent. This description makes use of the drawings and figures with examples, which represent tests and experiments, with tables of results where possible.

The section is usually ended with a general statement which is meant to indicate that, to one skilled in the art, other obvious applications and uses need not be described, e.g. "Since various modifications can be made in my invention as herein above described, and many apparently widely different embodiments of same made within the spirit and scope, it is intended that all matter contained in the accompanying specifications shall be interpreted as illustrative only and not in a limiting sense."

Claims

This section starts with: "What I claim as my invention is:"

1. The claims, which are numbered, are the main goal of the patent and the only parts which can be changed once the patent has been filed. The claims are classified as independent claims (limit of 3 – more cost extra) which stand alone, and dependent claims which refer to a previous claim. Extra claims (over 20 in all) add to the cost of the filing fee. Each country has different requirements concerning dependent and independent claims. If one or more of your claims are allowed by the examiner then the patent will be granted.

The first claim should cover the complete invention being broad and encompassing. Two or three independent claims are usually sufficient if supplemented by several dependent claims. These allow for variation in one or more of the parameters or components of the invention to be claimed.

Abstract

This is best written after the application is completed. It is a short paragraph which describes the invention, its purpose, operation and use. This abstract appears on the front page of the patent and in most collections and lists of patents and abstracts describing new issuances. It is this abstract which attracts potential buyers searching for new products and ideas to market and sell. Hence considerable care ought to be taken in its preparation.

The Provisional Patent

The best approach to writing a patent on your own is to obtain some previously issued patents on the subject and to acquire the jargon and style of the subject. Such patents can be ordered by number from the patent offices or can be downloaded from the WEB.

The response from the Patent Examiner is usually to deny the validity of all the claims based on previous patents which are included with his reply. The inventor would reply by changing the nature of the claims or point out to the Examiner that he/she is mistaken. A telephone call can do much to clarify the changes required to make the claims acceptable.

When there is no possibility of obtaining at least one claim, the patent will be finally denied. At this point it may be possible for the inventor to file a continuation-in-part in which new evidence and results are presented in the revised Embodiment to justify the new and altered claims. A patent agent at this time would be a great help in getting the patent approved.

Further Readings

1. Waller FJ (2011) Writing chemistry patents and intellectual property. A practical guide, Wiley, Hoboken
2. Kirk RE, Othmer DF (2006) The Kirk-Othmer encyclopedia of chemical technology, 5th edn, vol 18
3. Bryant JL (1999) Protecting your ideas: the inventor's guide to patents, Academic, San Diego
4. Pressman D (1986) Patent it yourself. Nolo Press, Berkeley
5. Flanagan JR (1983) How to prepare patent applications. Patent Educational Publishing, Troy

6. Richardson RO (1981) How to get your own patent. Sterling Publishing Company, New York

7. Hale AM (1981) Patenting manual, 2nd edn. S.P.I. Inc, New York

8. Buckles RA (1957) Ideas, inventions and patents. Wiley, New York

9. http://wikipedia.com

10. http://www.uspto.gov/web/offices/ac/qs/ope/fee092611.htm#comp (USA)

11. http://patent.womplex.ibm.com/ (USA)

12. http://www.uspto.gov/patft/index.html (USA)

13. http://www.cipo.ic.gc.ca/eic/site/cipointernet-internetopic.nsf/eng/wr00142.html (Canada)

14. http://www.lib.uwaterloo.ca/patents.html (Canada)

15. http://patents1.ic.gc.ca/intro-e.html (Canada)

Appendix E: Experiments

Experiment No. 1. The Vapor Pressure of a Substance

Introduction

The vapor pressure of a substance at a given temperature is a unique property of that substance and is a measure of its volatility. The standard boiling point of a substance is the temperature at which its vapor pressure is 760 mm of Hg at one atmosphere pressure. A substance would boil at a lower temperature when measured at a mountain top compared to measurements at the earth's surface or in underground mines where it would "boil" at a higher temperature. The object of this experiment is to measure the vapor pressure of a substance at several temperatures and to obtain its Standard Heat of Vaporization

One method used to determine the vapor pressure of a substance (a liquid in this case) is to measure the volume of its vapor. In a closed system, at different temperatures as the temperature of the substance is heated. The small sample of the substance is placed at the closed bottom of a tube. A short plug of mercury about 3–5 mm long is added to the tube (approximately 5 cm above the surface of the liquid sample) using an eye-dropper that is made by pulling apart two parts of a glass tube that has been soften in the flame of a Bunsen burner This is done by inclining the tube at 20–35° before inserting the eye-dropper with the mercury. The sample tube, a thermometer and a centimeter scale are attached as shown in Fig. exp. 1.1 and clamped to insert into a beaker of water that is slowly heated and stirred. As the water is heated the air in the sample tube expands and the liquid vaporized to a increasing extent pushing the mercury plug towards the open end. It can be shown that the measured movement of the mercury plug from a low temperature (where the vapor pressure of the sample liquid is negligible) to an elevated temperature can be related to the vapor pressure of the liquid.

Care with Mercury

Mercury vapor is toxic and it is important to avoid breathing it vapors. [J. Chem. Edu. Vol 12, page A529 (1965)] The equilibrium vapor pressure of mercury is 0.000185 mm, at 0°C, 0.00120 mm pressure of Hg at 20°C and 0.0127 mm at 50°C The vapors of mercury should therefore not be inhaled. Mercury amalgamates with gold and silver and finger rings should be removed during the laboratory period. Mercury is also absorbed through the skin and it should, therefore, not be handled unnecessarily. Used mercury must be recovered and a receptacle is provided for it.

O.V. Roussak and H.D. Gesser, *Applied Chemistry: A Textbook for Engineers and Technologists*, DOI 10.1007/978-1-4614-4262-2, © Springer Science+Business Media New York 2013

Fig. exp. 1.1 Mounted
sample tube

DO NOT HANDLE MERCURY
DO NOT THROW MERCURY DOWN THE SINK
DO NOT STIR WATER BATH WITH THERMOMETER
DO NOT HEAT BEAKER DIRECTLY WITH BURNER FLAME
DO NOT HEAT PLASTIC RULER ABOVE 56°c
DO NOT USE COMPRESED AIR FOR DRYING
REMOVE RINGS COMPOSED OF GOLD OR SILVER BECAUSE THESE METALS READILY
    AMALGAMATE WITH MERCURY

A porous plug is added to the top of the sample tube to prevent excessive return of mercury vapor into the laboratory.

A. Theory

The change in vapor pressure P of a liquid or solid as a function of temperature T is derived from thermodynamics and is given by the differential equation

$$\frac{dP}{dT} = \frac{\Delta H}{T(V_g - V_o)}$$

where V_g is the volume of 1 mole of vapor, V_o is the volume of 1 mole of either liquid or solid, and ΔH is the heat of vaporization or sublimation per mole of substance. If the volume of liquid is small in relation to the gaseous volume; if the vapor behaves as an ideal gas and if ΔH can be treated as a constant (independent of temperature) then $\dfrac{dP}{dT} = \dfrac{\Delta H}{TV_g} = \dfrac{\Delta HP}{nRT^2}$ but since $n = 1$ for one mole

$$\frac{dP}{P} = \frac{\Delta H}{R}\frac{dT}{T^2}$$

$$\text{Ln } P_2/P_1 = -\Delta\frac{H}{R}\ (1/T_2 - 1/T_1) = \Delta H(1/T_1 - 1/T_2)/R$$

$$\text{Log } P_2/P_1 = \Delta H(1/T_1 - 1/T_2)/(2.303\ R)$$

Hence if Log P is plotted as the ordinate against $1/T$ as the abscissa, then a straight line of negative slope $-\Delta H/2.303R$ will result. The Heat of Vaporization, ΔH, is given by $\Delta H = -2.303\ R \times$ Slope of line.

Experiment

The measurement of vapor pressure generally consists of an apparatus in which all air and foreign gases is evacuated from the sample. The compound is then kept at a constant temperature while equilibrium is obtained; the pressure in the apparatus is then measured. The simpler method to be used in this experiment assumes:

That the vapor and air behave as ideal gases.
The glass tubing has a uniform bore.
The thermal expansion coefficient of the glass and the plastic ruler are small and equal.

If a substance is placed in the bottom of a capillary it will exert its equilibrium vapor pressure when immersed in a constant temperature bath. When a droplet of mercury is then placed in the capillary tube, it isolates the substance, its vapor and some air. Hence as the temperature is increased, the air expands and the partial pressure of the vapor increases but the total pressure remains constant since the mercury plug moves to accommodate any pressure change.

Let P = the barometric pressure exerted on the system
E = the pressure exerted by the mercury plug
P_T = the total pressure exerted on the confined vapors

$$P_T = P + E \tag{E1.1}$$

If P_a = the partial pressure of air in volume V
and P_v = the vapor pressure of the liquid sample

$$\text{then }\ P_a + P_v = P_T \tag{E1.2}$$

$$\text{at }\ T_1, P_{a1} + P_{v1} = P_T \text{ or } P_T - P_{v1} = P_{a1} \tag{E1.3}$$

$$\text{at }\ T_2, P_{a2} + P_{v2} = P_T \text{ or } P_T - P_{v2} = P_{a2} \tag{E1.4}$$

V is the volume of vapor and is equal to $\pi r^2 L$ where r is the radius of the tube and L is the length between the sample and plug

$$\text{Then } P_{a1} = nRT_1/V_1 = nRT_1/V_1 = nRT_1/\pi r^2 L_1 \tag{E1.5}$$

$$\text{Similarly } P_{a2} = nRT_2/V_2 = nRT_2/V_2 = nRT_2/\pi r^2 L_2 \tag{E1.6}$$

Dividing (E1.4) by (E1.3) and combining (E1.5) and (E1.6)

$$(P_T - P_{V2})/(P_T - P_{V1}) = P_{a2}/P_{a1} = (nRT_2)/(\pi r^2 L_2)(nRT_1)/(\pi r^2 L_1) = (T_2/L_1)/(T_1 L_2) = \alpha \tag{E1.7}$$

$$(P_T - P_{V2}) = \alpha(P_T - P_{v1}) \text{ or } P_{v2} = P_T - \alpha(P_T - P_{v1}) \tag{E1.8}$$

If T_1 is sufficiently low then P_{v1} can be small and may be neglected, and

$$P_T - P_1 = P_T \tag{E1.9}$$

Hence

$$P_{v2} = P_T(1 - \alpha) \tag{E1.10}$$

Thus the vapor pressure can be calculated from a knowledge of the atmospheric pressure and the liquid-mercury distances at various temperatures. The approximation made in (E1.9) will be invalid when $P_{v1} \geq 50$ mm. However at higher temperatures the error introduced becomes less significant and it is possible by successive approximations to obtain very good results. See Question No. E6.

Apparatus

1. Three capillary tubes 2–3 mm I.D. 15 cm long
2. Plastic 6 in. ruler
3. Beaker (1,000) mL.
4. Bunsen burner, stand, wire gauze
5. Thermometer – 10° 110°C
6. Stirring rod
7. 2 Eyedroppers
8. Wall Barometer
9. Cork, rubber band, stopper, clamp and stand

NOTE. Do not heat the plastic ruler above 55°C; it will melt or deform.
   Do not use compressed air for drying since it generally contains some oil.

Part A

1. The sample tubing is of soft glass: seal one end and allow it to cool.

2. The ruler should be cut down the middle to separate the centimeter scale from the inch scale. The centimeter scale can then be conveniently tied to the thermometer and capillary tube by rubber bands, as shown in Fig. exp. 1.1.

3. Prepare two long thin capillary eyedroppers that can enter the above sample tubes. They can be prepared from a 6″ length of 6 mm OD soft glass tubing in the following manner. (a) Firepolish both ends and allow them to cool. (b) Place the center of the glass tube in the Bunsen flame while holding the ends of the tube with your finger tips and rotating it. (c) When the tube has softened in the center and becomes difficult to hold and rotate, remove it from the flame and gently pull it apart and cut in two. If the eyedropper can enter the sample tube, insert the fire polished ends into the rubber bulbs and store the finished eyedropper safely.

4. Put a cork or rubber stopper around the top end of the thermometer to facilitate clamping it in the apparatus.

5. Make a stirring rod from a soft glass rod.

Part B

1. By means of the eyedropper, deposit a 0.5–1 cm length of the liquid sample at the bottom of the sample tube. Do not wet the inside capillary wall above the liquid level: a wet wall will cause the mercury (introduced in the next step) to drop into the liquid.

2. With the second eyedropper, plug the sample tube with a thread of liquid mercury about 5 mm long, leaving a gap of about 3–4 cm between the upper level of the liquid sample and the mercury plug. It will be necessary to do this with the sample tube tilted 10–30° from the horizontal.

3. Bind the sample tube and thermometer to the centimeter scale and thermometer and set up the apparatus as shown in Fig. exp. 1.2. Be sure that the mercury is at least one inch below the level of the top of the water bath.
   NOTE: If the mercury thread should break or fall into the liquid sample, it will mean starting over again with a new capillary tube (Part B. Step 1). If continued difficulties are encountered, consult your instructor about having the sample tubes treated with Drifilm to prevent the glass from wetting. (Drifilm is sold in supermarkets to waterproof shoes and boots). Place all used tubing in the receptacle provided since most can be cleaned and reused.

4. Measure the length of the mercury thread, E, and take the barometer reading, recording the data.

5. Record both the temperature of the water and the distance from the top of the liquid sample to the bottom of the mercury plug at about 10 intervals from 0°C (T_1) to about 50°C (not beyond 55°C, because he plastic ruler may melt).
   NOTE: The value of L and T for the lowest temperature. T_1 should be obtained with care since these are used as L_1 and T_1 in each of the successive calculations of vapor pressure.

6. The temperature readings can be estimated to 0.1°C and the distance can be read to about 0.02 cm. Allow equilibrium to be established before noting the readings at each temperature.

Results

Tabulate the results and calculate the vapor pressure for the propylalcohol provided. Plot the results (log Vp vs. 1/T) and compare the value of the heat of vaporization with the literature values given at the end of this experiment. Extrapolating from the straight line, obtain a value for the boiling point of the liquid.

   Note: The plot can be done on semi-log graph paper that consists of a logarithm ordinate scale and a linear abscissa scale to represent the reciprocal of the absolute temperature. Plotting vapor pressure directly on the ordinate scale result in the logarithm of the vapor pressure being represented.

Fig. exp. 1.2 Apparatus for
the pressure – temperature
determination have been
scanned together on one sheet

Cork or rubber
stopper

Stirrer

Figure-2 assembly

Ring
stand

Discuss the results obtained with respect to the literature values listed in Table exp. 1.1 and comment on the sources of error. Then complete the assigned Exercises at the end of the section.

Questions

1. The vapor pressure of mercury at 20°C is 0.0012 mm of Hg. Calculate the equilibrium concentration of mercury in air in units of g/m^2.
   Note that the threshold toxic limit of mercury is 0.1 $mg.m^3$.
2. Show how the vapor pressure of a solution can be determined at a given temperature using the capillary-tube method.
3. It is possible to determine the vapor pressure of a liquid at temperatures above its boiling point. This can be done conveniently by sealing the top-end of the capillary tube and measuring the distance between the mercury thread and the sample and between the mercury thread and the top seal. If l is the distance between the sample and the mercury and λ is the distance between the mercury thread and the top seal, Δ is the change from l_1 to l_2 and λ_1 to λ_2 for the temperature interval T_1 to T_2, that is, $l_2 = l_1 + \Delta$, and $\lambda_2 = \lambda_1 - \Delta$. P_a is the partial pressure of air in the

Table exp. 1.1 Vapor pressure

| TemperatureC | | | | | | | | |
Pressure	1 mm	10 mm	40 mm	100 mm	400 mm	760 mm	MP°C	H_{vap} kJ/mol
Substance								
Water	17.3	11.3	34.1	51.6	83.0	100	0	40.66
Methanol	−44.0	−16.2	5.0	21.2	49.9	64.7	−97.8	35.2
Ethanol	−31.3	−2.3	19.0	34.9	63.5	78.4	−112	35.2
n-propyl alcohol	−15.0	14.7	36.4	52.8	82.0	97.8	−127	41.4
Isopropyl alcohol	−26.1	2.4	23.8	39.5	67.8	82.5	−85.8	39.9
Acetone	−59.4	−31.1	−9.4	7.7	39.5	56.5	−94.6	29.1

sample segment of the capillary tube, and P_v is the vapor pressure of the sample at temperature T. P_l is the initial pressure in the sealed upper part of the capillary tube at T_l where it is assumed that

$P_{v1} = 0$. Show that

$$P_{v2} = 7(P_l T_2)/T_1 ?7(\lambda_1/ \lambda_2) - (l_1/l_2)?\square =; P_l T_2/T_1 ?\square 7 \lambda_1/(\lambda_1 - \Delta)? - 7l_1/(l_1 + \Delta)?\square$$

Hint: $P_l + E = P_a + P_v$ and assume that $E = 0$, and $P_{v1} = 0$ when $T_1 = 0°C$

4. The vapor pressure of Hg has been neglected in these calculations. Discuss the significance of this assumption in the open-tube method and in the sealed-tube method at higher temperatures.
5. If the vapor pressure of a substance in the solid state and the liquid state is known at various temperatures, it is possible to calculate the heat of fusion.
   (A) explain this statement and (B) calculate the heat of fusion of benzene from the data given in Table exp. 1.1.
6. The vapor pressure of acetone was determined in an open tube and the data in the following table were obtained. Atmospheric pressure was 749.1 mm mercury, the length of the mercury thread was 6.2 mm. Determine the Heat of Vaporization of Acetone.

T (°C)	0.5	10.6	20.0	30.1	40.0	38.5	49.0	30.0
l (mm)	22.0	24.5	28.9	36.5	48.1	49.0	85.0	35.0

It must be realized that the assumption $P_{v1} = 0$ when $T_1 = 0°C$ is not valid. To obtain a more exact value of P_{v1} plot the logarithm of the vapor pressure against the reciprocale of the temperature (°K) and extrapolate the straight part of the line (at the higher temperatures) to 0°C (273 K). Using the extrapolated value for P_{v1} replot the data and calculate a heat of vaporization for acetone. A second extrapolation of the plot will improve the accuracy of the results.

References

1. Shen CT, Herrmamnn RA (1960) Anal Chem 32:418
2. Borrell P, Nyburg SC (1965) J Chem Ed 42:551
3. Gesser HD et al (1967) J Chem Ed 44:387
4. Shurvelll HF (1970) Can Chem Ed 5:12
5. Gesser HD et al (1971) Lab studies in Gen Chem. No 1092, W.H.Freeman & Co

Experiment No. 2. The Speed of a Boat in Water

Object: To determine how the surface properties of a boat affects its speed in water.

Introduction

Water is the "best" hydrophilic substance and it has often been asked whether an object moving in water should be hydrophilic or hydrophobic for reduced friction. Both types of coatings are available on the market but based on the hydrophilic nature of fish scales, it is most likely that a boat with a hydrophilic surface may move faster or consume less fuel than a boat with a hydrophobic surface, This can be tested by dropping a "torpedo" (with successive characteristics) in a water column and timing the fall between two points along the column.

The Experiment

A diagram of an apparatus is shown in Fig. exp. 2.1 for the measurement of the effect of a coating on a "torpedo" in a water column and timing its fall between two points. The effect of speed on the difference between the two types of surfaces properties of the "torpedo" helps us understand the difference The hollow "torpedo" can be loaded with lead shot to increase its speed as it falls through the column of water.

Procedure

1. Set up the apparatus as shown in Fig. exp. 2.1 and prepare two clean glass test-tube type "torpedoes" with different weights by adding different amounts of BB shot to the "torpedoes".

Fig. exp. 2.1 The glass tube (1), filled with water (2), with bob (3), sets off photo switch (4), that start the 4 digit timer (5), in electronic box (6). The photo switch (7) stops the timer. The polyester thread (8) is used to retrieve the bob bringing it to the start point (9) for the next test while the timer is reset. A and B represent the housing for the photo switches (4 and 7)

(The timed readings for the duration of the fall through the water column should be between about 5 and 15 s. and have four significant figures. The contact angle of a clean glass slide should not be more than 10 degrees).

2. Attach an appropriate length of fish twine to retrieve the "torpedo" from the bottom of the flow tube. Reset the timing unit after each run when the "torpedo" has been raised above the start line of the water column.

3. Repeat the drop between 5 and 10 times to obtain an average drop time and determine the mean deviation for the "torpedo" with the hydrophilic wetting (W) surface.

4. Repeat the drop-speed times as in step #3 and again calculate the average of the 5 or 10 drop times and mean deviation for the wetting "torpedo".

5. Repeat the drop-speed times as in step #3 and again calculate the average of the 10 or 5 drop times and mean deviation for the wetting "torpedo".

6. Report the results obtained in a Table listing the type of surface treatment and time of fall through the water and suggest further applications and experiments to verify your conclusions.

7. Set up a table for the results and calculate the significance of the wetting process on the speed of the "torpedos" through water and potential usefulness of hydrophilizing boats to save fuel and/or time in marine traffic.

8. If time permits repeat the above with another "torpedo" having a different mass to give a different fall speed.

Provide answers to the questions below.

Questions

1. Suggest how you can improve upon the results obtained, e.g. decrease the errors involved.
2. Suggest some other application and advantages of a hydrophilic surface.
3. Barnacles and other marine life interfere with a submerged surface in the sea. Suggest systems or methods that can be used on boats to keep surfaces clean and hydrophilic.
4. The human body has several hydrophilic surfaces. What are these surfaces and show how a water insoluble hydrophilic coating can be of benefit to us.
5. Examine the literature (from the WEB and US Patents) to determine what the speed enhancement systems have been reported for a boat moving in water.

Experiment No. 3. The Permeation of Lquids Through Solid Natural Beds

Introduction

The permeation of liquids through sand, gravel or soil is governed by the pressure head and wetting of the solid by the liquid (surface tension). It is the object of this experiment to show this effect and how it might be possible to circumvent the effect.

Procedure

The ground glass or glass beads (d = 0.5–5 mm) are cleaned by boiling in acid and rinsing in distilled water before being placed in a glass column fitted with a stopcock. A drawing of the apparatus is shown in Fig. exp. 3.1. To avoid air spaces in the glass column it is best to add the glass beads to the

Fig. exp. 3.1 Diagram of the apparatus used to measure the rate of flow of water (or other fluids) from a full reservoir (XY), through a column of packed solids (ZS), (glass beads) used to determine the effect of the solid surface properties (contact angle on the rage of flow. YZ is a variable length that is used to change the flow rate of water through the packed bed

column filled with water. The glass taper joint on the top of the glass beads are used to add a straight column that provides the pressure head that pushes the water through the column of beads. Columns of different lengths can be used to provide different speeds through the column and will show the effect of speed on the influence of the surface. A stopwatch is used to time the flow of water through the column to the top of the glass bead taper joint. The results of 10 experiments are averaged and the mean deviation is determined. The glass beads are then dried and treated with silanes to make them non-wetting (hydrophobic). The beads when dry, are then placed into the flow column and the flow experiments are repeated and the results are compared with the flow of water through the beads with a hydrophilic surface. Using statistical formulae determine the significance of the differences between the two averages and the surface effect on the speed of water through the packed beds of different surface wetting caractor.

It is also feasible to perform the experiment with oil as the liquid and to start with the oil and change the surface of the glass beads between the hydrophilic and hydrophobic states. The oil will show improved flow with hydrophobic beads.

NOTE: To avoid air being trapped in the packing it may be desirabale to fill the solid bed from the bottom up by attaching a rubber tube to the exit spout of the column and slowly raising the reservoir. Alternitly it is also best to add the beads slowly to the water filled column to avoid air pockets.

Questions

1. Why would you expect vapor holes to appear in the glass bead bed?
2. Under what conditions would the surface conditions become irrelevant?
3. What factors would you use to select a wall material for a flow system through which all types of liquids flow?
4. Explain why the wall effect disappears when the diameter of the flow tube is 5 cm or greater.
5. What wall material would you use to improve the flow of oil?

Experiment No. 4. The Molecular Weight of a Polymer by Viscosity Measurements

The viscosity of a gas is well understood. However the viscosity in the liquid state is more complex and readily influenced by many factors. Some of these factors will be examined in other laboratory experiments. At this time we restrict ourselves to simple well understood systems. The viscosity of a polymeric solution as discussed in Chap. 11.

The viscosity of a polymer solution is much higher than that of a simple solute and it increases as the molecular weights (M_n and M_w) a polymer increases. Figure 11.1 shows the relative position of these two quantities in terms of the changing average molecular weights of a typical polymer and its Degree of Polymerization (DP) Fig. 11.2. The average molecular weight of a polymer, M_v, as determined by viscosity measurements is given by

$$[\acute{\eta}] = KM_v \, P - 1$$

where $[\acute{\eta}]$ is the intrinsic viscosity; K and α are empirical constants.
$[\acute{\eta}]$ is determined by the limit $[\acute{\eta}] = \text{Limit } n_{sp}/c$

$$c \to 0$$

Procedure

(1) Prepare a solution of the polymer in the solvent. 500 mg in 100 mL (0.5%) and store in a glass bottle with a well fitting ground glass stopper. If necessary filter the solution through a course grade sintered glass filter. Prepare the water bath for the viscometer which must be supported in the bath and which must be visually observable in order to time the flow of the solution through the capillary part of the viscometer. Figure B.2 of Appendix B shows up how to set up an Ostwald viscometer.

Complete the following table by dilution of the prepared solution:

One method is to divide the 100 mL into 80 and 20 mL and dilute these each to 100 mL giving solutions **B** and **E** with 0.4 and 0.1 mg/100 mL respectively. The 0.3 mg/100 mL solution (**C**) can be made by taking 75 mL from the 80 mg/100 mL solution (**B**) and diluting it to 100 mL. Similarly, a volume of 66.7 mL of solution (**C**) diluted to 100 mL would give solution **D**. Viscosity measurements should be performed in a minimum of triplicates and averaged with mean deviations determined and plotted to indicate the possible error of each value.

Polymer Soln.	Flow time	Relative viscosity	Specific viscosity	Reduced viscosity
mg/100 mL	sec.	$\acute{\eta}_r = \acute{\eta}_s/\acute{\eta}_o = t_s/t_o$	$\acute{\eta}_{sp} = \acute{\eta}_r - 1$	$\acute{\eta}_{red} = \acute{\eta}_{sp}/C$
Pure solvent	$t_o =$			
: A 0.5	$t_5 =$			

(continued)

Table X Intrinsic viscosity–Molecular weight constants (K and a for equation P-1)

Polymer	Solvent	T (°C)	$K \times 10^4$	a
Polystyrene	Benzene	25	1.02	0.74
	Toluene	25	1.10	0.72
Polyvinylacetate	Acetone	25	1.88	0.69
PolymethylMethacrylate	Acetone	25	0.75	0.7
	Toluene	25	0.75	0.71
		30	0.7	0.72

Polymer Soln.		Flow time	Relative viscosity	Specific viscosity	Reduced viscosity
B	0.4	$t_4 =$			
C	0.3	$t_3 =$			
D	0.2	$t_2 =$			
E	0.1	$t_1 =$			

Plot the results of η_{red} against C (g/100 mL) and extrapolate the line to zero value of C. The ordinate value at C = 0 gives the value of Limiting Viscosity Number or Intrinsic Viscosity (η) = KM^a where estimated values of K and $\underline{a}$ for hydroxyethylmethacrylate are 0.75×10^{-4} and 0.71 respectively. Calculate the molecular weight of the polymer. Answer the assigned questions that follow:

Questions

1. Why is viscosity of a polymer a direct function of its molecular weight?
2. What can you do to reduce the viscosity of a polymer?
3. Under what conditions is the intrinsic viscosity, [ή], independent of the molecular weight of the solute?
4. A 1.232 g of Carboxy terminated polybutadiene (CTPB) is dissolved in a toluene-ethanol mixture and reacted with 2.9 ml of 0.0965 N KOH to the neutralization to a phenolphthalein end – point. Calculate the molecular weight of the sample polymer.
5. What is meant by the term "Degree of Polymerization"?
6. Why does a polymer have different "Molecular Weights"?
7. Comment on the variability of the intrinsic viscosity parameters in the Table X.

Experiment No. 5. Surface Chemistry

Introduction

The surface of a solid can be a very active and reactive substance. We are familiar with the Gas Mask which was initially used to remove poison chemicals from the air. The gas mask consists of a solid with a large active surface area that can trap poisonous vapors. Similarly the removal of unhealthy substances in water can be effected by passing the water through an active solid which has a large surface area that can trap the unwonted chemicals.

This experiment will illustrate the removal of an acid–oxalic acid (HOOCCOOH) from water using an activated form of carbon that has a large active surface area.

Background

Carbon prepared by various methods can be converted into an activated form that has a large surface. Such active carbons can have surface areas that range from about 50 m^2/g to over 1,000 m^2/g Molecules that encounter such surfaces tend to stick to the surface due to the attractive forces of the residual atom or molecules that constitute the material. The study of this adsorption process is called surface chemistry which was developed over 100 years ago and was initiated by the Freundlich equation

$$x/m = kc^{1/n} \tag{5.1}$$

where x = mass of substance adsorbed, m = the mass of the absorbent (charcoal), c is the equilibrium concentration of the unabsorbed substance, k and n are constants. The logarithmic equation is

$$\log x/m = \log k + (1/n) \log c \tag{5.2}$$

A plot of log x/m versus log c should be a straight line with a slope of 1/n. The Langmuir equation was developed about 1920 and is based on a kinetic principle concerned with the equilibrium between the adsorption and desorption occurring continuously.

$$F = kc/(1 + kc) \tag{5.3}$$

where F = the fraction (x/m) of solid surface covered by the adsorbed molecules, k is a constant at constant temperature and c is the concentration of the substance being adsorbed. The saturated surface is represented as F = S. Thus (B.13) becomes

$$c/F = c/S + 1/kS \tag{5.4}$$

A plot of c/F versus c will be a straight line with slope equal to 1/S.

Materials

Oxalic acid solution, 0.5 N (1.0 M); 1 l, Active charcoal 25 g; KMnO$_4$ 0.2 N (1.0 M); 5–250 mL Erlenmeyer flasks with stoppers.

Weigh out 2 g of the active charcoal to each of 5 flasks suitably labeled and add 100 mL of oxalic acid solution each diluted to give solutions of 0.4, 0.3, 0.2, 0.1 N and the undiluted solution 0.5 N

The equation:

$$5C_2O_4{}^{2-} + 2\ MnO_4{}^- + 16H^+ = 10\ CO_2 + 2\ Mn^{2+} + 8H_2O \tag{5.5}$$

The oxalic acid solution is used to standardize the permanganate solution (10 mL). 2.0 g of active charcoal is added to each of the five flasks. 100 mL solutions of different strengths oxalic acid is added to the flasks. The solutions are:

1. 0.5 N stock solution-100 mL;
2. 80 mL of stock solution + 20 mL of water = 0.4 N;
3. 60 mL stock solution + 40 mL of water = 0.3 N;
4. 40 mL of stock solution + 60 mL of water = 0.2 N,
5. 20 mL of stock solution + 80 mL of water = 0.1 N.

Thus the five flasks are prepared to contain each 2.0 g of active carbon, 100 mL of oxalic acid with concentrations of 0.1–0.5 N and stoppered and shaken to achieve equilibrium in the ½–1 h. Equilibrium is established when two successive titration of the oxalic acid with the permanganate solutions are close together indicating that equilibrium has been established.. This is done by removing 10 mL of the oxalic acid solution and by titrating it with the permanganate solution. Two successive equivalent titration values would indicate that equilibrium has been established. It is important to titrate the solutions of oxalic acid separated from the carbon which had adsorbed the oxalic acid from the solution. This can be done by filtering the solutions keeping them separate for titration of 25 mL portions with the standard permanganate solution.

Calculate the weight of oxalic acid used in each of the flasks and the weight adsorbed in the sample. Calculate the weight of oxalic acid adsorbed in each sample and determine the values of x/m, from the slope of the line. If the adsorption sites are independent of temperature then from the Langmuir equation the slope of c/F versus c is a straight line with slope = 1/S which would be independent of temperature though the intercepts would vary. Calculate the surface areas per unit weight of sample and comment on the differences determined by the two equations.

Questions

1. Two other methods have been reported more recently, The BET method that accounts for multilayers of adsorbed species and the Dubinin method. Comment on these two approaches in relation to the two described in this older approach.
2. Based on the size of molecules and assuming close-packing calculate the maximum area of a surface that a single layer of a mole of gas could cover.
3. A desorption curve of an adsorbent on a surface usually does not follow the adsorption curve. How this is usually explained?
4. Describe some of the interesting aspects of Surface Chemistry.
5. The different adsorption characteristics of substances are used in Gas Chromatography to separate and identify minute quantities of volatile substances. Explain how this is accomplished.

Experiment No. 6. Colorimetric Analysis–Filter Photometry

It is possible to determine the concentration of a liquid dissolved substance by optical methods. Prior calibration is essential. To enhance the visible perception of color as an indication of the concentration of the dissolved substance it is necessary to filter out the irrelevant components that interfere with the optical detection. Visual light consists of a continuum range of wavelengths from about 450 to 800 nm. Since most substances absorb light over a relatively narrow wavelength range it is helpful filter out the light that is not involved with the detection of the desired substance. Thus the substance with a red color will transmit at the high end of the wavelength spectrum between 600 and 800 nm and the stray or non-effective light that can be removed is from 600 nm down to the UV which is in the blue and green part of the spectrum which is shown in Table exp. 6.1.

Table exp. 6.1 Light and filters

Wavelength range	Color of light	Compliment color	Wavelengths
780–620	Red	Blue-Green	600–450 nm
580–520	Yellow	Purple-Violet	450–520, 580–780
530–500	Green	Red-Purple	450–530, 500–780
500–400	Green-Violet	Red-Yellow	500–780

The eye or a camera light detector can be used to detect differences in transmitted light intensity. Thus by removing the non-relevant light that passes through the unknown sample it is possible to use the eye as a detector for the degree of adsorption by the solution.

The optical adsorption usually follows the Beer's Law which states that the absorbed light at a specified frequency (or wavelength) is proportional to the concentration of the absorbing species. This can be expressed as $I_o - I \propto C$ where I_o is the incident light intensity and I is the intensity of light transmitted after passing through the material in question and C is the concentration of the absorbing substance. I_o is the same for the two solutions, (the known and the unknown) that have different concentrations of a light absorbing substance. I_t is the transmitted light that is adjusted in length l for one of the samples to make the transmitted light I_t equal for the two solutions.

$$\text{Since } Log(I_{t1}/I_o) = al_1C_1 \text{ and } Log(I_{t2}/I_o) = al_2C_2$$

Then $l_1C_1 = l_2C_2$ and since the lengths of the light path through the solution are known and the concentration of one of the solutions is known it is possible to determine the concentration of the remaining unknown solution.

The optical filters are selected to remove the light that is NOT absorbed by the solution.

Procedure

1. Test the equivalence of the two optical paths by examining the light passed through two identical solutions samples of the same path length and satisfy that $I_{t1} = I_{t2}$ when $l_1 = l_2$
   Make the necessary adjustments of light and solution length and location until this is achieved.
2. Replace one of the known sample tube with the unknown solution and adjust the heights of the solution until the transmitted light is of the same intensity. Measure the length of the two light-paths and calculate the concentration of the unknown solution. If it is not possible to match the colors of the two solutions, it may mean that the concentrations are too high. Try to dilute the samples by a factor of 2, 5, or 10 to achieve good visible contrast. See your instructor or demonstrator if the problem persists. NOTE: Some people are color-blind or have difficulty in determining the difference for a particular color.
3. Report the results.

Questions

1. Explain why long tubes are used. Would an meter long tube be better? Why?
2. What is the limit on the length of tubes that can be used?
3. Explain why the filters can improve the accuracy of the analysis
4. Why would black glass tubes be better than the clear glass tubes that are used?
5. The Duboscq Colorimeter is an instrument used to analyze samples by differences in color intensity. Explain how an accuracy of 2% is achieved.

Experiment No. 7. Characteristics of Geiger – Muller counters

A. Introduction

The Geiger-Miller (G.M.) counter is a simple device for detecting radiation. G.M. counters are of different sizes and design depending upon their usage. It consists of a chamber the inner surface of which is coated with an electrical conductor that acts as the cathode of the tube (Fig. exp. 7.1). The anode is a tungsten wire of about 0.1 mm diameter at the central axis of the chamber, which is insulated

Fig. exp. 7.1 Outline
diagram of the G.M. detector

from the cathode and is, made the anode. The cylindrical cathode is made vacuum tight at both ends. The chamber is filled with a monoatomic gas, usually argon or helium, at a pressure of 5–10 cm of Hg. Usually a quenching gas e.g., Butane or ethyl alcohol is filled at a pressure of 1–2 cm of Hg. Quenching is the termination of ionization current pulse in a G.M. tube. For accurate quantitative work, G.M tubes are contained on a lead block or "castle", which also surrounds the sample chamber. The lead serves to shield the tube and chamber from outside radiation (see Fig. exp. 7.1). The G.M. tube is connected to a high voltage power supply and a scaler that counts the pulses of emitted electrons.

Working

If a beta emitter is brought near the window of the tube, some of the beta particles penetrates the window and pass into the gas inside the tube. This results in the formation of positive ions and electrons. When a high potential difference is applied across the electrodes the ions move toward the electrode of opposite charge. The accelerated ions also react with the gaseous atoms in the tubes to produce more ions and this chain reaction continues resulting in great mass of ions an amplification of 10^6–10^8. On reaching the electrodes the mass of ions is neutralized to producing a flow of electrons in the external circuit and provide potential of 1–10 V (Figs exp. 7.2 and 7.3).

The above reaction is terminated by quenching the accelerated ions with organic or halogen gas. If this is not done, the chain reaction would continue for some time and during which the tube would not detect another beta particle. The circuit is designed to indicate the total number of counts which are dependent the disintegration rate of the radioactive sample and the potential applied across the electrodes. At low voltage count rate/voltage curve is exponential. A slight change in the voltage causes considerable change in count rate. At higher voltage the curves becomes almost linear and horizontal. This is termed as plateau region of the G.M. tube which now operates at its maximum efficiency.

The efficiency of the tube = (counts per second from the sample/disintegrations per second from the radioisotope) × 100%.

Fig. exp. 7.2 Electronic counting set up of the G.M. counter

Fig. exp. 7.3 Variation of the charge induced at the anode with the applied voltage in an ionization counter

Determination of the Geiger – Muller Plateau

Geiger – Muller tube must be operated at an acceptable voltage which has to be determined for each tube.

Source Preparation

1. Prepare a slurry from uranium oxide, acetone and a small amount of adhesive in a plastic beaker.
2. Transfer small amounts of the slurry by a pipette or a glass rod to a planchet.
3. Spread evenly and dry under lamp ensuring that the acetone does not boil.
4. Cover the planchet by an aluminum foil of thickness equivalent to 54 mg/cm^2 (0.0008^{II}).
5. Count the planchet and if a count rate of 15,000–20,000 counts per 100 s is not achieved remove the foil and build up additional layer of U_3O_8 by adding small amount of slurry.
6. Seal the foil with an adhesive and label it.

Fig. exp. 7.4 Characteristic voltage response of the G.M counter

Procedure

1. Insert the radioactive standard under the counter tube. Use shelf number two of a source holder for U_3O_8 and shelf number one for other weaker sources.
2. Set the operating voltage at 0 V. Increase the voltage until counts are registered. This will give fairly accurate indication of the starting voltage.
3. Starting from the threshold (starting) voltage perform 100 s counts at 25 V increments. Counts should be noted in each case.
4. Immediately beyond the threshold voltage a rapid rise in counting rate will occur until the plateau is reached. The termination of the plateau region will be noted by a second rapid increase in the counting rate as the voltage is further increased.
5. As soon as this second increase in the counting rate is noted decrease the voltage, as the G.M tube will be damaged if it is allowed to operate in this region.
6. Plot the counting rate (counts per minute) versus voltage as shown in Fig. exp. 7.4.

The plateau threshold voltage V_0 is the voltage at which the linear portion of the graph (the plateau) begins. The plateau slope can be calculated from the following equation:

$$(C_2 - C_1)/C_M \times 100/(V_2 - V_1)$$

Where, C_2 and C_1 = Two count values on the linear portion of the plateau (C_2 being greater than C_1). V_2 and V_1 = Respective voltages of C_2 and C_1.

As the G.M. tube ages, the slope increases and shortens. Therefore, the G.M. counter is usually at a voltage near the middle of the plateau or about 100 V above the threshold.

Determination of Counter Efficiency

Since all radioactivity detecting devices are not able to detect all the activities in a given sample, the efficiency of the counter must be determined so that the actual number of atomic disintegrations may be calculated.

Overall counter efficiency is determined by preparing standard sample sources and unknowns.

For measuring beta radioactivity of unknown composition, use a standard solution of Cs-137 or Sr-90 in equilibrium with its daughter. For alpha calibration use standard solution of natural uranium salt, Pu-239 or Am-241.

Determine the number of counts per minute in the standard sample by making three 1 min counts and three 10 min counts. Determine counting efficiency by comparing actual counts obtained with the known number of disintegration occurring per minute in the standard.

Counter efficiency = (counts per minute from the sample/disintegration per minute from radioisotope) × 100%.

Compare statistically 1 and 10 min counts by applying analysis of variance.

To Determine the Effect of Distance on Counting

The radiation emitted by a radioactive substance is scattered in all directions at random. Therefore, as the distance between the radioactive source and the G.M. tube becomes greater, less radioactivity is detected.

Procedure

Distance between the planchet and the tube is varied. Take three counts at each of the planchet positions for a period of 1 min. Tabulate the data.

Experimental data.

Voltage count rate	Voltage count rate	Voltage count rate

Calibrated standard —————————— —————————— ——————————
Observed Activity at V_0 ————— cpm ————————— cpm ————— cpm

Report

1. Using Microsoft Excel™ or similar, plot the count rate, R, against the voltage, V, and determine the operating voltage, V_0, of the G.M. detector.
2. Determine the efficiencies with which the calibrated reference sources were measured at V_0. Efficiencies.

Source	dpm	cpm	Efficiency
Source	dpm	cpm	Efficiency
Source	dpm	cpm	Efficiency
Source	dpm	cpm	Efficiency
Source	dpm	cpm	Efficiency
Source	dpm	cpm	Efficiency

3. Explain any differences observed in the efficiencies.
4. Determine the effect of distance on counting.

Distance	Counting per minute
Distance	Counting per minute
Distance	Counting per minute
Distance	Counting per minute
Distance	Counting per minute
Distance	Counting per minute
Distance	Counting per minute
Distance	Counting per minute
Distance	Counting per minute
Distance	Counting per minute

5. Explain the effect of distance on counting.

Questions

1. What is the standard for measuring radioactivity?
2. What activity in cpm is expected from a 0.035 μCi of P-32 when it is measured with 5.4% efficiency?
3. Would it be possible to determine the operating voltage if a source emitting a different type of radiation were used? For example, if a beta-emitting source were used in this experiment, would a gamma source give approximately the same result?
4. Why is it a good idea to periodically check the high voltage (HV) plateau for G.M. detector?
5. Make the following conversions:
   (a) From 1 pCi to **X** dpm;
   (b) From 1 nCi to **X** pCi;
   (c) From 1 mCi to **X** pCi.

Sources

1. Katz SA, Bryan JC (2011) Experiments in nuclear science. CRC Press/Taylor and Francis Group, Boca Raton, p 168
2. Aery NC (2010) Manual of environmental analysis. CRC Press/Taylor and Francis Group, Boca Raton/London/New York. Ane Books Pvt. Ltd. p 413

Experiment No. 8. Biofuel Ethanol

The century of inexpensive fuel automobiles and other vehicles is soon to end. Alternates are already in the market, but still very expensive. Some of these are worth examining: (1) alcohol from the juice of fruit plants, (2) glucose from corn or potatoes, and (3) hydrolyzed cellulose by (a) microwaves and (b) ultrasonics.

It must be pointed out that the paths (1) and (2) have been with us for several thousand years and do not need any explanation other than to consider such cost saving systems as continuous fermentation and production. The fuel (alcohol) from cellulose is still in the experimental stage or rather the economizing stage. However recent studies have shown that the exposure of complex cellulose to microwave heating or ultrasonics can liberate some of the bound glucose which can now be subject to fermentation and the formation of ethanol which has become a prominent candidate to replace gasoline.

The Fermentation Process

Most canned sweet fruit juices can be used directly to convert the glucose (sugar) to ethanol with little preparation. Select a bottle or can of fruit juice and pour 200 mL into a 250 mL Erlenmeyer and add 3 g of dry active yeast. Seal the opening with a one hole rubber stopper into which a glass tube is inserted and attached to a rubber hose that is immersed in a flask containing a solution of calcium hydroxide to exclude oxygen and to absorb the CO_2 emitted.

The fermentation is allowed to proceed for a week before being examined for the yield.

The Characterization of the Alcohol

1. Examine the solution and characterize the product by (a) density, (b) taste, (c) freezing point. When half of the solution is slowly frozen the liquid is separated and the above three tests repeated. (NOTE: freezing removes water preferentially leaving an enriched alcohol solution).
2. The Microwave/Ultrasonic Degradation of Cellulose.

Place two weighed samples of cellulose cotton (3 g) in separate beakers and add 50 mL of distilled water to each beaker. Soak and thoroughly wet the cotton and place the beakers in **M** a microwave oven and heat the sample for 5, 10 and 25 min and in **U** an ultrasonic bath and apply the ultrasonics to the sample for 5, 10 and 25 min. Ideally, it would be best to examine the water in the ultraviolet to determine if a part of the cotton had reacted to produce glucose or some other organic substance. Determine if a reaction had occurred and if so, what is the product.

Report the results and suggest other methods to convert cellulose into glucose and methods to test the process.

Questions

1. The present cost of producing ethanol from grain or farm sources is too expensive. Explain why this is the case and suggest changes that might reduce the costs.
2. The alcohol from the fermentation process can be enriched by freezing the solution or by separating the alcohol by distilling it. What differences can you expect in the quality of the wine from these two different processes?
3. It is claimed that making wine is more of an art than a science. Do you agree? Explain.
4. It is possible to convert cellulose into fermentable glucose. Comment on the consequence of an economical process being developed that can produce ethanol from cellulose based crops.
5. Comment on the difference between ethanol and methanol with regard to price and suitability as a "beverage" or "fuel".

Index

GPSR Compliance
The European Union's (EU) General Product Safety Regulation (GPSR) is a set
of rules that requires consumer products to be safe and our obligations to
ensure this.

If you have any concerns about our products, you can contact us on

ProductSafety@springernature.com

In case Publisher is established outside the EU, the EU authorized
representative is:

Springer Nature Customer Service Center GmbH
Europaplatz 3
69115 Heidelberg, Germany